Graduate Texts in Contemporary Physics

Series Editors:

R. Stephen Berry
Joseph L. Birman
Jeffrey W. Lynn
Mark P. Silverman
H. Eugene Stanley
Mikhail Voloshin

Springer
New York
Berlin
Heidelberg
Barcelona
Hong Kong
London
Milan
Paris
Singapore
Tokyo

Graduate Texts in Contemporary Physics

S.T. Ali, J.P. Antoine, and J.P. Gazeau: **Coherent States, Wavelets and Their Generalizations**

A. Auerbach: **Interacting Electrons and Quantum Magnetism**

B. Felsager: **Geometry, Particles, and Fields**

P. Di Francesco, P. Mathieu, and D. Sénéchal: **Conformal Field Theories**

J.H. Hinken: **Superconductor Electronics: Fundamentals and Microwave Applications**

J. Hladik: **Spinors in Physics**

Yu.M. Ivanchenko and A.A. Lisyansky: **Physics of Critical Fluctuations**

M. Kaku: **Introduction to Superstrings and M-Theory, 2nd Edition**

M. Kaku: **Strings, Conformal Fields, and M-Theory, 2nd Edition**

H.V. Klapdor (ed.): **Neutrinos**

J.W. Lynn (ed.): **High-Temperature Superconductivity**

H.J. Metcalf and P. van der Straten: **Laser Cooling and Trapping**

R.N. Mohapatra: **Unification and Supersymmetry: The Frontiers of Quark-Lepton Physics, 2nd Edition**

H. Oberhummer: **Nuclei in the Cosmos**

G.D. Phillies: **Elementary Lectures in Statistical Mechanics**

R.E. Prange and S.M. Girvin (eds.): **The Quantum Hall Effect**

B.M. Smirnov: **Clusters and Small Particles: In Gases and Plasmas**

M. Stone: **The Physics of Quantum Fields**

F.T. Vasko and A.V. Kuznetsov: **Electronic States and Optical Transitions in Semiconductor Heterostructures**

(continued following index)

Michael Stone

The Physics of Quantum Fields

With 67 Illustrations

 Springer

Michael Stone
Loomis Laboratory
Department of Physics
University of Illinois
Urbana, IL 61801
USA

Series Editors

R. Stephen Berry
Department of Chemistry
University of Chicago
Chicago, IL 60637
USA

Mark P. Silverman
Department of Physics
Trinity College
Hartford, CT 06106
USA

Joseph L. Birman
Department of Physics
City College of CUNY
New York, NY 10031
USA

H. Eugene Stanley
Center for Polymer
 Studies
Physics Department
Boston University
Boston, MA 02215
USA

Jeffrey W. Lynn
Department of Physics
University of Maryland
College Park, MD 20742
USA

Mikhail Voloshin
Theoretical Physics Institute
Tate Laboratory of Physics
University of Minnesota
Minneapolis, MN 55455
USA

Library of Congress Cataloging-in-Publication Data
Stone, Michael, Ph.D.
 The physics of quantum fields / Michael Stone.
  p. cm. — (Graduate texts in contemporary physics)
 Includes bibliographical references and index.
 ISBN 0-387-98909-9 (hc : alk. paper)
 1. Quantum field theory. I. Title. II. Series.
QC174.45.S66 2000
530.14′3—dc21   99-39802

Printed on acid-free paper.

© 2000 Springer-Verlag New York, Inc.
All rights reserved. This work may not be translated or copied in whole or in part without the written permission of the publisher (Springer-Verlag New York, Inc., 175 Fifth Avenue, New York, NY 10010, USA), except for brief excerpts in connection with reviews or scholarly analysis. Use in connection with any form of information storage and retrieval, electronic adaptation, computer software, or by similar or dissimilar methodology now known or hereafter developed is forbidden.
The use of general descriptive names, trade names, trademarks, etc., in this publication, even if the former are not especially identified, is not to be taken as a sign that such names, as understood by the Trade Marks and Merchandise Marks Act, may accordingly be used freely by anyone.

Production managed by Allan Abrams; manufacturing supervised by Jacqui Ashri.
Photocomposed copy prepared from the author's LaTeX files.
Printed and bound by R.R. Donnelley and Sons, Harrisonburg, VA.
Printed in the United States of America.

9 8 7 6 5 4 3 2 1

ISBN 0-387-98909-9 Springer-Verlag New York Berlin Heidelberg   SPIN 10740660

Dedicated to the memory of my father, Thomas Alfred Stone.

Preface

This book is intended to provide a general introduction to the physics of quantized fields and many-body physics. It is based on a two-semester sequence of courses taught at the University of Illinois at Urbana-Champaign at various times between 1985 and 1997. The students taking all or part of the sequence had interests ranging from particle and nuclear theory through quantum optics to condensed matter physics experiment.

The book does not cover as much ground as some texts. This is because I have tried to concentrate on the basic conceptual issues that many students find difficult. For a computation-method oriented course an instructor would probably wish to suplement this book with a more comprehensive and specialized text such as Peskin and Schroeder *An Introduction to Quantum Field Theory*, which is intended for particle theorists, or perhaps the venerable *Quantum Theory of Many-Particle Systems* by Fetter and Walecka.

The most natural distribution of the material if the book is used for a two-semster course is as follows:

1st Semester: Chapters 1-11.

2nd semester: Chapters 12-18.

The material in the first 11 chapters is covered using traditional quantum mechanics operator language. This is because the text is intended for people with a wide range of interests. Were I writing for particle-theory students only, I would start with path integrals from chapter one. For a broader readership, it seems useful to maintain continuity with traditional hamiltonian quantum mechanics for as long as one as there is no penalty in ease of comprehension — and this is the case with the simple field theories discussed in the earlier chapters.

In the second half of the book the path integral comes into its own. It is seen as an efficient generator Feynman rules and Ward identities, and is, of course, indispensible for understanding the connection between renormalization and critical phenomena, as well as non-perturbative phemomena such as tunneling.

Although the book is not intended primarily for students of condensed matter physics, many of the examples discussed in the text are drawn from that field. This choice partly reflects my own interests, which have, over the years, wandered from high energy physics through lattice gauge theories to systems with real crystal lattices. There is, however, another reason: condensed matter systems can been seen and felt. I believe that it is far easier to acquire a visceral understanding of spontaneous symmetry breaking in a superfluid, than it is to grasp chiral symmetry breaking in QCD. Furthermore, condensed matter systems have mathematically well-defined hamiltonians and any field theory method applied to them has to reproduce the measured properties. This is usually not the case in relativistic systems, where additional principles have to be applied in order to decide which of several regularization-dependendent answers is correct. While there is nothing wrong with this, it is frequently disturbing to the beginner. Also it was only when Ken Wilson used field theory to address the concrete condensed matter problem of critical phenomena that the origin of the perturbation theory divergences was understood.

Michael Stone
Urbana, Illinois
October, 1999

Contents

Preface		vi
1	**Discrete Systems**	**1**
1.1	One-Dimensional Harmonic Crystal	1
	1.1.1 Normal Modes	1
	1.1.2 Harmonic Oscillator	4
	1.1.3 Annihilation and Creation Operators for Normal Modes	5
1.2	Continuum Limit	7
	1.2.1 Sums and Integrals	7
	1.2.2 Continuum Fields	8
2	**Relativistic Scalar Fields**	**12**
2.1	Conventions	12
2.2	The Klein-Gordon Equation	13
	2.2.1 Relativistic Normalization	14
	2.2.2 An Inner Product	16
	2.2.3 Complex Scalar Fields	17
2.3	Symmetries and Noether's Theorem	18
	2.3.1 Internal Symmetries	18
	2.3.2 Space-Time Symmetries	21
3	**Perturbation Theory**	**25**
3.1	Interactions	25
3.2	Perturbation Theory	26

		3.2.1	Interaction Picture	26
		3.2.2	Propagators and Time-Ordered Products	27
	3.3	Wick's Theorem		30
		3.3.1	Normal Products	30
		3.3.2	Wick's Theorem	30
		3.3.3	Applications	32

4 Feynman Rules — 37

	4.1	Diagrams		37
		4.1.1	Diagrams in Space-time	37
		4.1.2	Diagrams in Momentum Space	41
	4.2	Scattering Theory		43
		4.2.1	Cross-Sections	44
		4.2.2	Decay of an Unstable Particle	47

5 Loops, Unitarity, and Analyticity — 48

	5.1	Unitarity of the S Matrix		48
	5.2	The Analytic S Matrix		51
		5.2.1	Origin of Analyticity	51
		5.2.2	Unitarity and Branch Cuts	52
		5.2.3	Resonances, Widths, and Lifetimes	54
	5.3	Some Loop Diagrams		56
		5.3.1	Wick Rotation	56
		5.3.2	Feynman Parameters	58
		5.3.3	Dimensional Regularization	59

6 Formal Developments — 62

	6.1	Gell-Mann Low Theorem		62
	6.2	Lehmann-Källén Spectral Representation		64
	6.3	LSZ Reduction Formulae		67
		6.3.1	Amputation of External Legs	67
		6.3.2	In and Out States and Fields	68
		6.3.3	Borcher's Classes	71

7 Fermions — 72

	7.1	Dirac Equation		72
	7.2	Spinors, Tensors, and Currents		74
		7.2.1	Field Bilinears	74
		7.2.2	Conservation Laws	75
	7.3	Holes and the Dirac Sea		75
		7.3.1	Positive and Negative Energies	75
		7.3.2	Holes	78
	7.4	Quantization		79
		7.4.1	Normal and Time-Ordered Products	81

8 QED — 83
8.1 Quantizing Maxwell's Equations — 83
8.1.1 Hamiltonian Formalism — 83
8.1.2 Axial Gauge — 84
8.1.3 Lorentz Gauge — 85
8.2 Feynman Rules for QED — 88
8.2.1 Møller Scattering — 90
8.3 Ward Identity and Gauge Invariance — 92
8.3.1 The Ward Identity — 92
8.3.2 Applications — 93

9 Electrons in Solids — 97
9.1 Second Quantization — 97
9.2 Fermi Gas and Fermi Liquid — 100
9.2.1 One-Particle Density Matrix — 100
9.2.2 Linear Response — 103
9.2.3 Diagram Approach — 104
9.2.4 Applications — 106
9.3 Electrons and Phonons — 114

10 Nonrelativistic Bosons — 117
10.1 The Boson Field — 117
10.2 Spontaneous Symmetry Breaking — 118
10.3 Dilute Bose Gas — 122
10.3.1 Bogoliubov Transfomation — 122
10.3.2 Field Equations — 125
10.3.3 Quantization — 126
10.3.4 Landau Criterion for Superfluidity — 128
10.3.5 Normal and Superfluid Densities — 129
10.4 Charged Bosons — 131
10.4.1 Gross-Pitaevskii Equation — 131
10.4.2 Vortices — 132
10.4.3 Connection with Fluid Mechanics — 134

11 Finite Temperature — 136
11.1 Partition Functions — 136
11.2 Worldlines — 137
11.3 Matsubara Sums — 140

12 Path Integrals — 143
12.1 Quantum Mechanics of a Particle — 143
12.1.1 Real Time — 143
12.1.2 Euclidean Time — 146
12.2 Gauge Invariance and Operator Ordering — 148
12.3 Correlation Functions — 150

	12.4	Fields	152
	12.5	Gaussian Integrals and Free Fields	153
		12.5.1 Real Fields	153
		12.5.2 Complex Fields	155
	12.6	Perturbation Theory	156

13 Functional Methods — 158
- 13.1 Generating Functionals — 158
  - 13.1.1 Effective Action — 161
- 13.2 Ward Identities — 166
  - 13.2.1 Goldstone's Theorem — 167

14 Path Integrals for Fermions — 171
- 14.1 Berezin Integrals — 171
  - 14.1.1 A Simple Supersymmetry — 174
- 14.2 Fermionic Coherent States — 177
- 14.3 Superconductors — 179
  - 14.3.1 Effective Action — 181

15 Lattice Field Theory — 185
- 15.1 Boson Fields — 185
- 15.2 Random Walks — 189
- 15.3 Interactions and Bose Condensation — 191
  - 15.3.1 Rotational Invariance — 192
- 15.4 Lattice Fermions — 195
  - 15.4.1 No Chiral Lattice Fermions — 200

16 The Renormalization Group — 201
- 16.1 Transfer Matrices — 202
  - 16.1.1 Continuum Limit — 204
  - 16.1.2 Two-Dimensional Ising Model — 205
- 16.2 Block Spins and Renormalization Group — 206
  - 16.2.1 Correlation Functions — 212

17 Fields and Renormalization — 213
- 17.1 The Free-Field Fixed Point — 213
- 17.2 The Gaussian Model — 215
- 17.3 General Method — 219
- 17.4 Nonlinear σ Model — 220
  - 17.4.1 Renormalizing — 225
  - 17.4.2 Solution of the RGE — 228
- 17.5 Renormalizing $\lambda \varphi^4$ — 229

18 Large N Expansions — 233
- 18.1 $O(N)$ Linear σ-Model — 233

	18.2	Large N Expansions	237
		18.2.1 Linear *vs.* Nonlinear σ-Models	241
A	**Relativistic State Normalization**		**246**
B	**The General Commutator**		**248**
C	**Dimensional Regularization**		**250**
	C.1	Analytic Continuation and Integrals	250
	C.2	Propagators	252
D	**Spinors and the Principle of the Sextant**		**254**
	D.1	Constructing the γ-Matrices	254
	D.2	Basic Theorem	255
	D.3	Chirality	256
	D.4	$Spin(2N)$, $Pin(2N)$, and $SU(N) \subset SO(2N)$	257
E	**Indefinite Metric**		**258**
F	**Phonons and Momentum**		**261**
G	**Determinants in Quantum Mechanics**		**264**
Index			**267**

1
Discrete Systems

1.1 One-Dimensional Harmonic Crystal

We begin with the quantum mechanics of a vibrating crystal. To the naked eye the crystal appears to be a continuous elastic solid. We know, however, that, when viewed through a sufficiently powerful microscope it will be revealed to be composed of individual atoms held together by chemical bonds. For our purpose the atoms and bonds can be thought of as "balls and springs," and the crystal as an assembly of coupled harmonic oscillators. If you understand the quantum mechanics of harmonic oscillators, it will not be difficult to apply this understanding to study the effectively continuous crystal. This is our task in this chapter.

1.1.1 Normal Modes

To avoid the complexities of real crystals with their plethora of elastic constants and modes, we will consider a simple one-dimensional model of a crystal.

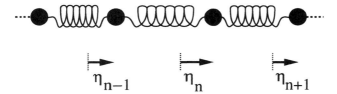

Fig 1. *A one-dimensional crystal.*

We will take a line of atoms of unit mass whose equilibrium positions are at a set of sites on the x axis labeled by the integer n, and separated by a distance a. We will assume the atoms are free to vibrate only in the x direction, so we are dealing with longitudinal waves, and denote the displacement of the atom at site n by η_n.

The quickest route to the dynamics uses the lagrangian. As always in mechanics this is the difference of the kinetic energy T and the potential energy V. For a *harmonic crystal* V is a sum of terms of the form $\frac{1}{2}\lambda(\eta_n - \eta_{n+1})^2$, where λ is the spring constant. Thus

$$L = T - V = \sum_n \left\{ \frac{1}{2}\dot{\eta}_n^2 - \frac{\lambda}{2}(\eta_n - \eta_{n+1})^2 \right\}. \tag{1.1}$$

From Lagrange's equations, one for each η_n,

$$\frac{d}{dt}\left(\frac{\partial L}{\partial \dot{\eta}_n}\right) - \frac{\partial L}{\partial \eta_n} = 0, \tag{1.2}$$

we find the classical equations of motion

$$\ddot{\eta}_n = \lambda(\eta_{n+1} + \eta_{n-1} - 2\eta_n). \tag{1.3}$$

These have solutions in the form of complex traveling waves

$$\eta_n = e^{ikn - i\omega t}, \tag{1.4}$$

where

$$\omega^2 = 2\lambda(1 - \cos k). \tag{1.5}$$

In the long-wavelength limit $k \ll 1$, this dispersion relation reduces to

$$\omega^2 = \lambda k^2, \tag{1.6}$$

which means that the long-wavelength sound waves have velocity $\sqrt{\lambda}$.

In the next chapter we will have cause to consider an additional term in the lagrangian, which corresponds to a harmonic potential $\frac{1}{2}\Omega^2 \eta_n^2$ pinning each of the particles to the vicinity of its initial location. Including this, L becomes

$$L = \sum_n \left\{ \frac{1}{2}\dot{\eta}_n^2 - \frac{\lambda}{2}(\eta_n - \eta_{n+1})^2 - \frac{1}{2}\Omega^2 \eta_n^2 \right\}. \tag{1.7}$$

The dispersion relation is now

$$\omega^2 \to 2\lambda(1 - 1\cos k) + \Omega^2 \approx \lambda k^2 + \Omega^2. \tag{1.8}$$

The additional potential therefore creates a *gap* in the spectrum, so there are no solutions corresponding to any frequency below Ω.

To determine the normal modes we must impose boundary conditions. Suppose we take periodic boundary conditions by identifying atom $n + N$ with atom n. This means that η_n must equal η_{n+N}. Consequently, we require e^{ikN} to be unity and the allowed values of k are therefore

$$k_m = \frac{2\pi m}{N}, \quad m = 0, 1, \ldots, N - 1. \tag{1.9}$$

We can now write a normal-mode expansion

$$\eta_n(t) = \sum_{m=0}^{N-1} \left\{ A_m e^{ik_m n - i\omega_m t} + A_m^* e^{-ik_m n + i\omega_m t} \right\}. \tag{1.10}$$

Because the total displacement is a real number, we have added to each original complex exponential solution its complex conjugate.

From (1.1) we read off the momentum canonically conjugate to the dispacement η_n

$$\pi_n \stackrel{def}{=} \frac{\partial L}{\partial \dot{\eta}_n} = \dot{\eta}_n. \tag{1.11}$$

In quantum mechanics the displacement η_n and its canonical conjugate π_n become operators $\hat{\eta}_n$ and $\hat{\pi}_n$ with commutation relations

$$[\hat{\eta}_n, \hat{\pi}_m] = i\hbar \delta_{nm}. \tag{1.12}$$

From (1.10) we find that

$$\pi_n(t) = \dot{\eta}_n(t) = \sum_{m=0}^{N-1} \left\{ -i\omega_m A_m e^{ik_m n - i\omega_m t} + i\omega_m A_m^* e^{-ik_m n + i\omega_m t} \right\}. \tag{1.13}$$

We have a choice as to how to include time evolution in the quantum mechanics formalism. In the Schrödinger[1] picture we put the time dependence in the Hilbert-space states and leave the operators time independent. This is the customary approach in elementary quantum mechanics courses, and is what we usually have in mind when we write equations like (1.12). In field theory it turns out to be more convenient to use the *Heisenberg*[2] picture where the operators are explicitly time dependent. For any operator $\hat{O}$ we have

$$\hat{O}(t) = e^{\frac{i}{\hbar}\hat{H}t} \hat{O}(0) e^{-\frac{i}{\hbar}\hat{H}t}, \tag{1.14}$$

and

$$\frac{d\hat{O}}{dt} = \frac{i}{\hbar}[\hat{H}, \hat{O}]. \tag{1.15}$$

When we use the Heisenberg picture, we must specify the times at which the fields in the commutation relation are to be evaluated. To retain its simple form (1.12) must be replaced by an equal-time commutator

$$[\hat{\eta}_n(t), \hat{\eta}_m(t)] = i\hbar \delta_{nm}. \tag{1.16}$$

Finding the commutator with the operators evaluated at two different times requires solving the dynamics of the system.

---

[1] Erwin Schrödinger. Born August 12, 1887, Vienna. Died January 4, 1961, Vienna. Nobel Prize for Physics 1933.

[2] Werner Karl Heisenberg. Born December 5, 1901. Died February 1, 1976, Munich. Nobel Prize for Physics 1932.

1.1.2 Harmonic Oscillator

Let us recall how the Heisenberg picture works for the harmonic oscillator.

For a unit mass oscillator with angular frequency ω, the hamiltonian is

$$\hat{H} = \frac{1}{2}(\hat{p}^2 + \omega^2 \hat{x}^2). \tag{1.17}$$

Here the operators $\hat{x}(t)$ and $\hat{p}(t)$ obey the equal-time commutation relation

$$[\hat{x}(t), \hat{p}(t)] = i\hbar. \tag{1.18}$$

The equations of motion are

$$\frac{d\hat{x}(t)}{dt} = \frac{i}{\hbar}[\hat{H}, \hat{x}] = \hat{p}(t), \tag{1.19}$$

$$\frac{d\hat{p}(t)}{dt} = \frac{i}{\hbar}[\hat{H}, \hat{p}] = -\omega^2 \hat{x}(t). \tag{1.20}$$

Differentiating the first equation with repect to t, and substituting for $\frac{d\hat{x}}{dt}$ from the second shows that

$$\frac{d^2\hat{x}}{dt^2} + \omega^2 \hat{x} = 0. \tag{1.21}$$

The Heisenberg operator $\hat{x}(t)$ therefore satisfies exactly the same equation of motion as the classical variable $x(t)$ it replaces.

We could write down the solution to (1.21) in terms of sines and cosines, but it is more productive to introduce the operators $\hat{a}(t)$ and $\hat{a}^\dagger(t)$ by writing

$$\hat{x}(t) = \sqrt{\frac{\hbar}{2\omega}}(\hat{a}(t) + \hat{a}^\dagger(t)) \tag{1.22}$$

$$\hat{p}(t) = \sqrt{\frac{\hbar}{2\omega}}(-i\omega \hat{a}(t) + i\omega \hat{a}^\dagger(t)). \tag{1.23}$$

Equivalently,

$$\hat{a}(t) = \sqrt{\frac{\omega}{2\hbar}}\left(\hat{x}(t) + i\frac{\hat{p}(t)}{\omega}\right), \tag{1.24}$$

$$\hat{a}^\dagger(t) = \sqrt{\frac{\omega}{2\hbar}}\left(\hat{x}(t) - i\frac{\hat{p}(t)}{\omega}\right). \tag{1.25}$$

Their equal-time commutation relations are found from those of $\hat{x}$, $\hat{p}$, to be

$$[\hat{a}(t), \hat{a}^\dagger(t)] = 1. \tag{1.26}$$

We also see that

$$\hat{H} = \hbar\omega(\hat{a}^\dagger(t)\hat{a}(t) + \frac{1}{2}). \tag{1.27}$$

So

$$\frac{d\hat{a}(t)}{dt} = \frac{i}{\hbar}[\hat{H}, \hat{a}(t)] = -i\omega\hat{a}(t) \Rightarrow \hat{a}(t) = \hat{a}(0)e^{-i\omega t}, \quad (1.28)$$

$$\frac{d\hat{a}^\dagger(t)}{dt} = \frac{i}{\hbar}[\hat{H}, \hat{a}^\dagger(t)] = +i\omega\hat{a}^\dagger(t) \Rightarrow \hat{a}^\dagger(t) = \hat{a}^\dagger(0)e^{+i\omega t}. \quad (1.29)$$

From now on we will write $\hat{a}$ for $\hat{a}(0)$, and similarly for $\hat{a}^\dagger(0)$. In field theory these are called the *annihilation* and *creation* operators, respectively.

The time dependence of $\hat{x}(t)$ and $\hat{p}(t)$ is now explicit:

$$\hat{x}(t) = \sqrt{\frac{\hbar}{2\omega}}(\hat{a}e^{-i\omega t} + \hat{a}^\dagger e^{+i\omega t}), \quad (1.30)$$

$$\hat{p}(t) = \sqrt{\frac{\hbar}{2\omega}}(-i\omega\hat{a}e^{-i\omega t} + i\omega\hat{a}^\dagger e^{+i\omega t}). \quad (1.31)$$

If we substitute these expressions into the hamiltonian, we find that it is time independent

$$\hat{H} = \hbar\omega(\hat{a}^\dagger\hat{a} + \frac{1}{2}), \quad (1.32)$$

just as it is in classical mechanics.

1.1.3 Annihilation and Creation Operators for Normal Modes

Inspired by the harmonic oscillator, let us try setting

$$\hat{\eta}_n(t) = \sum_{m=0}^{N-1} \sqrt{\frac{\hbar}{2\omega_m}} \frac{1}{\sqrt{N}} \{\hat{a}_m e^{ik_m n - i\omega_m t} + \hat{a}_m^\dagger e^{-ik_m n + i\omega_m t}\}, \quad (1.33)$$

$$\hat{\pi}_n(t) = \sum_{m=0}^{N-1} \sqrt{\frac{\hbar}{2\omega_m}} \frac{1}{\sqrt{N}} \{-i\omega_m \hat{a}_m e^{ik_m n - i\omega_m t} + i\omega_m \hat{a}_m^\dagger e^{-ik_m n + i\omega_m t}\}, \quad (1.34)$$

where $[\hat{a}_m, \hat{a}_n^\dagger] = \delta_{mn}$, and computing the equal-time commutator, $[\hat{\eta}_n(t), \hat{\pi}_m(t)]$, to see if it comes out right. We have some hope that this will work since the $\sqrt{\frac{\hbar}{2\omega}}$'s are suggested by the harmonic-oscillator case, and the $\frac{1}{\sqrt{N}}$'s serve to normalize the normal modes.

In dealing with these sorts of sums it is useful to remember the finite Fourier series identity

$$\sum_{m=0}^{N-1} e^{ik_m(n-n')} = N\delta_{nn'}, \quad (1.35)$$

which is easily proved from the formula for the sum of a geometric progression.

A short calculation shows that everything works, and

$$[\hat{\eta}_n(t), \hat{\pi}_m(t)] = i\hbar \delta_{nm} \tag{1.36}$$

as it should.

We can also express the hamiltonian in terms of $\hat{a}_m$, $\hat{a}_m^\dagger$. We find

$$\hat{H} = \sum_n \left\{ \frac{1}{2}\hat{\pi}_n^2 + \frac{\lambda}{2}(\hat{\eta}_n - \hat{\eta}_{n+1})^2 \right\} = \sum_{m=0}^{N-1} \hbar\omega_m (\hat{a}_m^\dagger \hat{a}_m + \frac{1}{2}). \tag{1.37}$$

We now know essentially everything about our model crystal. We only need to remember how to construct the Hilbert space for the harmonic oscillator by acting with $\hat{a}^\dagger$ on the ground state, and then generalize this to the crystal. Then we are home. Recall that in constructing the harmonic-oscillator operator representation we postulate the existence of a ground state $|0\rangle$ such that $\hat{a}|0\rangle = 0$, and then the states

$$|n\rangle = \frac{(\hat{a}^\dagger)^n}{\sqrt{n!}} |0\rangle \tag{1.38}$$

are normalized, $\langle n|n \rangle = 1$, energy eigenstates

$$\hat{H}|n\rangle = \hbar\omega(n + \frac{1}{2})|n\rangle. \tag{1.39}$$

Von Neumann[3] proved that this representation of the harmonic oscillator operator algebra is unique.

The Hilbert space for the crystal is a tensor product of N copies of the harmonic oscillator space. This may sound complicated, but all we need is to assume that there is a state $|0\rangle$ that obeys

$$\hat{a}_m|0\rangle = 0, \quad \forall m. \tag{1.40}$$

Then

$$|n_0, n_1, \ldots, n_{N-1}\rangle = \frac{(\hat{a}_0^\dagger)^{n_0}}{\sqrt{n_0!}} \frac{(\hat{a}_1^\dagger)^{n_1}}{\sqrt{n_1!}} \cdots \frac{(\hat{a}_{N-1}^\dagger)^{n_{N-1}}}{\sqrt{n_{N-1}!}} |0\rangle \tag{1.41}$$

is a normalized eigenstate of

$$\hat{H} = \sum_{m=0}^{N-1} \hbar\omega_m (\hat{a}_m^\dagger \hat{a}_m + \frac{1}{2}), \tag{1.42}$$

with eigenvalue

$$E = E_0 + n_0 \hbar\omega_0 + n_1 \hbar\omega_1 + \cdots + n_{N-1} \hbar\omega_{N-1}. \tag{1.43}$$

Here

$$E_0 = \sum \frac{1}{2} \hbar\omega_m. \tag{1.44}$$

---

[3] Johann (John) von Neumann. Born December 3, 1903, Budapest, Hungary. Died February 8, 1957, Washington DC.

The Hilbert space spanned by these states is called *Fock Space*.

We call the excited states *phonons*. We say that there are n_1 phonons in the first mode, n_2 in the second, and so on. They obey Bose statistics because the *occupation numbers* n_n may be as large as we wish. We think of the phonons as elementary "particles" that possess definite energy and momentum and may, when suitable interaction terms are included in the hamiltonian, scatter off one another just as any of the other "-ons" (mesons, photons, and so on) known to physics. The duality of the field [here $\hat{\eta}(x,t)$] and particle is the heart of quantum field theory.

1.2 Continuum Limit

1.2.1 Sums and Integrals

Now we stand back and blur our vision so that the atomic crystal appears as an elastic continuum. Viewed without a microscope the displacements η_n become a field $\eta(x)$, where x can be any real number. Naturally N must be taken very large so that we have some macroscopic size to our system.

Of course $x = na$ still, but we will be interested in slowly varying functions so that $f(an)$ can be regarded as a smooth function of x. The basic "rule" for this blurring is

$$a \sum_n f(an) \to \int f(an)a\,dn \to \int f(x)\,dx, \tag{1.45}$$

Now

$$a \sum_n f(na)\frac{1}{a}\delta_{nm} = f(ma) \to \int f(x)\delta(x-y)\,dx = f(y). \tag{1.46}$$

The Dirac delta function therefore corresponds to

$$\frac{\delta_{nn'}}{a} \to \delta(x-x'). \tag{1.47}$$

The divergent quantity $\delta(0)$ (in x space) is obtained by setting $n = n'$ and is thus to be understood as the reciprocal of the lattice spacing, or, equivalently, the number of normal modes per unit volume.

For Fourier sums we recall that

$$\sum_{m=0}^{N-1} e^{ik_m(n-n')} = N\delta_{nn'}, \tag{1.48}$$

where $k_m = \frac{2\pi m}{N}$. Since we want $k_m n = (k_m/a)na \to kx$, we must scale k so the continuum wavenumber is $k_m/a \to k$. In (1.48) the dimensionless k_m runs between 0 and 2π. We can equally well have the sum go symmetrically between $-\pi$ and $+\pi$, so the continuum k ranges between $-\frac{\pi}{a}$ and $+\frac{\pi}{a}$. Thus

$$\delta(x-x') \leftarrow \frac{\delta_{nn'}}{a} = \frac{1}{Na}\sum_m e^{ik_m(n-n')} \to \int_{-\infty}^{\infty} \frac{dk}{2\pi} e^{ik(x-x')}. \tag{1.49}$$

The limits on the last integral should be $\pm\frac{\pi}{a}$, but, if we are only interested in functions varying slowly on the scale of a, we can take the limits on the integral to be infinite. This then is the usual Fourier-integral representation of the delta function.

It is good practice when doing Fourier transforms in field theory to treat x and k asymmetrically. Always put the 2π's with the dk's. This is because $\frac{dk}{2\pi}$ has the physical meaning of the number of normal modes per unit (spatial) volume with wavenumber between k and $k + dk$. In other words,

$$\sum_m F(k_m/a) = \sum_m F(k) \leftrightarrow Na \int \frac{dk}{2\pi} F(k) = (Volume) \int \frac{dk}{2\pi} F(k). \qquad (1.50)$$

The Fourier integral for $\delta(k - k')$ is

$$\int dx\, e^{i(k-k')x} = 2\pi \delta(k - k'), \qquad (1.51)$$

so $2\pi \delta(0)$ (in k space), although again mathematically divergent, has the physical meaning $\int dx = V$, the volume of the system. Again it is good practice to put a 2π with each $\delta(k)$, because this combination has a direct physical interpretation.

Note that the symbol $\delta(0)$ has a very different physical interpretation depending on whether δ is a delta function in x or in k space.

1.2.2 Continuum Fields

We might take the continuum version of our crystal lagrangian to be

$$L = T - V = \int dx \left\{ \frac{1}{2}\rho_0 \dot{\eta}(x)^2 - \frac{\kappa}{2}(\partial_x \eta)^2 \right\}, \qquad (1.52)$$

where ρ_0 is the equilibrium mass density and $\eta(x)$ is the displacement field. The elastic constant κ is the one-dimensional equivalent of the bulk modulus. It is, however, common in field theory to absorb constants (in this case ρ_0) into the fields to in order to make the coefficient of the kinetic term simply $\frac{1}{2}$. We can do this by defining $\varphi(x) = \sqrt{\rho_0}\eta(x)$. Then, after defining $\kappa/\rho_0 = c^2$, and adding a pinning term $\propto \varphi^2$, we will write

$$L = \int dx \left\{ \frac{1}{2}\dot{\varphi}^2 - \frac{c^2}{2}(\partial_x \varphi)^2 - \frac{m^2 c^4}{2}\varphi^2 \right\}. \qquad (1.53)$$

The equation of motion can be found directly from the principle of least action. Here, as usual, the action S is defined by $S = \int dt\, L$. We can express the equation of motion as

$$\frac{\delta S}{\delta \varphi(x, t)} = 0, \qquad (1.54)$$

where the functional (sometimes called the Fréchet) derivative $\delta/\delta\varphi$ is defined by

$$\delta S = \int dx\, dt\, \frac{\delta S}{\delta \varphi(x, t)} \delta\varphi(x, t). \qquad (1.55)$$

In the present case we find, after integrating by parts and discarding the boundary terms, that

$$\delta S = \int dx\, dt\, (-\partial_{tt}^2 \varphi + c^2 \partial_{xx}^2 \varphi - m^2 c^4 \varphi)\delta\varphi, \tag{1.56}$$

so the equation of motion is

$$\partial_{tt}^2 \varphi = c^2 \partial_{xx}^2 \varphi - m^2 c^4 \varphi. \tag{1.57}$$

In the absence of the pinning term this is just the wave equation for wave speed c.
There are solutions

$$\varphi(x, t) = e^{ikx - i\omega_k t} \tag{1.58}$$

where ω_k is the positive square root of

$$\omega_k^2 = c^2 k^2 + m^2 c^4. \tag{1.59}$$

To quantize, we need the corresponding hamiltonian. We first define the canonically conjugate momentum field by

$$\pi(x) \equiv \frac{\delta L}{\delta \dot\varphi(\mathbf{x})} = \dot\varphi(x). \tag{1.60}$$

In evaluating this functional derivative, we regarded φ and $\dot\varphi$ as independent variables, as we always do in lagrangian mechanics. We then use the conjugate field to define the hamiltonian as $H = \sum p\dot q - L$, except that now we need to integrate, rather than sum, over the continuous variable x. We find

$$H = \int dx (\pi(x)\dot\varphi(x) - L)$$
$$= \int dx \left\{ \frac{1}{2}\dot\varphi^2 + \frac{c^2}{2}(\partial_x \varphi)^2 + \frac{m^2 c^4}{2}\varphi^2 \right\}. \tag{1.61}$$

We now write down the quantum fields

$$\hat\varphi(x) = \int \frac{dk}{2\pi} \sqrt{\frac{\hbar}{2\omega_k}} \left\{ \hat a_k e^{+ikx - i\omega_k t} + \hat a_k^\dagger e^{-ikx + i\omega_k t} \right\} \tag{1.62}$$

and

$$\hat\pi(x) = \int \frac{dk}{2\pi} \sqrt{\frac{\hbar}{2\omega_k}} \left\{ -i\omega_k \hat a_k e^{+ikx - i\omega_k t} + i\omega_k \hat a_k^\dagger e^{-ikx + i\omega_k t} \right\}, \tag{1.63}$$

with

$$[\hat a_k, \hat a_{k'}^\dagger] = 2\pi \delta(k - k'). \tag{1.64}$$

Their equal-time commutator comes out to be

$$[\hat\varphi(x, t), \hat\pi(x', t)] = i\hbar \delta(x - x'). \tag{1.65}$$

This is exactly what is needed for the Heisenberg equations of motion to coincide with the classical ones.

The quantum hamiltonian $\hat{H}$ can be written in terms of $\hat{a}_k$ and $\hat{a}_k^\dagger$ as

$$\hat{H} = \int \frac{dk}{2\pi} \frac{1}{2} \hbar \omega_k (\hat{a}_k^\dagger \hat{a}_k + \hat{a}_k \hat{a}_k^\dagger). \tag{1.66}$$

In writing this expression we have taken care to keep the $\hat{a}_k^\dagger$, $\hat{a}_k$'s in the ordering that they appear when we expand out the $\hat{\varphi}$'s. If we use (1.55) and $Vol = 2\pi \delta(k=0)$, we see that

$$\hat{H} = E_0 + \int \frac{dk}{2\pi} \hbar \omega_k (\hat{a}_k^\dagger \hat{a}_k), \tag{1.67}$$

where the ground-state energy is

$$E_0 = (Vol) \int \frac{dk}{2\pi} \frac{1}{2} \hbar \omega_k = \sum_{modes} \frac{1}{2} \hbar \omega_k. \tag{1.68}$$

For a strictly continuous system there is no cut-off in the k integral and the zero point energy density is divergent. This is not necessarily a problem because this energy is only experimentally accessible when we have some control over either the ω_k or the density of states. We can, for example, obtain this control by confining the field in a resonant cavity whose size is variable. We may then measure *changes* in E_0, which is then known as the *Casimir* energy. Another case where only changes in the energy are important occurs when one part of the system can modify the parameters of another part. Taking advantage of such coupling to lower the ground-state energy drives many examples of *spontaneous symmetry breaking*.

The divergence in the total zero-point energy *is* important when we consider quantum fields coupled to gravity. Any energy density acts a source for the gravitational field, and a uniform, divergent, vacuum energy density should give rise to a large cosmological constant. The smallness of the observed cosmological constant has prompted much theoretical speculation.

Of course everything we have done here can be extended to three dimensions. For a scalar field φ we have action

$$S = \int d^4x \, \mathcal{L}, \tag{1.69}$$

where

$$\mathcal{L} = \frac{1}{2} \dot{\varphi}^2 - \sum_{i=1}^{3} \frac{c^2}{2} (\partial_i \varphi)^2 - \frac{m^2 c^4}{2} \varphi^2 \tag{1.70}$$

is the *lagrangian density*, and a mode expansion

$$\hat{\varphi}(x) = \int \frac{d^3k}{(2\pi)^3} \sqrt{\frac{\hbar}{2\omega_\mathbf{k}}} \left\{ \hat{a}_\mathbf{k} e^{+i\mathbf{k}\cdot\mathbf{x} - i\omega_k t} + \hat{a}_\mathbf{k}^\dagger e^{-i\mathbf{k}\cdot\mathbf{x} + i\omega_k t} \right\}, \tag{1.71}$$

with

$$[\hat{a}_\mathbf{k}, \hat{a}_{\mathbf{k}'}^\dagger] = (2\pi)^3 \delta^3(\mathbf{k} - \mathbf{k}'). \tag{1.72}$$

This gives
$$[\hat{\varphi}(\mathbf{x}, t), \partial_t \hat{\varphi}(\mathbf{x}', t)] = i\hbar \delta^3(\mathbf{x} - \mathbf{x}'), \tag{1.73}$$
and so on.

2
Relativistic Scalar Fields

It is easiest to start our study of field theory with relativistic systems. Lorentz invariance provides many simplifications and guiding principles, as well as the allure of a compact notation.

2.1 Conventions

For the next few chapters we are going to focus on applications to particle physics. In this field it is customary to simplify our life by selecting *natural units*, where both $\hbar$ and the speed of light are set to unity. The latter choice means that both time and distance can be measured in *fermi*,[1] where 1 fm = 10^{-15} m. The fermi (which is also one *femtometer*, so the abreviation is apposite) is roughly the size of a nucleon (proton or neutron) so 1 fm $\approx 3^{-1} \times 10^{-23}$ s is the time light takes to cross a such a particle. This is a typical strong-interaction time scale. We also equate mass m with its corresponding energy, mc^2, and measure both in GeV.

Having set $\hbar = 1$, we may also measure mass in units of fm^{-1} by making use of the corresponding Compton wavelength, $\hbar/mc$. The conversion factor is 1 GeV ≈ 0.20 fm^{-1}. If we want to state the dimensions $[x]$ of a quantity x, we can do so either in terms of mass, M, or length, L, since $[M] = [L]^{-1}$. For example lagrangian densities have dimension M^4, so that the *action* is dimensionless.

---

[1] Named for Enrico Fermi. Born September 29, 1901, Rome. Died November 28, 1954, Chicago. Nobel Prize in Physics 1938.

We must now select a convention for the metric in Minkowski space. We will take $g_{00} = 1$, $g_{11} = g_{22} = g_{33} = -1$, so that for a particle of mass m and $p^\mu = (E_\mathbf{p}, \mathbf{p})$ we have $p^2 \equiv p^\mu p^\nu g_{\mu\nu} = E_\mathbf{p}^2 - \mathbf{p}^2 = m^2$. The metric is used to raise and lower Lorentz tensor indices. For example $p_\mu = g_{\mu\nu} p^\nu = (E_\mathbf{p}, -\mathbf{p})$.

2.2 The Klein-Gordon Equation

The formula $E_\mathbf{p}^2 = \mathbf{p}^2 + m^2$ is of course the reason we introduced the pinning term in the previous chapter. We interpret the dispersion relation obtained there, $\omega_\mathbf{p}^2 = c^2 \mathbf{p}^2 + m^2 c^4$, as the energy-momentum relation for relativistic particles of mass m, and from now on in our discussions of such particles we will replace $\omega_\mathbf{k}$ by $E_\mathbf{k} = \sqrt{\mathbf{k}^2 + m^2}$.

The wave equation that gives us this energy-momentum relation

$$\left(\partial_t^2 - \sum_1^3 \partial_i^2 + m^2 \right) \varphi = 0, \tag{2.1}$$

is named after Klein and Gordon because they were the first to publish it. The Klein-Gordon (KG) equation was in fact investigated and abandoned by Schrödinger even before he discovered his Schrödinger equation. Schrödinger had been exploring the behavior of de Broglie[2] waves when their associated particles were in Kepler orbits about a nucleus. Debye[3] remarked that in order to do this properly it was necessary to have a wave equation. Schrödinger found one, and because de Broglie's waves were relativistic, the one he found was the relativistic KG equation. He wrote a paper but then withdrew it because the KG equation did not give the correct fine structure for the hydrogen spectrum. He published only the nonrelativistic wave equation that now bears his name.

We now understand that the KG equation failed to give the correct fine structure because it takes no account of *spin*. The spin of the particles associated with a field is determined by the Lorentz indices it carries. Being a Lorentz scalar, the KG field $\varphi(x)$ describes spin-0 particles. To describe spin-$\frac{1}{2}$ electrons we need a *spinor* field, ψ_α, and this will be discussed in Chapter 7. For the moment we will stay with the scalar field φ and explore as many concepts as we can without cluttering the page with the added complications of spin or other indices.

---

[2]Louis-Victor (-Pierre-Raymond) de Broglie (7th duc de Broglie). Born August 15, 1892, Dieppe, France. Died March 19, 1987, Paris. Nobel Prize for Physics 1929.

[3]Peter (Joseph William) Debye. (Petrus Josephus Wilhelmus Debije.) Born March 24, 1884, Maastricht, The Netherlands. Died November 2, 1966, Ithaca, NY. Nobel Prize for Chemistry 1936.

2.2.1 Relativistic Normalization

When we adopt the formulae from Chapter 1 to Lorentz-invariant systems, the factors of $1/\sqrt{2E_\mathbf{k}}$ appearing in the mode expansion (1.62) tend to make life awkward. They mean that simple expressions such as

$$\langle \mathbf{k}|\hat{\varphi}|0\rangle = \frac{1}{\sqrt{2E_\mathbf{k}}} e^{-i\mathbf{k}\cdot\mathbf{x}+iE_\mathbf{k}t}, \tag{2.2}$$

the matrix element of the Lorentz scalar field $\hat{\varphi}$ between the vacuum and the one-particle state $|\mathbf{k}\rangle$, are not themselves Lorentz scalars. This defect can be remedied by some cosmetic redefinitions.

We rewrite the mode expansion as

$$\hat{\varphi}(x) = \int \frac{d^3k}{(2\pi)^3} \frac{1}{2E_\mathbf{k}} \left\{ \hat{a}_\mathbf{k} e^{+i\mathbf{k}\cdot\mathbf{x}-iE_\mathbf{k}t} + \hat{a}_\mathbf{k}^\dagger e^{-i\mathbf{k}\cdot\mathbf{x}+iE_\mathbf{k}t} \right\}. \tag{2.3}$$

Here we have multiplied and divided by factors of $\sqrt{2E_\mathbf{k}}$ and rescaled $\hat{a}$ and $\hat{a}^\dagger$ so that now

$$[\hat{a}_\mathbf{k}, \hat{a}_{\mathbf{k}'}^\dagger] = 2E_\mathbf{k}(2\pi)^3 \delta^3(\mathbf{k}-\mathbf{k}'). \tag{2.4}$$

This means that the state $|\mathbf{k}\rangle \equiv \hat{a}_\mathbf{k}^\dagger|0\rangle$ is normalized as

$$\langle \mathbf{k}|\mathbf{k}'\rangle = 2E_\mathbf{k}(2\pi)^3 \delta^3(\mathbf{k}-\mathbf{k}'). \tag{2.5}$$

With this rescaling we now have

$$\langle \mathbf{k}|\hat{\varphi}|0\rangle = e^{-i\mathbf{k}\cdot\mathbf{x}+iE_\mathbf{k}t}, \tag{2.6}$$

where the right-hand side is now a Lorentz scalar. We now find that matrix elements of operators with specified Lorentz-transformation properties will transform in the same way as the operators themselves.

To achieve this desirable state of affairs we seem to have paid the price of some awkward looking $1/2E_\mathbf{k}$'s in the mode expansion, and an overly complicated expression for the inner product. Actually the various $2E_\mathbf{k}$ factors are now in their natural locations. For example, by using the identity

$$\delta(k^2 - m^2) = \frac{1}{2k^0}\delta(k^0 - E_\mathbf{k}) + \frac{1}{2k^0}\delta(k^0 + E_\mathbf{k}), \tag{2.7}$$

we see that the combination

$$\int \frac{d^3k}{(2\pi)^3} \frac{1}{2E_\mathbf{k}} f(\mathbf{k}, E_\mathbf{k}) \tag{2.8}$$

can be written

$$\int \frac{d^4k}{(2\pi)^4} 2\pi\delta(k^2 - m^2)\theta(k_0) f(\mathbf{k}, k_0) \tag{2.9}$$

(here $k^2 = k_0^2 - \mathbf{k}^2$ means the lorentzian inner product) which now contains a manifestly Lorentz-invariant measure, a manifestly Lorentz-invariant delta function, and a Heaviside step function, which is also left unchanged by orthochronous (no time-reversing) Lorentz transformations.

The inner product $\langle \mathbf{k} | \mathbf{k}' \rangle = 2E_\mathbf{k} (2\pi)^3 \delta^3(\mathbf{k} - \mathbf{k}')$ is also sensible. The state $|\mathbf{k}\rangle$ has continuum normalization, and, just as the continuum normalized state $|k\rangle$ with wavefunction e^{ikx} in ordinary quantum mechanics, represents a *beam* of particles with common momentum $\mathbf{k}$. The new normalization of $|\mathbf{k}\rangle$ means that the particle density in the beam is equal to twice its energy. Suppose we make this normalization in some frame and then Lorentz transform to a frame where the particle is moving faster so its energy is higher. Then the $2E$ normalization means that the particle density in this frame is higher also – but this is exactly as it should be! Lorentz contraction makes what was a cubic meter in the first frame become smaller in the new one, while the number of particles in the volume is unchanged by the transformation. The density of particles therefore increases just as the energy does. A formal description of how this works is given in appendix A.

The states in (2.5) are said to have *Lorentz-invariant* normalization, and integrals such as

$$I = \int \prod_i \left(\frac{d^3 k_i}{(2\pi)^3} \frac{1}{2E_{\mathbf{k}_i}} \right) f(\mathbf{k}_i, E(\mathbf{k}_i)), \tag{2.10}$$

are said to be over *Lorentz-invariant Phase Space* (often abbreviated LIPS).

Note that with $x_0 = t$ and the lorentzian inner product $kx = k_0 x_0 - \mathbf{k} \cdot \mathbf{x}$ we can write

$$\hat{\varphi}(x) = \int \frac{d^3 k}{(2\pi)^3} \frac{1}{2E_\mathbf{k}} \left\{ \hat{a}_\mathbf{k} e^{-ikx} + \hat{a}_\mathbf{k}^\dagger e^{+ikx} \right\}, \tag{2.11}$$

and

$$\langle \mathbf{k} | \hat{\varphi} | 0 \rangle = e^{ikx}. \tag{2.12}$$

Often in quantum mechanics we need to insert a *complete set of states* into an expression. This is equivalent to inserting an expression for the identity operator, or a *resolution of unity*, in terms of the states in the Hilbert space. For example if $|n\rangle$ is a complete orthonormal set for our Hilbert space, then

$$1 = \sum_n |n\rangle \langle n|. \tag{2.13}$$

In field theory, the Hilbert space decomposes into sectors each having a different number of particles of various types. (In a free theory no matrix elements of the hamiltonian connect these sectors, but this is not the case when interactions create and destroy particles.) Each sector has its own resolution of unity. With our normalization we have

$$1_N = \frac{1}{N!} \int \prod_{i=1}^N \left(\frac{d^3 k_i}{(2\pi)^3} \frac{1}{2E_{\mathbf{k}_i}} \right) |\mathbf{k}_1, \ldots, \mathbf{k}_N\rangle \langle \mathbf{k}_1, \ldots, \mathbf{k}_N|. \tag{2.14}$$

Here the N-particle state is defined as

$$|\mathbf{k}_1, \ldots, \mathbf{k}_N\rangle = \hat{a}^\dagger_{\mathbf{k}_1} \ldots \hat{a}^\dagger_{\mathbf{k}_N} |0\rangle. \tag{2.15}$$

We find

$$\begin{aligned}&\langle \mathbf{p}_1, \ldots, \mathbf{p}_N | \mathbf{k}_1, \ldots, \mathbf{k}_N \rangle \\ &= \langle \mathbf{p}_1 | \mathbf{k}_1 \rangle \langle \mathbf{p}_2 | \mathbf{k}_2 \rangle \ldots \langle \mathbf{p}_N | \mathbf{k}_N \rangle \\ &+ \text{permutations}.\end{aligned} \tag{2.16}$$

The permutations are needed because the **p**'s need only equal the **k**'s as a *set*. It is the overcounting created by selecting an order for the momenta that mandates the $N!$ in (2.14).

A word of caution: Strictly speaking each factor d^3k in the density of states should come with a factor of the volume, V, of the system. These are cancelled by the factors of $1/\sqrt{V}$ that should similarly be included in the states if we want to have a single-particle normalization rather than a beam. The reader should bear this in mind when interpreting some of the formulae.

2.2.2 An Inner Product

With our metric conventions, we can write the KG equation as

$$(-p^2 + m^2)\varphi \equiv (\partial_\mu \partial^\mu + m^2)\varphi \equiv (\partial^2 + m^2)\varphi = 0. \tag{2.17}$$

There are solutions of the form $\varphi_\mathbf{p} = e^{-ipx} = e^{i\mathbf{p}\cdot\mathbf{x} - iE_\mathbf{p}t}$, where $E_\mathbf{p} = \pm\sqrt{\mathbf{p}^2 + m^2}$. We say the solutions have *positive frequency* (associated with annihilation operators) if we have selected the positive root $E_\mathbf{p} > 0$.

Provided that φ satisfies the KG equation, we have

$$\partial^\mu(\varphi^* \partial_\mu \varphi - (\partial_\mu \varphi^*)\varphi) = 0. \tag{2.18}$$

We define the current

$$J_\mu = i(\varphi^* \partial_\mu \varphi - (\partial_\mu \varphi^*)\varphi), \tag{2.19}$$

so that $\partial_\mu J^\mu = 0$

With $J^\mu = (J^0, \mathbf{J})$ this reads

$$\partial_\mu J^\mu = \partial_t J^0 + \nabla \cdot \mathbf{J} = 0, \tag{2.20}$$

showing that J^0 is the density of some conserved "charge." In other words,

$$Q = \int_{t=\text{const.}} J^0 \, d^3x \tag{2.21}$$

is time independent and indeed Lorentz invariant so

$$Q = \int_\Sigma J^\mu \, d\Sigma_\mu \tag{2.22}$$

for any spacelike hyperplane.

2.2 The Klein-Gordon Equation

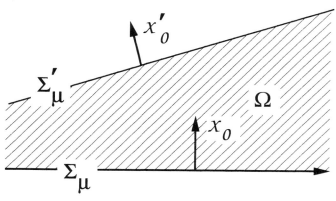

Fig 1. The difference between $\int_\Sigma J_\mu \, d\Sigma^\mu$ and $\int_{\Sigma'} J_\mu \, d\Sigma^\mu$ is $\int_\Omega \partial^\mu J_\mu = 0$.

This charge gives us a natural inner product on the space of solutions of the KG equation. Define

$$(\varphi_{\mathbf{p}}, \varphi_{\mathbf{p}'}) = i \int_{x_0=const.} \left(\varphi_{\mathbf{p}}^* \partial_0 \varphi_{\mathbf{p}'} - (\partial_0 \varphi_{\mathbf{p}}^*) \varphi_{\mathbf{p}'} \right) d^3x. \quad (2.23)$$

For two positive-energy solutions we find

$$(\varphi_{\mathbf{p}}, \varphi_{\mathbf{p}'}) = 2E_{\mathbf{p}}(2\pi)^3 \delta^3(\mathbf{p} - \mathbf{p}'). \quad (2.24)$$

The product of a positive with a negative frequency solution vanishes.

As with any complex inner product, (a, b) is conjugate symmetric, $(a, b) = (b, a)^*$, and sesquilinear,

$$\begin{aligned}(a, \lambda b) &= \lambda(a, b), \\ (\lambda a, b) &= \lambda^*(a, b),\end{aligned} \quad (2.25)$$

but it is *not* positive definite. This is because the inner product of a negative frequency solution with itself is a negative number. One consequence of this failure of positive definiteness is that whatever the charge Q represents it is *not* a probability density or the number of particles, so we cannot use the inner product to interpret the KG equation as a relativistic version of the Schrödinger equation.

2.2.3 Complex Scalar Fields

The inner product vanishes identically when the field φ is real. After quantizing, real fields become hermitian field operators and correspond to particles, such as the photon or π^0, which are their own antiparticles. Complex, nonhermitian, fields are needed for particles such as $\pi^\pm$, which are not their own antiparticles.

Suppose that $\varphi = \frac{1}{\sqrt{2}}(\varphi_1 + i\varphi_2)$, where $\varphi_{1,2}$ are real fields, then

$$S = \sum_{i=1}^{2} \int d^4x \left(\frac{1}{2} \varphi_i (\partial^2 + m^2) \varphi_i \right)$$

$$= \int d^4x \left(\varphi^*(\partial^2 + m^2)\varphi \right). \tag{2.26}$$

From the equal-time commutators

$$\delta(x_0 - x'_0)[\hat{\varphi}_i(x), \partial_0 \hat{\varphi}_j(x')] = i\delta_{ij}\delta^4(x - x'), \tag{2.27}$$

we find

$$\delta(x_0 - x'_0)[\hat{\varphi}(x), \partial_0 \hat{\varphi}^\dagger(x')] = i\delta^4(x - x'). \tag{2.28}$$

The normal-mode expansion is

$$\hat{\varphi}(x) = \int \frac{d^3k}{(2\pi)^3} \frac{1}{2E_\mathbf{k}} \left\{ \hat{a}_\mathbf{k} e^{-ikx} + \hat{b}^\dagger_\mathbf{k} e^{+ikx} \right\}, \tag{2.29}$$

where now we need an independent coefficient for the second term, since it is no longer related to the first by the hermiticity of $\hat{\varphi}$.

We use the inner product to pick off the coefficients $\hat{a}, \hat{b}$ as

$$\begin{aligned} \hat{a}_\mathbf{k} &= (e^{-ikx}, \hat{\varphi}), & \hat{b}_\mathbf{k} &= (e^{-ikx}, \hat{\varphi}^\dagger), \\ \hat{a}^\dagger_\mathbf{k} &= (\hat{\varphi}, e^{-ikx}), & \hat{b}^\dagger_\mathbf{k} &= (\hat{\varphi}^\dagger, e^{-ikx}). \end{aligned} \tag{2.30}$$

It is then straightforward to use (2.28) to compute

$$[\hat{a}_\mathbf{k}, \hat{a}^\dagger_{\mathbf{k}'}] = (e^{ikx}, e^{ik'x}) = 2E_\mathbf{k}(2\pi)^3 \delta^3(\mathbf{k} - \mathbf{k}'). \tag{2.31}$$

Similarly,

$$[\hat{b}_\mathbf{k}, \hat{b}^\dagger_{\mathbf{k}'}] = 2E_\mathbf{k}(2\pi)^3 \delta^3(\mathbf{k} - \mathbf{k}'). \tag{2.32}$$

We also find that the $\hat{a}$'s and $\hat{a}^\dagger$'s commute with the $\hat{b}$'s and $\hat{b}^\dagger$'s.

2.3 Symmetries and Noether's Theorem

2.3.1 Internal Symmetries

From both classical and quantum mechanics we know that symmetries imply conservation laws. We also know that the conserved quantities *generate* the symmetry operations that gave rise to them. Probably the most familiar example of this is angular momentum. Rotational symmetry leads to angular momentum conservation and the three components $\hat{J}_i$ of the angular-momentum vector can be exponentiated to give the unitary operator $U = \exp -i\theta \mathbf{n} \cdot \hat{\mathbf{J}}$, which rotates states through an angle θ about the axis $\mathbf{n}$.

In addition to rotations in physical space, we can have rotations in an *internal space*. Isospin is an example of this. The three pions, $\pi^\pm$, π^0, transform into each other under isospin rotations, forming an $I = 1$ representation of the isospin algebra $su(2)$. This algebra is mathematically isomorphic to the Lie algebra of the angular momentum $\hat{J}_i$'s, but its physical meaning is quite distinct. The associated Lie group $SU(2)$ is (almost) a symmetry of the strong interactions.

2.3 Symmetries and Noether's Theorem

Our complex scalar field φ has an action (2.26) whose integrand, the lagrangian density $\mathcal{L}$, is left unchanged by the substitution

$$\varphi(x) \to e^{-i\alpha}\varphi(x), \quad \varphi^*(x) \to e^{i\alpha}\varphi^*(x), \tag{2.33}$$

or, equivalently,

$$\begin{aligned}\varphi_1 &\to \cos\alpha\,\varphi_1 + \sin\alpha\,\varphi_2,\\ \varphi_2 &\to -\sin\alpha\,\varphi_1 + \cos\alpha\,\varphi_2.\end{aligned} \tag{2.34}$$

In these expressions the angle α is independent of position.

Emmy Noether,[4] one of the creators of modern abstract algebra, made her one foray into mathematical physics in 1915 by pointing out that the lagrangian formalism gives a simple way to extract the conserved quantities from this sort of symmetry. In the present case we simply promote α to a position-dependent parameter $\alpha(x)$ and use $\varphi(x) \to e^{-i\alpha(x)}\varphi(x)$, with small $\alpha(x)$ as a variation in the action principle. Substituting in (2.26) we find

$$\delta S = \int i(\varphi^*\partial_\mu\varphi - (\partial_\mu\varphi^*)\varphi)\partial^\mu\alpha(x)\,d^4x. \tag{2.35}$$

That there is no term here with α undifferentiated is due to the symmetry. If we take $\alpha(x)$ to vanish at infinity, then we can integrate by parts to find

$$\delta S = \int -\alpha(x)\partial^\mu(\varphi^*\partial_\mu\varphi - (\partial_\mu\varphi^*)\varphi)\,d^4x. \tag{2.36}$$

Now the equations of motion are equivalent to the statement that $\delta S = 0$ for *any* variation. *A fortiori* it must be zero for variations of the restricted type generated by $\alpha(x)$. Thus

$$0 = \partial^\mu(\varphi^*\partial_\mu\varphi - (\partial_\mu\varphi^*)\varphi). \tag{2.37}$$

We have known this since (2.12), but now we understand that the current conservation law is not the result of some mere coincidence.

The operator that generates the transformations is the total charge

$$\hat{Q} = \int_{t=\text{const.}} \hat{J}_0\,d^3x = i\int_{t=\text{const.}} d^3x(\hat{\varphi}^\dagger\partial_0\hat{\varphi} - (\partial_0\hat{\varphi}^\dagger)\hat{\varphi}). \tag{2.38}$$

From the commutation relation

$$[\hat{\varphi}(\mathbf{x},t), \partial_0\hat{\varphi}^\dagger(\mathbf{x}',t)] = i\delta^3(\mathbf{x}-\mathbf{x}'), \tag{2.39}$$

we find that

$$[\hat{Q}, \hat{\varphi}(x)] = -\hat{\varphi}(x), \quad [\hat{Q}, \hat{\varphi}^\dagger(x)] = \hat{\varphi}^\dagger(x). \tag{2.40}$$

---

[4] Amalie (Emmy) Noether. Born March 23, 1882, Erlangen, Germany. Died April 14, 1935, Bryn Mawr, PA.

20   2. Relativistic Scalar Fields

Exponentiating via the Campbell-Baker-Hausdorff formula gives

$$e^{i\alpha\hat{Q}}\hat{\varphi}(x)e^{-i\alpha\hat{Q}} = e^{-i\alpha}\hat{\varphi}(x). \tag{2.41}$$

In terms of the $\hat{a}$' and $\hat{b}$'s, we find that

$$\hat{Q} = \int \frac{d^3k}{(2\pi)^3} \frac{1}{2E_k} (\hat{a}_k^\dagger \hat{a}_k - \hat{b}_k^\dagger \hat{b}_k). \tag{2.42}$$

The eigenvalues of $\hat{Q}$ are therefore the difference between the number of $\hat{a}$ quanta and the number of $\hat{b}$ quanta in the eigenstate. I must confess here that if one takes care to preserve the ordering of the $\hat{b}$ and $\hat{b}^\dagger$ terms, what one actually finds as the integrand in (2.38) is $(\hat{a}_k^\dagger \hat{a}_k - \hat{b}_k \hat{b}_k^\dagger)$. In this form the commutator term $\propto \delta^3(0)$ would make the ground state $|0\rangle$ contain infinitely many $\hat{b}$ quanta. I have discarded this commutator term in order to treat the $\hat{a}$ and $\hat{b}$ particles on equal footing. As we will see in the next chapter I should therefore write the symbol $N(\hat{Q})$ instead of $\hat{Q}$ in (2.38).

From (2.40) we see that if $\hat{Q}|state\rangle = N|state\rangle$, then $\hat{Q}\hat{\varphi}|state\rangle = (N-1)\hat{\varphi}|state\rangle$. Thus $\hat{\varphi}(x)$ creates a negative Q charge at the space-time point x, or annihilates a positive Q charge at x. Similarly $\hat{\varphi}^\dagger(x)$ creates a positive Q charge at x, or annihilates a negative Q charge.

Diagramatically we represent this is as:

Fig 2. *Creation and annihilation. Note that the arrows run in the direction of the charge.*

A more general formulation of Noether's theorem applies when there is a lagrangian density $\mathcal{L}$ depending on a set of fields φ_i, and their first derivatives $\partial_\mu \varphi_i$, which is invariant under the infinitesimal transformations

$$\varphi_\alpha \to \varphi_\alpha + i\epsilon_i \lambda^i_{\alpha\beta} \varphi_\beta, \tag{2.43}$$

where ϵ_i is a set of small parameters and the hermitian matrices λ^i form a Lie algebra with commutation relations

$$[\lambda^i, \lambda^j] = if^{ij}{}_k \lambda^k. \tag{2.44}$$

Again we assume that $\mathcal{L}$ is invariant only when the parameters ϵ_i are independent of x. These are called *global* transformations (as opposed to *local* transformations) and the resulting symmetry is called a *global* symmetry.

2.3 Symmetries and Noether's Theorem

We have

$$\delta S = \int d^4x \left[\epsilon_i \left\{ \frac{\partial \mathcal{L}}{\partial \varphi_\alpha} \lambda^i_{\alpha\beta} \varphi_\beta + \frac{\partial \mathcal{L}}{\partial(\partial_\mu \varphi_\alpha)} \lambda^i_{\alpha\beta} \partial_\mu \varphi_\beta \right\} + \partial_\mu \epsilon_i \left\{ \frac{\partial \mathcal{L}}{\partial(\partial_\mu \varphi_\alpha)} \lambda^i_{\alpha\beta} \varphi_\beta \right\} \right]. \quad (2.45)$$

Now the group of terms proportional to ϵ_i in (2.36) vanishes by assumption. Integrating by parts and using the stationarity of the action shows that the equation of motion for the fields implies that

$$0 = \partial_\mu J^{(i)\mu} = \partial_\mu \left\{ \frac{\partial \mathcal{L}}{\partial(\partial_\mu \varphi_\alpha)} \lambda^i_{\alpha\beta} \varphi_\beta \right\}. \quad (2.46)$$

This suggests defining a current

$$J^{(i)\mu} = -i \frac{\partial \mathcal{L}}{\partial(\partial_\mu \varphi_\alpha)} \lambda^i_{\alpha\beta} \varphi_\beta. \quad (2.47)$$

The charge densities $J^{(i)0}$ can be written

$$J^{(i)0} = -i\pi_\alpha(x) \lambda^i_{\alpha\beta} \varphi_\beta(x), \quad (2.48)$$

and from $[\hat{\varphi}_\alpha(\mathbf{x}, t), \hat{\pi}_\beta(\mathbf{x}', t)] = i\delta^3(\mathbf{x} - \mathbf{x}')\delta_{\alpha\beta}$ we find that their quantum equivalents obey

$$[\hat{J}^{(i)0}(\mathbf{x}), \hat{J}^{(j)0}(\mathbf{x}')] = -i\hat{\pi}_\alpha(x)[\lambda^i, \lambda^j]_{\alpha\beta} \hat{\varphi}_\beta(x) \delta^3(\mathbf{x} - \mathbf{x}')$$
$$= if^{ij}_{\;\;k} \hat{J}^{(k)0} \delta^3(\mathbf{x} - \mathbf{x}'). \quad (2.49)$$

Equation (2.49) is an example of a *current algebra*. Integrating the charge densities over all space gives us a set of charges $\hat{Q}^{(i)}$, and we immediately see that the commutator of these reproduces the Lie algebra.

2.3.2 Space-Time Symmetries

Noether's theorem can also be used to derive the energy-momentum tensor. For a translationally invariant system, the lagrangian density $\mathcal{L}$ must be invariant under

$$\varphi(x^\mu) \to \varphi(x^\mu + \alpha^\mu) \approx \varphi(x^\mu) + \alpha^\mu \partial_\mu \varphi(x^\mu). \quad (2.50)$$

We derive the consequence of this by adopting the same strategy as before, and allowing α^μ to be position dependent. Then

$$0 = \delta S = \int d^4x \left\{ \frac{\partial \mathcal{L}}{\partial \varphi}(\alpha^\mu \partial_\mu \varphi) + \frac{\partial \mathcal{L}}{\partial(\partial_\nu \varphi)} \partial_\nu(\alpha^\mu \partial_\mu \varphi) \right\} \quad (2.51)$$

$$= \int d^4x \left[\left\{ \frac{\partial \mathcal{L}}{\partial \varphi} \partial_\mu \varphi + \frac{\partial \mathcal{L}}{\partial(\partial_\nu \varphi)} \partial^2_{\mu\nu} \varphi \right\} \alpha^\mu + \left\{ \frac{\partial \mathcal{L}}{\partial(\partial_\nu \varphi)} \partial_\mu \varphi \right\} \partial_\nu \alpha^\mu \right] \quad (2.52)$$

$$= \int d^4x\, \alpha^\mu \left[\partial_\mu \mathcal{L} - \partial_\nu \left\{ \frac{\partial \mathcal{L}}{\partial(\partial_\nu \varphi)} \partial_\mu \varphi \right\} \right]. \quad (2.53)$$

2. Relativistic Scalar Fields

If we define the *canonical energy-momentum tensor* by

$$T^\mu_{\ \nu} = \frac{\partial \mathcal{L}}{\partial(\partial_\mu \varphi)} \partial_\nu \varphi - \delta^\mu_\nu \mathcal{L}, \tag{2.54}$$

then $\partial_\mu T^\mu_{\ \nu} = 0$. We see that $T^0_{\ 0} = T^{00}$ is just the usual definition of the energy density. Other components have the interpretation:

- $T^i_{\ 0}$, $i = 1, 2, 3$ is the *energy flux*.
- $T^{0i} = -T^0_{\ i}$, $i = 1, 2, 3$, is the *momentum density*.

For a real scalar field we find that the doubly covariant tensor $T_{\mu\nu} = g_{\mu\sigma} T^\sigma_{\ \nu}$ is symmetric

$$T_{\mu\nu} = \partial_\mu \varphi \partial_\nu \varphi - g_{\mu\nu} \mathcal{L}. \tag{2.55}$$

When $T_{\mu\nu}$ (and hence $T^{\mu\nu}$) has this property, then the tensor

$$M^{\alpha\mu\nu} = x^\mu T^{\alpha\nu} - x^\nu T^{\alpha\mu} \tag{2.56}$$

obeys $\partial_\alpha M^{\alpha\mu\nu} = 0$. The charges $L^{\mu\nu} = \int d^3x M^{0\mu\nu}$ form a four-dimensional generalization of angular momentum and generate the Lorentz transformations. From the asymmetric treatment of μ and ν in the definition (2.54), however, it is not surprising that in many field theories the canonical energy-momentum tensor is *not* symmetric.

This lack of symmetry is not a disaster for Lorentz invariance. The energy-momentum tensor is not unique because we can always add terms such as $\partial_\alpha B^{\alpha\mu\nu}$, where B is antisymmeric in the first two indices, without spoiling $\partial_\mu T^{\mu\nu} = 0$. Such additional terms in the Lorentz-transformation generators are usually related to the intrinsic angular momentum of the particles, and serve to take care of the transformation properties of any tensor or spinor indices that the fields may be carrying.

A powerful method of generating a conserved, and automatically symmetric, energy-momentum tensor is to introduce an arbitrary background metric, $g_{\mu\nu}$, and define

$$T_{\mu\nu} = \frac{2}{\sqrt{g}} \frac{\delta S}{\delta g^{\mu\nu}}, \tag{2.57}$$

where $\sqrt{g} = \sqrt{\det g_{\mu\nu}}$ is the volume element.

For a real scalar field we would set

$$S = \int d^4x \sqrt{g}\, \frac{1}{2} \left(g^{\mu\nu} \partial_\mu \varphi \partial_\nu \varphi - m^2 \varphi^2 \right). \tag{2.58}$$

The only tricky bit in evaluating (2.57) is knowing the variation of $\sqrt{g}$. Employing the identity $\ln \det g = \operatorname{tr} \ln g$ we find that this is given by

$$\delta \sqrt{g} = -\frac{1}{2} \sqrt{g}\, g_{\mu\nu} \delta g^{\mu\nu}. \tag{2.59}$$

Using this to vary (2.58) we recover (2.55).

2.3 Symmetries and Noether's Theorem

I leave as an exercise for those who are familar with covariant derivatives the application of Noether's theorem to prove that $\nabla^\mu T_{\mu\nu} = 0$ in an arbitrary spacetime.

The quantity

$$P^i = \int d^3x \, T^{0i} = -P_i, \quad i = 1, 2, 3 \tag{2.60}$$

is the momentum density. For the complex scalar field we find that

$$\hat{P}^i = \int d^3x \left(\partial^0 \hat{\varphi}^\dagger \partial^i \hat{\varphi} + \partial^0 \hat{\varphi} \partial^i \hat{\varphi}^\dagger \right)$$

$$= \int \frac{d^3k}{(2\pi)^3} \frac{1}{2E_\mathbf{k}} k^i (\hat{a}_\mathbf{k}^\dagger \hat{a}_\mathbf{k} + \hat{b}_\mathbf{k}^\dagger \hat{b}_\mathbf{k}) \tag{2.61}$$

and

$$[\hat{P}^i, \hat{\varphi}] = [\int d^3x \left(\partial^0 \hat{\varphi}^\dagger \partial^i \hat{\varphi} + \partial^0 \hat{\varphi} \partial^i \hat{\varphi}^\dagger \right), \hat{\varphi}]$$

$$= -i \partial^i \hat{\varphi} = +i \partial_i \hat{\varphi}. \tag{2.62}$$

If we define

$$U(\mathbf{a}) = e^{-ia^i \hat{P}^i} = e^{-i\mathbf{a} \cdot \hat{\mathbf{P}}}, \tag{2.63}$$

we have

$$U(\mathbf{a}) \hat{\varphi}(\mathbf{x}) U^\dagger(\mathbf{a}) = \hat{\varphi}(\mathbf{x}) - ia^i[\hat{P}^i, \varphi(\mathbf{x})] + \ldots$$
$$= \hat{\varphi}(\mathbf{x}) - ia^i(i\partial_i \hat{\varphi}(\mathbf{x})) + \ldots$$
$$= \hat{\varphi}(\mathbf{x}) + a^i \partial_i \hat{\varphi}(\mathbf{x}) + \ldots$$
$$= \hat{\varphi}(\mathbf{x} + \mathbf{a}). \tag{2.64}$$

[The reader should verify from the Campbell-Baker-Hausdorff theorem that the terms represented by $+\ldots$ are indeed the rest of the Taylor series exansion for $\hat{\varphi}(\mathbf{x} + \mathbf{a})$.]

Let us confirm that we have the signs correct by looking at the expression

$$\langle \mathbf{k} | \hat{\varphi}(\mathbf{x}) | 0 \rangle = e^{-i \mathbf{k} \cdot \mathbf{x}} \tag{2.65}$$

from this point of view. We write

$$\langle \mathbf{k} | \hat{\varphi}(\mathbf{x}) | 0 \rangle = \langle \mathbf{k} | e^{-i\mathbf{x} \cdot \hat{\mathbf{P}}} \hat{\varphi}(0) e^{+i\mathbf{x} \cdot \hat{\mathbf{P}}} | 0 \rangle = e^{-i \mathbf{k} \cdot \mathbf{x}}, \tag{2.66}$$

where at the last step we have used the fact that $\langle \mathbf{k} | e^{-i\mathbf{x} \cdot \hat{\mathbf{P}}} = \langle \mathbf{k} | e^{-i\mathbf{k} \cdot \mathbf{x}}$ and $\hat{\mathbf{P}} | 0 \rangle = 0$. We see that the manipulations with $\hat{\mathbf{P}}$ do lead to the expression that we already knew.

After all this formalism I should point out some real physics. In relativistic systems with no "Æther," our field momentum $\mathbf{P}$ is what is usually called momentum.

It is a scalar-field analog of the quantity

$$\mathbf{P} = \frac{1}{c^2} \int d^3x \, \mathbf{E} \times \mathbf{H} \tag{2.67}$$

familiar from classical electrodynamics. When our fields are describing phonons in a medium such as the crystal in Chapter 1, or in a fluid, then caution is necessary. There are two distinct symmetries that we might call "translations", and therefore two distinct notions of "momentum." First we have the translations we have been discussing in this section where we translate the location of the wave with respect to the uniform background medium. The second one is where we translate the medium and the wave together. Only the latter gives rise to "momentum" in the Newtonian sense. In fluid mechanics the former symmetry corresponds to what is called *pseudomomentum*. It is interesting to note that phonons in a crystal carry only pseudomomentum, while phonons in a fluid carry both types of momentum. Since in a fluid the momentum density $\rho \mathbf{v}$ coincides with the mass current, this means that there is a phonon "wind." We will discuss this more in later.

3
Perturbation Theory

3.1 Interactions

The simplest interaction is with a c-number external source. If we add to the lagrangian density for a real scalar field a term $J(x)\varphi(x)$, the equation of motion becomes

$$(\partial^2 + m^2)\varphi(x) = J(x). \tag{3.1}$$

In terms of the model crystal of Chapter 1, J is an external force acting on the atoms. In the particle interpretation J acts as a source of $\hat{\varphi}$ quanta. This is not unreasonable: hitting the crystal produces phonons.

If we want interactions between the particles represented by the theory, then we need nonlinear field equations, and therefore terms of higher order than quadratic in the lagrangian. For the scalar fields we have considered the simplest interactions are simply polynomials in the fields. For example, the real $(\varphi^4)_4$ model has interaction terms

$$\mathcal{L}_{int} = \frac{\lambda}{4!}\varphi^4. \tag{3.2}$$

A variant of this has N real fields with

$$\mathcal{L}_{int} = \frac{\lambda}{2}\left(\sum_i \varphi_i^2\right)^2. \tag{3.3}$$

This has $O(N)$ as a symmetry group, and there are conserved quantities associated with the charges that generate the group.

3. Perturbation Theory

In particle physics the interactions are largely determined by symmetries and their associated conservation laws. One writes down the most general interaction with the required symmetries. Usually the interaction terms with the least number of derivatives are the most important. Too many derivatives render the interaction *nonrenormalizable*, or, in equivalent modern parlance, *irrelevant*. Such nonrenormlizable terms should not affect physics at energies far below any high-energy cut-off. In this chapter we will only deal with the simplest interactions.

3.2 Perturbation Theory

3.2.1 Interaction Picture

Although the general theory of fields is best investigated using the Heisenberg picture, the fastest route to perturbation theory is via the *interaction picture*. This was first used for this purpose by Dyson.[1] In the interaction picture we use the Heisenberg picture for the free-field time evolution, but allow the interactions to make transitions between the free-field Heisenberg states.

The Schrödinger-picture time evolution gives

$$i\partial_t |\psi(t)\rangle_S = \hat{H}(t)|\psi(t)\rangle_S, \qquad (3.4)$$

so if $\hat{H}$ is time independent,

$$|\psi(t)\rangle_S = \exp(-i\hat{H}t)|\psi(0)\rangle_S. \qquad (3.5)$$

If $\hat{H} = \hat{H}_0 + \hat{V}$, we define the interaction picture by

$$|\psi(t)\rangle_I = e^{i\hat{H}_0 t}|\psi(t)\rangle_S. \qquad (3.6)$$

In other words we peel off the free-theory Schrödinger time evolution so as to be able to concentrate on the effect of the interactions. Then

$$\begin{aligned} i\partial_t |\psi(t)\rangle_I &= -\hat{H}_0|\psi(t)\rangle_I + e^{i\hat{H}_0 t}\left(i\partial_t|\psi(t)\rangle_S\right) \\ &= -\hat{H}_0|\psi(t)\rangle_I + e^{i\hat{H}_0 t}(\hat{H}_0 + \hat{V})e^{-i\hat{H}_0 t}|\psi(t)\rangle_I \\ &= \hat{V}_I(t)|\psi(t)\rangle_I, \end{aligned} \qquad (3.7)$$

where

$$\hat{V}_I(t) = e^{i\hat{H}_0 t}\hat{V}(t)e^{-i\hat{H}_0 t}. \qquad (3.8)$$

Note that for a time-independent $\hat{V}$, the time dependence of $\hat{V}_I$ is that of a free Heisenberg field.

Recall, from time-dependent perturbation theory in quantum mechanics, that the solution to the matrix equation

$$i\partial_t |\psi(t)\rangle = \hat{H}(t)|\psi(t)\rangle \qquad (3.9)$$

---

[1] Freeman John Dyson. Born December 15, 1923, Crowthorne, Berkshire, UK.

is

$$|\psi(t)\rangle = T \exp\left\{-i \int_0^t \hat{H}(t') dt'\right\} |\psi(0)\rangle, \quad (3.10)$$

where the time-ordered exponential is defined by

$$T \exp\left\{-i \int_0^t \hat{H}(t') dt'\right\}$$
$$= 1 - i \int_0^t \hat{H}(t') dt' - \frac{1}{2!} \int_0^t \int_0^t T\left\{\hat{H}(t')\hat{H}(t'')\right\} dt' dt'' + \cdots. \quad (3.11)$$

The time-ordering symbol T means $T\left\{\hat{H}(t')\hat{H}(t'')\right\} = \hat{H}(t')\hat{H}(t'')$ if $t' > t''$ and $\hat{H}(t'')\hat{H}(t')$ if $t'' > t'$, and so on.

Note the difference between

$$e^{\int_a^b A(t) dt} \approx e^{A(t_n = t_b)\delta t + A(t_{n-1})\delta t + \ldots + A(t_2)\delta t + A(t_1 = t_a)\delta t} \quad (3.12)$$

and

$$T e^{\int_a^b A(t) dt} \approx e^{A(t_n = t_b)\delta t} e^{A(t_{n-1})\delta t} \ldots e^{A(t_2)\delta t} e^{A(t_1 = t_a)\delta t}, \quad (3.13)$$

where $t_n > t_{n-1} \ldots$. Only if the A's at different times commute are the two expressions the same.

We will also write

$$|\psi(t_2)\rangle = U(t_2, t_1)|\psi(t_1)\rangle, \quad (3.14)$$

where the unitary evolution operator is

$$U(t_2, t_1) = T \exp\left\{-i \int_{t_1}^{t_2} \hat{H}(t) dt\right\}. \quad (3.15)$$

3.2.2 Propagators and Time-Ordered Products

The key ingredients in the Dyson approach to perturbation theory are matrix elements of time-ordered products $\langle m|T\{\hat{H}(t_1)\hat{H}(t_2)\ldots\hat{H}(t_n)\}|n\rangle$, where the time dependence of the fields composing the $\hat{H}(t_i)$ is that of free fields.

We will build these complicated time-ordered products up from vacuum expectation values of time-ordered products of pairs of fields such as $\langle 0|T\{\hat{\varphi}(x)\hat{\varphi}(x')\}|0\rangle$ or $\langle 0|T\{\hat{\varphi}(x)\hat{\varphi}^\dagger(x')\}|0\rangle$ together with *normal products* that are combinations of fields with easily calculable matrix elements. These vacuum expectation values of time-ordered products of pairs of fields are called Feynman *propagators*, and are the principal feature of the diagrammatic approach to computation.

Diagrammatically we represent the propagators by

3. Perturbation Theory

Fig 1. The propagators $\langle 0|T\,\hat{\varphi}(x)\hat{\varphi}(y)|0\rangle$ and $\langle 0|T\,\hat{\varphi}^\dagger(x)\hat{\varphi}(y)|0\rangle$. The arrow on the latter runs from the $\hat{\varphi}^\dagger$ field to the $\hat{\varphi}$ field.

We begin by discussing

$$iG(x,x') = \langle 0|T\{\hat{\varphi}(x)\hat{\varphi}(x')\}|0\rangle$$
$$= \theta(x'_0 - x_0)\langle 0|\hat{\varphi}(x')\hat{\varphi}(x)|0\rangle + \theta(x_0 - x'_0)\langle 0|\hat{\varphi}(x)\hat{\varphi}(x')|0\rangle. \tag{3.16}$$

We can insert complete sets of intermediate states to get, for example,

$$\langle 0|\hat{\varphi}(x')\hat{\varphi}(x)|0\rangle = \int \frac{d^3k}{(2\pi)^3}\frac{1}{2k_0}\langle 0|\hat{\varphi}(x')|\mathbf{k}\rangle\langle\mathbf{k}|\hat{\varphi}(x)|0\rangle, \tag{3.17}$$

where

$$\langle 0|\hat{\varphi}(x')|\mathbf{k}\rangle = e^{-ikx'}, \quad \langle\mathbf{k}|\hat{\varphi}(x)|0\rangle = e^{ikx}. \tag{3.18}$$

Thus

$$iG(x,x') = \int \frac{d^3k}{(2\pi)^3}\frac{1}{2k_0}\left\{e^{-ik(x'-x)}\theta(x'_0-x_0) + \theta(x_0-x'_0)e^{+ik(x'-x)}\right\}. \tag{3.19}$$

Now the Heaviside step function, $\theta(x'_0 - x_0)$, may be written as

$$\theta(x'_0 - x_0) = i\int \frac{d\omega}{2\pi}\frac{e^{-i\omega(x'_0-x_0)}}{\omega + i\epsilon}. \tag{3.20}$$

Using this in addition to the partial fraction decomposition

$$\frac{i}{2k_0}\left(\frac{1}{\omega - k_0 + i\epsilon} - \frac{1}{\omega + k_0 - i\epsilon}\right) = \frac{i}{\omega^2 - (\mathbf{k}^2 + m^2) + i\epsilon}, \tag{3.21}$$

where k_0^2 has been expanded out as $(\mathbf{k}^2 + m^2)$, we can assemble the two terms in (3.19) to get

$$iG(x,x') = \int \frac{d\omega}{2\pi}\frac{d^3k}{(2\pi)^3}\frac{i}{\omega^2 - (\mathbf{k}^2+m^2) + i\epsilon}e^{-ik(x'-x)}. \tag{3.22}$$

We now use the lorentzian inner product to rewrite this as the simple-seeming expression

$$iG(x,x') = \int \frac{d^4k}{(2\pi)^4}\frac{i}{k^2 - m^2 + i\epsilon}e^{-ik(x'-x)}. \tag{3.23}$$

Here the new k_0 is an independent variable (the old ω) and is no longer a function of $\mathbf{k}$. We say that that the 4-momentum in (3.23) is *off-shell*, that is, it does not satisfy the mass-shell condition $k^2 = m^2$.

3.2 Perturbation Theory

Note that the integrand has poles at $\omega = \pm\left(\sqrt{\mathbf{k}^2 + m^2} - i\epsilon\right)$. The $i\epsilon$ term can be omitted if we understand that we are to route the ω integral round the poles on the real ω axis as shown in the Fig.2.

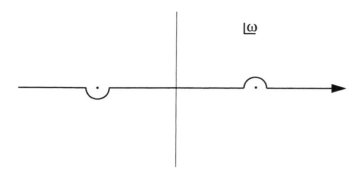

Fig 2. The Feynman contour.

This integral defining $G(x, x')$ can be evaluated in closed form as a modified Bessel function, but we have no immediate use for the expression so we will defer discussing it until later.

Let us confirm that we have our signs correct. Starting from the definition of time ordering we have

$$(\partial_{x'}^2 + m^2)\langle 0|T\hat{\varphi}(x')\hat{\varphi}(x)|0\rangle$$
$$= \langle 0|T(\partial_{x'}^2 + m^2)\hat{\varphi}(x')\hat{\varphi}(x)|0\rangle + \delta(x_0' - x_0)\langle 0|[\dot{\hat{\varphi}}(x'), \hat{\varphi}(x)]|0\rangle$$
$$= \delta(x_0' - x_0)\left(-i\delta^3(\mathbf{x}' - \mathbf{x})\right) = -i\delta^4(x' - x). \qquad (3.24)$$

Here we have used the fact that $(\partial_{x'}^2 + m^2)\hat{\varphi}(x') = 0$ and that $\partial_{x_0}\theta(x_0) = \delta(x_0)$.

Applying the Klein-Gordon operator to our integral expression we have

$$(\partial_{x'}^2 + m^2)iG(x', x) = -i\int \frac{d^4k}{(2\pi)^4}\frac{k^2 - m^2}{k^2 - m^2 + i\epsilon}e^{-ik(x'-x)} = -i\delta^4(x' - x), \qquad (3.25)$$

which is happily the same.

We see that the Feynman propagator is a Green function[2] for the KG equation. The choice of contour for the ω integration selects a particular set of initial conditions for this Green function so that it is neither a retarded nor an advanced Green function, but a peculiar sum of both. It is a remarkable part of the formalism that time-ordered products, which are not the first things one would have thought of calculating, play such an important role — and are indeed the only quantities for which there exists a simple perturbation theory.

---

[2]George Green. Born July 14, 1793, Sneinton, Nottinghamshire, England. Died March 31, 1841, Sneinton.

3.3 Wick's Theorem

3.3.1 Normal Products

We next define the *normal products*. The present definition only makes sense for free fields, but that is all we will need.

Decompose

$$\hat{\varphi}(x) = \hat{\varphi}^{(+)}(x) + \hat{\varphi}^{(-)}(x), \qquad (3.26)$$

where $\hat{\varphi}^{(+)}$ is the positive frequency part of the field, i.e. the part with the $e^{-i\omega t}$, $\omega > 0$. This is the part containing the annihilator $\hat{a}$'s and $\hat{b}$'s. $\hat{\varphi}^{(-)}$ is the negative frequency part, which contains the creation operators.

We now define the normal-product of a monomial in the $\hat{\varphi}^{(\pm)}$ to be the expression obtained by placing the annihilation operators to the right and the creation operators to the left. For example,

$$N\left(\hat{\varphi}^{(+)}(x_1)\hat{\varphi}^{(-)}(x_2)\hat{\varphi}^{(-)}(x_3)\hat{\varphi}^{(+)}(x_4)\right)$$
$$= \hat{\varphi}^{(-)}(x_2)\hat{\varphi}^{(-)}(x_3)\hat{\varphi}^{(+)}(x_1)\hat{\varphi}^{(+)}(x_4). \qquad (3.27)$$

The ordering of the $\hat{\varphi}^{(-)}(x)$ factors among themselves is unimportant since all $\hat{\varphi}^{(-)}(x)$'s commute with one another, similarly for the $\hat{\varphi}^{(+)}(x)$'s. We next extend the definition to sums of products by linearity and the distributive law.

The basic, and obvious, property of normal products is that $\langle 0|N(\hat{O})|0\rangle$ is zero — unless $\hat{O}$ is the identity operator, when $\langle 0|N(1)|0\rangle = 1$.

Clearly

$$T\left(\hat{\varphi}(x_1)\hat{\varphi}(x_2)\right) = N\left(\hat{\varphi}(x_1)\hat{\varphi}(x_2)\right) + \text{c-number}. \qquad (3.28)$$

We find the c-number by taking the vacuum expectation value

$$\langle 0|T\left(\hat{\varphi}(x_1)\hat{\varphi}(x_2)\right)|0\rangle = 0 + \text{c-number}, \qquad (3.29)$$

so

$$T\left(\hat{\varphi}(x_1)\hat{\varphi}(x_2)\right) = N\left(\hat{\varphi}(x_1)\hat{\varphi}(x_2)\right) + \langle 0|T\left(\hat{\varphi}(x_1)\hat{\varphi}(x_2)\right)|0\rangle \qquad (3.30)$$

3.3.2 Wick's Theorem

Wick's theorem is a generalization of (3.30) to an arbitrary number of factors. The theorem states that

$$T\left(\hat{\varphi}(x_1)\hat{\varphi}(x_2)\ldots\hat{\varphi}(x_N)\right) =$$
$$N\left(\hat{\varphi}(x_1)\hat{\varphi}(x_2)\ldots\hat{\varphi}(x_N)\right) +$$
$$+ \langle 0|T\left(\hat{\varphi}(x_1)\hat{\varphi}(x_2)\right)|0\rangle N\left(\hat{\varphi}(x_3)\hat{\varphi}(x_4)\ldots\hat{\varphi}(x_N)\right) + \text{perms.}$$
$$+ \langle 0|T\left(\hat{\varphi}(x_1)\hat{\varphi}(x_2)\right)|0\rangle \langle 0|T\left(\hat{\varphi}(x_3)\hat{\varphi}(x_4)\right)|0\rangle N\left(\hat{\varphi}(x_5)\hat{\varphi}(x_6)\ldots\hat{\varphi}(x_N)\right) + \text{perms.}$$
$$+ \cdots . \qquad (3.31)$$

The general pattern should be clear: we *contract* pairs of fields into propagators in all possible ways and leave the rest in the normal product. One can prove this result by induction, but this is tedious and unenlightening. It is more instructive to prove a generating function form of the theorem.

Theorem:

Let $J(x)$ be a c-number function, then

$$T\left\{e^{-i\int d^4x J(x)\hat{\varphi}(x)}\right\} =$$
$$N\left\{e^{-i\int d^4x J(x)\hat{\varphi}(x)}\right\} e^{-\frac{1}{2}\int d^4x\, d^4y J(x)\langle 0|T(\hat{\varphi}(x)\hat{\varphi}(y))|0\rangle J(y)}. \quad (3.32)$$

COMMENT: By replacing $J(x) = iK(x)$, expanding out, and comparing coefficients we recover the previous form of the theorem. The motivation for inserting the i's is that the left-hand side is a unitary operator when $J(x)$ is real.

PROOF: Recall that $e^A e^B = e^{A+B+\frac{1}{2}[A,B]}$ provided A, B commute with $[A, B]$. This condition holds for A, B sums of $\hat{\varphi}$'s since the commutators of free fields are c-numbers. Therefore, if $t_n > t_{n-1} > \ldots > t_1$, and we write $\hat{H}$ for $J(x)\hat{\varphi}(x)$, we can break up the time-ordered product in an approximate way as

$$T\left\{e^{-i\int \hat{H}(t)\, dt}\right\} \approx e^{-i\Delta t \hat{H}(t_n)} e^{-i\Delta t \hat{H}(t_{n-1})} \ldots e^{-i\Delta t \hat{H}(t_1)}$$
$$= e^{-i\Delta t \sum_1^n \hat{H}(t_i) - \frac{1}{2}(\Delta t)^2 \sum_{k>l}[\hat{H}(t_k),\hat{H}(t_l)]}. \quad (3.33)$$

Taking the limit of many t_i's so the sums become integrals we get

$$T\left\{e^{-i\int d^4x J(x)\hat{\varphi}(x)}\right\}$$
$$= e^{-i\int d^4x J(x)\hat{\varphi}(x)} e^{-\frac{1}{2}\int d^4x\, d^4y J(x)J(y)\theta(x_0-y_0)[\hat{\varphi}(x),\hat{\varphi}(y)]}. \quad (3.34)$$

This is a nice result in itself — but not yet quite what we need.

We now note that

$$N\left\{e^{-i\int d^4x J(x)\hat{\varphi}(x)}\right\}$$
$$\equiv e^{-i\int d^4x J(x)\hat{\varphi}^{(-)}(x)} e^{-i\int d^4x J(x)\hat{\varphi}^{(+)}(x)}$$
$$= e^{-i\int d^4x J(x)\hat{\varphi}(x)} e^{-\frac{1}{2}\int d^4x\, d^4y J(x)J(y)[\hat{\varphi}^{(-)}(x),\hat{\varphi}^{(+)}(y)]}, \quad (3.35)$$

so

$$T\left\{e^{-i\int d^4x J(x)\hat{\varphi}(x)}\right\} = N\left\{e^{-i\int d^4x J(x)\hat{\varphi}(x)}\right\}$$
$$\times e^{\frac{1}{2}\int d^4x\, d^4y J(x)J(y)([\hat{\varphi}^{(-)}(x),\hat{\varphi}^{(+)}(y)] - \theta(x_0-y_0)[\hat{\varphi}(x),\hat{\varphi}(y)])}. \quad (3.36)$$

The expression $[\hat{\varphi}^{(-)}(x), \hat{\varphi}^{(+)}(y)] - \theta(x_0 - y_0)[\hat{\varphi}(x), \hat{\varphi}(y)]$ looks a trifle complicated, but it *is* only a c-number, so we can evaluate it by taking its vacuum expectation value. We find

$$[\hat{\varphi}^{(-)}(x), \hat{\varphi}^{(+)}(y)] - \theta(x_0 - y_0)[\hat{\varphi}(x), \hat{\varphi}(y)]$$
$$= \langle 0|[\hat{\varphi}^{(-)}(x), \hat{\varphi}^{(+)}(y)]|0\rangle - \theta(x_0 - y_0)\langle 0|[\hat{\varphi}(x), \hat{\varphi}(y)]|0\rangle$$
$$= \langle 0| - \hat{\varphi}(y)\hat{\varphi}(x) - \theta(x_0 - y_0)[\hat{\varphi}(x), \hat{\varphi}(y)]|0\rangle$$
$$= -\langle 0|T(\hat{\varphi}(x)\hat{\varphi}(y))|0\rangle. \quad (3.37)$$

Q.E.D

3.3.3 Applications

Yukawa Potential

Let us consider the specific example

$$J(x) = g_1 \delta^3(\mathbf{x} - \mathbf{x}_1) + g_2 \delta^3(\mathbf{x} - \mathbf{x}_2). \tag{3.38}$$

Taking this for our interaction hamiltonian corresponds to two immobile sources for the $\hat{\varphi}$ field. These might be two heavy nucleons that can emit and absorb $\hat{\varphi}$ field quanta.

Let us find the force between them arising from the exchange of virtual $\hat{\varphi}$ quanta. To do this we use the *vacuum persistence amplitude*

$$\langle 0|S|0 \rangle = \langle 0|U(+\infty, -\infty)|0 \rangle. \tag{3.39}$$

We imagine that we make g_1 and g_2 time dependent. We slowly switch them on so that the initial vacuum evolves adiabatically into the true vacuum ground state, $|0_J\rangle$, of $\hat{H}$ with the sources present. Then, after a very long time, we slowly switch them off so the state returns to the original $|0\rangle$. Remember that $U(t_2, t_1) = e^{i\hat{H}_0 t_2} e^{-i\hat{H}(t_2 - t_1)} e^{-i\hat{H}_0 t_1}$ for a time-independent hamiltonian. When $\hat{H}$ is time dependent, this is replaced by $U(t_2, t_1) = e^{iH_0 t_2} T\{e^{-i \int_{t_1}^{t_2} \hat{H}(t) dt}\} e^{-i\hat{H}_0 t_1}$ where the central factor is the Heisenberg time-evolution operator. The vacuum persistence amplitude will therefore equal

$$\langle 0|U(+\infty, -\infty)|0 \rangle = \langle 0|T\{e^{-i \int_{t_1}^{t_2} \hat{H}(t) dt}\}|0\rangle e^{iE_0(t_2 - t_1)}. \tag{3.40}$$

Here we have used $\langle 0|\hat{H}_0 = \langle 0|E_0$ and $\hat{H}_0|0\rangle = E_0|0\rangle$. Now the adiabatically slow variation in $\hat{H}(t)$ evolves $|0\rangle$ to the current eigenstate of $H(t)$, therefore, to an arbitrarily good approximation,

$$\langle 0|T\{e^{-i \int_{t_1}^{t_2} \hat{H}(t) dt}\}|0\rangle = e^{-i \int_{t_1}^{t_2} E(g(t)) dt}. \tag{3.41}$$

For most of the time $E(g(t))$ will equal E_J, the energy with the sources present. Therefore

$$\langle 0|S|0 \rangle = \langle 0|U(+\infty, -\infty)|0 \rangle = e^{-i(E_J - E_0)T}, \tag{3.42}$$

where T is the total time that the sources were present.

Our theorem tells us that $e^{-i(E_J - E_0)T}$ is equal to

$$\exp\left\{-\frac{1}{2} \int d^4x \, d^4y \, J(x) \langle 0|T(\hat{\varphi}(x)\hat{\varphi}(y))|0\rangle J(y)\right\} \tag{3.43}$$

$$= A \exp -g_1 g_2 \int dt_1 \, dt_2 \int \frac{d^4k}{(2\pi)^4} \frac{i}{k^2 - m^2 + i\epsilon} e^{-ik_0(t_1 - t_2)} e^{i\mathbf{k}\cdot(\mathbf{x}_1 - \mathbf{x}_2)}. \tag{3.44}$$

The coefficient A contains the divergent *self-energy* that comes from terms with both arguments in $G(x, y)$ on the same source world-line. These correspond to virtual $\hat{\varphi}$ particles that are emitted and absorbed by the same source. We are not interested in these, but only in the variation in the vacuum energy as a function of the separation of the sources.

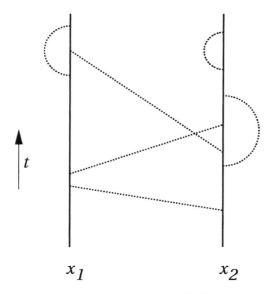

Fig 3. One of the diagrams being summed to get the Yukawa interaction.

The integrals over t_1 and t_2 provide a factor of $2\pi\delta(k_0)T$. Therefore,

$$e^{-i\Delta E T} = A \exp -ig_1 g_2 T \int \frac{d^3 k}{(2\pi)^3} \frac{e^{i\mathbf{k}\cdot(\mathbf{x}_1 - \mathbf{x}_2)}}{-\mathbf{k}^2 - m^2}. \tag{3.45}$$

Now

$$\int \frac{d^3 k}{(2\pi)^3} \frac{e^{i\mathbf{k}\cdot(\mathbf{x}_1 - \mathbf{x}_2)}}{\mathbf{k}^2 + m^2} = \frac{1}{4\pi} \frac{e^{-m|\mathbf{x}_1 - \mathbf{x}_2|}}{|\mathbf{x}_1 - \mathbf{x}_2|}, \tag{3.46}$$

which is the *Yukawa potential*.[3] It is also the Green function for the three-dimensional equation

$$(-\partial^2 + m^2) G(\mathbf{x}, \mathbf{y}) = \delta(\mathbf{x} - \mathbf{y}). \tag{3.47}$$

We see that the energy of interaction is

$$\Delta E = -g_1 g_2 \frac{1}{4\pi} \frac{e^{-m|\mathbf{x}_1 - \mathbf{x}_2|}}{|\mathbf{x}_1 - \mathbf{x}_2|}, \tag{3.48}$$

which corresponds to an *attractive* interaction. It is a fact that the exchange of scalar (or any even integer spin) particles always leads to attractive forces.

---

[3] Hideki Yukawa, Born January 23, 1907, Tokyo, Japan. Died September 8, 1981, Kyoto. Nobel Prize for Physics 1949.

Mössbauer Effect

The operator part of $S = U(\infty, -\infty)$ is embedded in the normal products. S can be written

$$S(J) = \text{c-number } e^{-i\int d^4x\, J(x)\hat{\varphi}^{(-)}} e^{-i\int d^4x\, J(x)\hat{\varphi}^{(+)}}. \tag{3.49}$$

Define

$$\begin{aligned}\hat{a}(J) &= \int d^4x\, J(x)\hat{\varphi}^{(+)} \\ &= \int d^4x \int \frac{d^3k}{(2\pi)^3}\frac{1}{2E_\mathbf{k}} J(x)\hat{a}_\mathbf{k} e^{-iE_\mathbf{k}t+i\mathbf{k}\cdot\mathbf{x}} \\ &= \int \frac{d^3k}{(2\pi)^3}\frac{1}{2E_\mathbf{k}} \hat{a}_\mathbf{k} \tilde{J}(k).\end{aligned} \tag{3.50}$$

Here $\tilde{J}(k)$ is the Fourier transform of $J(x)$ always evaluated with the frequency on-shell i.e, $k_0 = E_\mathbf{k}$. Since J is real, $\tilde{J}(-k)^* = \tilde{J}(k)$.

We find that

$$\begin{aligned}[\hat{a}(J), \hat{a}^\dagger(J)] &= \int \frac{d^3k}{(2\pi)^3}\frac{1}{2E_\mathbf{k}} |\tilde{J}(k)|^2 \\ &= \int \frac{d^4k}{(2\pi)^4} 2\pi\, |\tilde{J}(k)|^2\, \delta(k^2 - m^2)\theta(k_0).\end{aligned} \tag{3.51}$$

We can think of $\hat{a}(J), \hat{a}^\dagger(J)$ as oddly normalized annihilation and creation operators. In particular, if we set μ equal to the positive quantity $\langle 0|[\hat{a}(J), \hat{a}^\dagger(J)]|0\rangle$, the states

$$|N\rangle = \mu^{-\frac{N}{2}} \frac{(\hat{a}^\dagger(J))^N}{\sqrt{N!}} |0\rangle \tag{3.52}$$

are normalized,

$$\langle N|N\rangle = 1. \tag{3.53}$$

The quantity μ also appears as the real part of

$$K = \int d^4x\, d^4y\, J(x)J(y)\langle 0|T\left(\hat{\varphi}(x)\hat{\varphi}(y)\right)|0\rangle. \tag{3.54}$$

If we write this as

$$\int d^4x\, d^4y\, J(x)J(y) \int \frac{d^4k}{(2\pi)^4} \frac{i}{k^2 - m^2 + i\epsilon} e^{-ik(x-y)}, \tag{3.55}$$

and use

$$\frac{1}{x + i\epsilon} = \frac{P}{x} - i\pi\delta(x), \tag{3.56}$$

we find that

$$K = \int \frac{d^4k}{(2\pi)^4} \pi\, |\tilde{J}(k)|^2\, \delta(k^2 - m^2) + \text{imaginary}$$

$$= \int \frac{d^4k}{(2\pi)^4} 2\pi \, |J(k)|^2 \, \delta(k^2 - m^2)\theta(k_0) + \text{imaginary}. \tag{3.57}$$

In the last step we have used $\tilde{J}(-k)^* = \tilde{J}(k)$ to observe that the integrand is even in k_0.

With these results in mind, we can write

$$S(J) = e^{-i\hat{a}^\dagger(J)} e^{-i\hat{a}(J)} e^{-\frac{1}{2}\mu} \times \text{(phase)}. \tag{3.58}$$

We see that

$$|\langle N|S|0\rangle|^2 = \frac{\mu^N}{N!} e^{-\mu}, \tag{3.59}$$

therefore

$$\sum_N |\langle N|S|0\rangle|^2 = e^{-\mu} \sum_N \frac{\mu^N}{N!} = 1. \tag{3.60}$$

Now (3.59) is the probability that the action of J on the system produces exactly N quanta. We see that this probablity sums to unity as it should, and that it has a Poisson distribution. In particular, provided μ is finite, we see that there is a probablity $e^{-\mu}$ that we disturb the system but produce no particles whatsoever.

The *Mössbauer effect* exploits this occasional lack of particle emission for precise metrology. When a free nucleus emits a γ-ray, it recoils. The photon has less energy than the difference between the nuclear energy levels by the amount of energy taken by the recoiling nucleus. If it meets another nucleus of the same isotope, but in its ground state, the γ-ray photon does not have enough energy to excite this nucleus to its excited state. If, however, we embed the nucleus in a crystal, then the decay corresponds to a sharp "rap" on the nucleus — an impulsive $J(x) \propto \delta^3(\mathbf{x} - \mathbf{x}_{nucleus})\delta(t - t_{rap})$. The recoil energy is carried away by the emitted phonons. Sometimes no phonons are emitted and the γ-ray has its full energy $E_\gamma = E_{N^*} - E_{N^0}$ and so is able to excite another, similarly bound, nucleus.

This points out the distinction between momentum and pseudomomentum alluded to in the previous chapter. The crystal phonons carry *real* energy but only *pseudo*momentum. Even if no phonon is emitted, newtonian momentum $|p| = E_\gamma/c^2$ is still imparted to the crystal. The crystal, though, has a macroscopically large mass, so the amount of energy lost, $\Delta E = |p|^2/2M_{crystal}$, is small compared to the width of the $N^0 \to N^*$ resonance.

It is amusing to contemplate how the entire crystal is able to respond as a rigid body, even though one would think that the news of the impulsive rap can travel across it no faster than the speed of sound. The paradox is resolved by an application of the uncertainty principle. Processes where no phonons are created when a single atom out of the N atoms in the crystal receives an impulse $\mathbf{I}$ correspond to those relatively rare occasions when, just prior to the impulse being delivered, a zero-point fluctuation has each of the $N - 1$ other atoms in the crystal moving together as a crystal with mean momentum $\mathbf{I}/N$ per atom. Since momentum is a good quantum number for a crystal that is free to move, the single about-to-be-struck atom will then have momentum $-(N - 1)(\mathbf{I}/N)$. After the impulse the single atom has momentum $\mathbf{I}/N$ like the rest. The other atoms do not have to be told where to go therefore. They are already going there. The impulse merely sends the single atom chasing off after its brethren.

4

Feynman Rules

In this chapter we will introduce Feynman diagrams and Feynman rules. By now the reader will have become familiar with the basic ideas of operator-valued fields. I will therefore be less punctilious about putting hats on quantities that are operator equivalents of classical quantities, doing so only when there is a possibility of confusion.

4.1 Diagrams

4.1.1 Diagrams in Space-time

Most experiments in high-energy physics involve scattering one particle off another. When we consider such a process, the evolution operator $U(\infty, -\infty)$ tells us how an interaction-picture state evolves from the far distant past, where it represents the well-separated and therefore noninteracting incoming particles, into the far future where it describes the no-longer-interacting collision debris. Because of this important role, the operator $U(\infty, -\infty)$ is called the S (for *scattering*) matrix. We have

$$S = U(\infty, -\infty) = T\left\{\exp -i \int H_I d^4x\right\}, \tag{4.1}$$

where the integration is over all space-time.

Let us consider the simplest case where $H_I = \lambda \varphi^4/4!$ We expand out the time-ordered exponential and use Wick's theorem to break the resultant expressions up into contracted propagators and normal products containing the annihilation and

38   4. Feynman Rules

creation operators. The latter determine to which matrix elements of S the diagram corresponds when sandwiched between initial and final states.

We find the following diagrams contributing to S at first order:

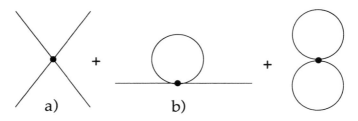

Fig 1. Diagrams contributing to S in first order.

Each dot in the diagrams corresponds to an *interaction vertex* at which we have a factor of λ.

For example diagram a) corresponds to

$$-i\lambda \frac{1}{4!} N\left\{(\varphi(x))^4\right\}. \tag{4.2}$$

The factor of $\frac{1}{4!}$ will disappear when we take matrix elements, as it is the number of ways we can soak up the external states into the normal products.

Diagram b) is

$$-i\lambda \frac{1}{2} \int d^4x\, (iG(x,x)) \frac{1}{2} N\left\{\varphi(x)\varphi(x)\right\}. \tag{4.3}$$

Again the factor of 1/2 in front of the the normal product will disappear when we take matrix elements.

At second order we have:

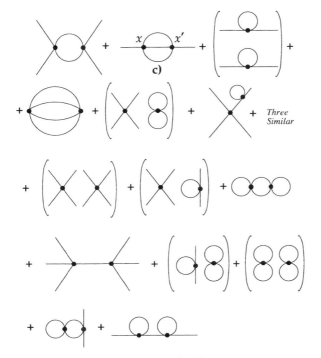

Fig 2. Diagrams contributing to S in second order.

Diagram c) is

$$(-i\lambda)^2 \frac{1}{3!} \int d^4x \, d^4x' \, (iG(x,x'))^3 \frac{1}{2} N\{\varphi(x)\varphi(x')\}. \tag{4.4}$$

Where everything comes from should be clear, except perhaps for the 1/3! in the expression for diagram c). This is an example of a *symmetry factor*.

Symmetry factors are best computed by counting the diffrent ways of attaching lines to the vertices. An alternative method is to observe that the symmetry factor is found by dividing by the order of the group of automorphisms of the graph. For example, the 1/48 in the "gooseberry" graph of Fig. 3 arises because the graph is unchanged if we permute any of the four lines, and if we interchange the two vertices. The automorphism group is therefore $S_4 \otimes Z_2$ whose order is $4! \times 2 = 48$.

Diagrams with no external legs have no normal products associated with them, and are not involved in the description of scattering processes. These are *vacuum bubbles* and relate to how the interactions modify the energy and structure of the vacuum state.

Factorization of Vacuum Bubbles

A vacuum diagram or bubble is any piece of a diagram that is not connected to any external line.

4. Feynman Rules

Suppose any diagram has $n + p$ vertices, so it is of order λ^{n+p}. Suppose n of the vertices are in diagrams connected to external lines and p of them are in bubbles. Then this diagram contains a factor

$$\frac{1}{(n+p)!} \binom{n+p}{p} \times \text{(externally connected parts)} \times \text{(bubbles)}. \tag{4.5}$$

The $1/(n+p)!$ comes from the $1/n!$ in the definition of the time-ordered exponential, and the combinatorial factor from the number of ways of assigning the vertices to the two sets. Since

$$\binom{n+p}{p} = \frac{(n+p)!}{n!p!}, \tag{4.6}$$

the sum over diagrams factorizes,

$$\left\{ \sum_n \frac{1}{n!} \text{(externally connected parts)} \right\} \times \left\{ \sum_p \frac{1}{p!} \text{(bubbles)} \right\}. \tag{4.7}$$

Therefore

$$\langle a|S|b\rangle = \langle a|S'|b\rangle \langle 0|U(\infty, -\infty)|0\rangle, \tag{4.8}$$

where S' is evaluated by ommiting any vacuum bubbles, and the last term, the vacuum persistence amplitude, is something we have met before. Provided the vacuum is stable against decay (not always true, but required for a quiet life), it only contributes an overall phase and therefore does not affect cross sections that depend only on $|\langle a|S|b\rangle|^2$.

Actually we can extract some physics from this factorization. We know from Chapter 3 that the vacuum persistence amplitude, $\langle 0|U(\infty, -\infty)|0\rangle$, can be evaluated as

$$\begin{aligned} \langle 0|U(\infty, -\infty)|0\rangle &= \langle 0|e^{iH_0 T} e^{-iHT}|0\rangle \\ &= e^{-i(E-E_0)T}, \end{aligned} \tag{4.9}$$

where E and E_0 are the ground-state energies of H_0 and H, respectively, and T is the duration of the system. If we look at the vacuum bubbles, we see that each disconnected bubble diagram has an overall integration over x and then integrations over the relative coordinates of its vertices. Each disconnected diagram therefore has a factor of VT, where T we have already defined and V is the volume of the system. Diagrams with two bubbles are proportional to $(VT)^2$, with three to $(VT)^3$, and so on. By arguments similar to those above one can convince oneself that the bubble sum exponentiates

$$\langle 0|U(\infty, -\infty)|0\rangle = e^{-iW}, \tag{4.10}$$

where

$$W = iVT \text{(connected diagrams)}. \tag{4.11}$$

In evaluating the diagrams in (4.11) we fix one vertex so we do not integrate over the overall position of the diagram in space-time.

The vacuum energy density $\mathcal{E}_0$ is therefore given by a sum of diagrams in which each distinct vacuum bubble appears once and once only.

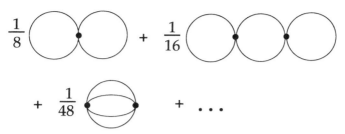

Fig 3. Connected diagrams contributing to the vacuum energy-density, together with their symmetry factors.

This result is often called the *Brueckner* (or *Bethe*) *Goldstone* theorem. It is a worthwhile exercise to show that the resulting energy shift is a real number.

Self-Energy Insertions on External Legs

Another set of diagrams we can and should ignore are those where *self-energy* insertions appear in the external legs. These are diagrams such as the last family in Fig 2. We ignore these because interactions which do not involve any of the other incoming or outgoing particles serve only to evolve the "bare" noninteracting single particle eigenstate into the exact eigenstate.

Along with evolving the state, these diagrams renormalize the mass of the particle and modify the efficiency with which the bare φ couples to the state. With this modification the matrix element of the full Heisenberg field between the vacuum and the one-particle states is given by

$$\langle \mathbf{k}|\varphi(x)|0\rangle = \sqrt{Z}e^{-ikx}. \tag{4.12}$$

We know its x dependence has to be of this form from Chapter 2. The coefficient Z, which is always less than unity, is called the *wave function renormalization constant*. We will discuss these topics more in later chapters.

4.1.2 Diagrams in Momentum Space

Look at

$$\begin{aligned}
\langle \mathbf{k}'_1, \mathbf{k}'_2|S|\mathbf{k}_1, \mathbf{k}_2\rangle &= -i\frac{\lambda}{4!}\int d^4x \langle \mathbf{k}'_1, \mathbf{k}'_2|N(\varphi(x)^4)|\mathbf{k}_1, \mathbf{k}_2\rangle \\
&= -i\lambda \int d^4x\, e^{ik'_2 x} e^{ik'_1 x} e^{-ik_1 x} e^{-ik_2 x} \\
&= -i\lambda(2\pi)^4 \delta^4(k'_1 + k'_2 - k_1 - k_2).
\end{aligned} \tag{4.13}$$

Note how the 4! was absorbed by the choice of which of the states to eat with which of the four field operators. Each integration over the position of a vertex will yield a similar energy-momentum conserving delta function.

42   4. Feynman Rules

Here is the general procedure, or *Feynman Rules*,[1] for a general $\lambda \varphi^N / N!$ interaction:

- Draw diagrams appropriate to the process being considered. Ignore vacuum bubbles and self-energies on external legs.
- Put a $-i\lambda(2\pi)^4 \delta(p_{out} - p_{in})$ for each vertex
- Put a propagator $\frac{i}{p^2 - m^2 + i\epsilon}$ for each line. Use a different p_i for each line.
- Multiply by the symmetry factor.
- Perform a $\int \frac{d^4 p_i}{(2\pi)^4}$ integration for each of the lines.

The energy-momentum delta functions help a great deal at the last step. One can do most of the integrals by simply conserving momentum at each vertex. If there are V vertices and L lines, there will only be $N_{loop} = V - L + 1$ integrations left to do — the "1" in this expression coming from an overall energy-momntum conservation delta function for the external lines. A little playing around will soon convince you that N_{loop} is the number of independent loops in the diagram. It is the same as the number of equations you would get if you regarded the graph as an electric circuit and applied Kirchhoff's rules. Indeed it is convenient to use a "loop momentum" in the same way that one would use a "loop current" in such a circuit.

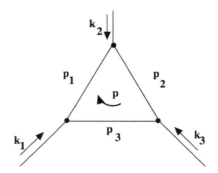

Fig 4. Momentum routing.

For example the graph of Fig. 4, which occurs in $\lambda\varphi^3$ theory, is expressed by running momentum p around the loop

$$(-i\lambda)^3 (2\pi)^4 \delta^4(k_1 + k_2 + k_3) \int \frac{d^4 p}{(2\pi)^4} \frac{i}{p_1^2 - m^2} \frac{i}{p_2^2 - m^2} \frac{i}{p_3^2 - m^2}. \quad (4.14)$$

Here

$$p_1 = p + k_1, \quad p_2 = p + k_1 + k_2, \quad p_3 = p + k_1 + k_2 + k_3 \equiv p. \quad (4.15)$$

Similarly the graph of Fig. 5

---

[1] Richard Phillips Feynman. Born May 11, 1918, New York, NY. Died February 15, 1988, Los Angeles, CA. Nobel Prize for Physics 1965.

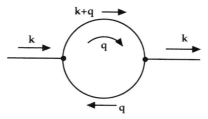

Fig 5. A self-energy graph.

has loop momentum q:

$$(-i\lambda)^2(2\pi)^4\delta^4(k-k')\int \frac{d^4q}{(2\pi)^4}\frac{i}{(k+q)^2-m^2}\frac{i}{q^2-m^2}. \qquad (4.16)$$

4.2 Scattering Theory

For $\lambda\varphi^3$ theory three diagrams contribute at lowest order to the two-body $p_1, p_2 \to p_1', p_2'$ scattering process.

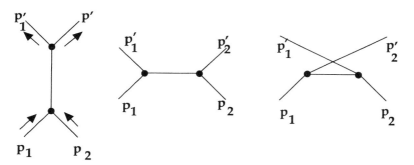

Fig 6. Three tree diagrams.

They give

$$\langle \mathbf{p}_1', \mathbf{p}_2'|S|\mathbf{p}_1, \mathbf{p}_2\rangle = (-i\lambda)^2(2\pi)^4\delta^4(k-k')$$
$$\times \left\{\frac{i}{(p_1+p_2)^2-m^2}+\frac{i}{(p_1-p_1')^2-m^2}+\frac{i}{(p_1-p_2')^2-m^2}\right\}$$
$$=(-i\lambda)^2(2\pi)^4\delta^4(k-k')$$
$$\times \left\{\frac{i}{s-m^2}+\frac{i}{t-m^2}+\frac{i}{u-m^2}\right\}. \qquad (4.17)$$

Here the symbols s, t, u are standard notation for the the three Lorentz-invariant kinematic invariants

$$s = (p_1+p_2) = \text{squared center of mass energy.}$$
$$t = (p_1-p_1') = \text{squared momentum transfer.}$$
$$u = (p_1-p_2'). \qquad (4.18)$$

These quantities are not independent of one another. For a general amplitude with different particles on each external leg, we find that

$$s + t + u = (p_1)^2 + (p_2)^2 + (p_1')^2 + (p_2')^2 + 2p_1(p_2 - p_1' - p_2' + p_1)$$
$$= m_1^2 + m_2^2 + m_3^2 + m_4^2, \quad (4.19)$$

where the m_i are the masses of the particles on the four legs.

When all four masses are the same these invariants take a simple form. Written in terms of the center-of-mass scattering angle θ and momentum $\mathbf{k}^2$ we have

$$s = 4(m^2 + \mathbf{k}^2),$$
$$t = -2\mathbf{k}^2(1 - \cos\theta),$$
$$u = -2\mathbf{k}^2(1 + \cos\theta). \quad (4.20)$$

In the physical region where θ is real and $\mathbf{k}^2 > 0$, the invariant s is always positive and $> 4m^2$, while t is always negative.

We see that the scattering amplitude has a pole whenever any of s, t, u is equal to the square of the mass of the particle represented by the propagator. These poles are said to be due to contributions from the s, t, and u *channels*. When the intermediate state particles are unstable, the poles will move to complex value of s, t, u and the amplitude will be large, but not infinite. We then have *resonances* in these channels. We will discuss this extensively in the next chapter.

4.2.1 Cross-Sections

Suppose

$$\langle \mathbf{p}_1', \ldots, \mathbf{p}_n' | S | \mathbf{p}_1, \mathbf{p}_2 \rangle = \langle p_f | S | p_i \rangle \quad (4.21)$$

is a general two-body to n'-body amplitude. Define T_{fi} by extracting the overall energy-momentum delta function from S:

$$\langle p' | S | p \rangle = (2\pi)^4 i \delta^4(p_f - p_i) T_{fi}. \quad (4.22)$$

Here $p_f = \sum p'$ and $p_i = p_1 + p_2$. The probability of making the transition $i \to f$ is

$$|\langle p' | S | p \rangle|^2 = \left((2\pi)^4 \delta^4(p_f - p_i)\right)^2 |T_{fi}|^2. \quad (4.23)$$

What sense are we to make of the square of a delta function? Well, we can proceed heuristically and use the fact that $\delta(x) f(x) = \delta(x) f(0)$ for any function f and guess that $\delta(x)^2$ should be thought of as $\delta(0)\delta(x)$. We apply this idea to (4.23) and follow Chapter 1 in interpreting $(2\pi)^4 \delta^4(0)$ as the volume of our universe time its duration. Thus we conjecture that

$$|\langle p' | S | p \rangle|^2 = (Volume)(Time)(2\pi)^4 \delta^4(p_f - p_i) |T_{fi}|^2. \quad (4.24)$$

In other words, we are claiming that the transition rate $i \to f$ per unit volume is

$$\Gamma_{fi} = (2\pi)^4 \delta^4(p_f - p_i) |T_{fi}|^2. \quad (4.25)$$

Equation (4.25) is indeed correct, and may be recognizable as a form of Fermi's "Golden Rule." You should have met a derivation of the Golden Rule in your quantum mechanics classes. Recall that time-dependent perturbation theory generically says that the amplitude to be in a new state growa linearly with the elapsed time t. The *probablility* to be in a particular state therefore seems to grow as t^2. When we are making transitions to a continuum of levels, however, the energy-time uncertainty principle restricts the range of available energy levels to be inversely proportional to t. The net effect is a transition probability to allowed levels that grows linearly with time.

The rate of transition to a restricted range of final states is

$$\Gamma = \sum_{states} (2\pi)^4 \delta^4(p_f - p_i) |T_{fi}|^2. \quad (4.26)$$

Now the connection between rates and scattering cross-sections is

$$\Gamma = \sigma \rho_1 \rho_2 |v_{12}|, \quad (4.27)$$

where σ is the cross-section, ρ_1 is the target number-density, ρ_2 is the beam number-density, and $|v_{12}|$ is the relative speed between the beam and target. (Note dimensions: $[\Gamma] = L^{-4}$, $[\sigma] = L^2$, $[\rho] = L^{-3}$, $[v_{12}] = L^0$.) Therefore, the cross section is

$$\sigma = \frac{1}{\rho_1 \rho_2 |v_{12}|} \frac{1}{n!} \int \prod_{i=1}^{n} \left(\frac{d^3 p_i}{(2\pi)^3} \frac{1}{2E_{\mathbf{p}_i}} \right) (2\pi)^4 \delta^4(p_f - p_i) |T_{fi}|^2. \quad (4.28)$$

The integration is over the required range of momentum space (perhaps restricted by some detector acceptance). For the total cross-section we sum over all of accessible phase space. The $1/n!$ in (4.28) is for the case that there are n *identitical* particles in the final state.

The cross-section σ is a Lorentz invariant quantity, and so is the integral of $|T_{fi}|^2$ over the LIPS. Consequently, the quantity $\rho_1 \rho_2 |v_{12}|$ must also be Lorentz invariant.

Theorem:

The flux factor is given by the Lorentz invariant expression

$$\begin{aligned} \rho_1 \rho_2 |v_{12}| &= 4\sqrt{(p_1 p_2)^2 - p_1^2 p_2^2} \\ &= 8i \Delta(\sqrt{s}, m_1, m_2), \end{aligned} \quad (4.29)$$

where $\Delta(a, b, c)$ is the area of the triangle with sides a, b, c.

COMMENT The i in this expresion comes from the fact that the area of a triangle with sides $\sqrt{s}, m_1, m_2$ is imaginary. No real triangle has the length of any of its sides longer than the sum of the other two, but for real momenta $\sqrt{s}$, the center of mass energy, is always greater than the sum of the rest masses.

PROOF: We begin by proving the famous theorem of Heron[2] giving the area of a triangle in terms of the lengths of its sides.

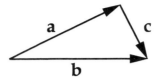

For the triangle in any number of dimensions with the vectors **a**, **b**, **c** making up the sides we have

$$a^2 b^2 - (\mathbf{a} \cdot \mathbf{b})^2 = a^2 b^2 (1 - \cos^2 \theta) = a^2 b^2 \sin^2 \theta = 4(Area)^2$$
$$= a^2 b^2 - (\frac{1}{2}(\mathbf{a} - \mathbf{b})^2 - a^2 - b^2)^2 = a^2 b^2 - \frac{1}{4}(c^2 - a^2 - b^2)^2$$
$$= a^2 b^2 - \frac{1}{4}(c^4 + a^4 + b^4 - 2a^2 c^2 - 2a^2 b^2 + 2a^2 b^2)$$
$$= -\frac{1}{4}(a^4 + b^4 + c^4 - 2a^2 b^2 - 2a^2 c^2 - 2b^2 c^2)$$
$$= 4(\Delta(a, b, c))^2. \tag{4.30}$$

Here the expression $\Delta(a, b, c)$ is defined by

$$\sqrt{\frac{1}{16}(a+b+c)(a+b-c)(a-b+c)(-a+b+c)}$$
$$= \Delta(a, b, c) \equiv \sqrt{s(s-a)(s-b)(s-c)}, \tag{4.31}$$

where $s \equiv (a+b+c)/2$ is the *semiperimeter* of the triangle.

This expression for the area of a triangle in terms of s and the three sides was a standard high-school formula in the days before pocket calculators. I prove it here because it seems less familiar to today's students.

Now look at

$$I = (p_1 p_2)^2 - p_1^2 p_2^2. \tag{4.32}$$

If we take particle 1 to be at rest, we have $p_1 = (E_0, 0)$ and $p_2 = (E_2, \mathbf{p}_2)$

$$I = E_1^2 E_2^2 - E_1^2(-\mathbf{p}_2^2 + E_2^2) = \mathbf{p}_2^2 E_1^2$$
$$= \left|\frac{\mathbf{p}_2}{E_2^2}\right|^2 E_1^2 E_2^2$$
$$= \left(\frac{1}{4}\right)(\rho_1 \rho_2 |v_{12}|)^2. \tag{4.33}$$

---

[2]Heron of Alexandria, Born: *circa* 65 in (possibly) Alexandria, Egypt. Died *circa* 125, location unknown.

In the last line we have used the fact that the velocity of a particle is $|\mathbf{p}|/E_\mathbf{p}$ and that, with our relativistic normalization for the plane-wave states $|\mathbf{p}\rangle$, the number of particles per unit volume is $2E_\mathbf{p}$.

Thus

$$\rho_1 \rho_2 |v_{12}| = 8i\,\Delta(\sqrt{s}, m_1, m_2) \tag{4.34}$$

as was required.

4.2.2 Decay of an Unstable Particle

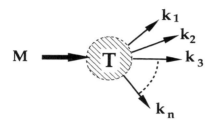

Fig 7. A decaying particle.

Suppose a stationary (and hence completely delocalized) particle of mass M, and 4-momentum $K = (M, 0)$, decays to a set of n identical particles with momenta $\mathbf{k}_i$, $i = 1, \ldots, n$. The number of decay events per unit time per unit volume will be given by integrating the final momenta over LIPS

$$\Gamma = \frac{1}{n!} \int \prod_{i=1}^{n} \left(\frac{d^3 k_i}{(2\pi)^3} \frac{1}{2E_{\mathbf{k}_i}} \right) (2\pi)^4 \delta^4 \left(\sum k_i - K \right) |T_{fi}|^2. \tag{4.35}$$

Since the number-density of the unstable particle is $2M$, we have the following expression for the lifetime τ of each individual particle:

$$\frac{1}{\tau} = \frac{1}{2M} \frac{1}{n!} \int \prod_{i=1}^{n} \left(\frac{d^3 k_i}{(2\pi)^3} \frac{1}{2E_{\mathbf{k}_i}} \right) (2\pi)^4 \delta^4 \left(\sum k_i - K \right) |T_{fi}|^2. \tag{4.36}$$

[The lifetime is defined by the survival probability being $P(t) \propto e^{-t/\tau}$.]

Equation (4.36) is useful formula in itself, but we will soon have an opportunity to derive it in a completely different manner, which will cast considerable light on the effect of higher loops on particle properties.

5
Loops, Unitarity, and Analyticity

In this chapter we will explore how unitarity, the requirement that probability be conserved in quantum time evolution, interacts with the perturbation expansion.

5.1 Unitarity of the S Matrix

It is convenient to separate off the parts of the S matrix where nothing happens and so to write

$$S = 1 + iT. \tag{5.1}$$

Then unitarity, the requirement that $S^\dagger S = 1$, tells us that

$$1 = 1 - i(T^\dagger - T) - T^\dagger T. \tag{5.2}$$

In other words

$$T^\dagger T = 2\,\mathrm{Im}\,T. \tag{5.3}$$

A consequence of this identity is

$$2\,\mathrm{Im}\,\langle n|T|n\rangle = \sum_m |\langle m|T|n\rangle|^2. \tag{5.4}$$

Equation (5.4) can be summarized by the gnomic assertion that the probability for something (a transition from $|n\rangle$ to any other state) to occur is the imaginary part of the amplitude for it *not* to occur. This is a form of the *optical theorem*.

Let us look at how (5.4) works diagramatically by computing the imaginary part of the following bubble graph.

5.1 Unitarity of the S Matrix

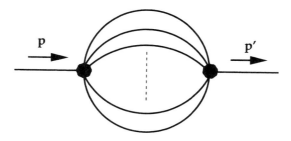

Fig 1. A bubble graph.

The bubble has n internal lines. The symmetry factor for the diagram is $1/n!$. There are $n-1$ loops so this is a messy diagram in p space, but it is easy to evalute in x space where it is simply an integral over x_1 and x_2 of $G(x_1, x_2)^n$. Breaking the propagators up into their time-ordered parts gives us

$$\int d^4x_1 d^4x_2\, e^{-ipx_1+ip'x_2} \left[\theta(x_2^0 - x_1^0) \int \prod_{i=1}^n \left(\frac{d^3k_i}{(2\pi)^3} \frac{1}{2E_{\mathbf{k}_i}} \right) e^{-i(k_1+\ldots+k_n)(x_2-x_1)} \right.$$

$$\left. +\theta(x_1^0 - x_2^0) \int \prod_{i=1}^n \left(\frac{d^3k_i}{(2\pi)^3} \frac{1}{2E_{\mathbf{k}_i}} \right) e^{+i(k_1+\ldots+k_n)(x_2-x_1)} \right] (-i\lambda)^2. \quad (5.5)$$

Put $x_2 - x_1 = z$ and do the x_1 integral to get

$$(2\pi)^4 \delta^4(p - p') \int d^4z\, e^{ipz} [\ldots] (-i\lambda)^2. \quad (5.6)$$

Since we are calculating T, we drop this energy-momentum delta function.
Finally we do the z integrals to get $(-i\lambda)^2$ times

$$-i\left[\int \prod_{i=1}^n \left(\frac{d^3k_i}{(2\pi)^3} \frac{1}{2E_{\mathbf{k}_i}} \right) (2\pi)^3 \delta^3(\mathbf{p} - \sum \mathbf{k}) \frac{i}{p_0 - (E_1 + \ldots E_n) + i\epsilon} \right.$$

$$\left. + \int \prod_{i=1}^n \left(\frac{d^3k_i}{(2\pi)^3} \frac{1}{2E_{\mathbf{k}_i}} \right) (2\pi)^3 \delta^3(\mathbf{p} - \sum \mathbf{k}) \frac{i}{-p_0 - (E_1 + \ldots E_n) + i\epsilon} \right].$$

$$(5.7)$$

We recognize these as the two diagrams of "old fashioned" Rayleigh-Schrödinger perturbation theory where we sum over intermediate states with energy denominators.

5. Loops, Unitarity, and Analyticity

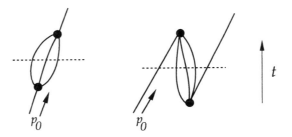

Fig 2. Time-ordered Rayleigh-Schrödinger diagrams.

In the left-hand figure the intermediate states (at the time shown by the dotted line) have energy $E_1 + E_2 + E_3$ so the energy denominator is

$$\frac{1}{p_0 - E_1 - E_2 - E_3}. \tag{5.8}$$

In the right-hand figure the intermediate states have energy $2p_0 + E_1 + E_2 + E_3$ so the denominator is

$$\frac{1}{p_0 - 2p_0 - E_1 - E_2 - E_3} = \frac{1}{-p_0 - E_1 - E_2 - E_3}. \tag{5.9}$$

In the old fashioned approach momentum is conserved at each vertex, but energy is not. Indeed each diagram of Feynman perturbation theory at order N can be decomposed into $N!$ diagrams of the old fashioned perturbation theory, one for each of the $N!$ time orderings of the vertices. The notorious "counterterms" in Rayleigh-Schrödinger perturbation theory are just the instructions to omit the vacuum bubbles.

Now let's find the imaginary part. Assume that $p^0 > 0$, then only the first term can contribute. We use the familar mantra

$$\frac{1}{\omega + i\epsilon} = \frac{P}{\omega} - i\pi\delta(\omega), \tag{5.10}$$

to find

$$2\,\text{Im}\,T = \frac{1}{n!} \int \prod_{i=1}^{n} \left(\frac{d^3 k_i}{(2\pi)^3} \frac{1}{2E_{k_i}}\right) (2\pi)^4 \delta^4(p - \sum k) |\lambda|^2, \tag{5.11}$$

which is the final-states summation in the cross section

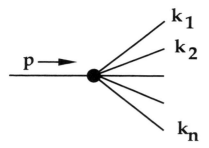

Fig 3. Final states.

Thus to this order

$$2 \operatorname{Im} \langle \mathbf{p}|T|\mathbf{p}\rangle = \sum_\mathbf{k} |\langle \{\mathbf{k}\}|T|\mathbf{p}\rangle|^2 . \tag{5.12}$$

Sometimes the imaginary part of the forward amplitude is referred to as the *Absorbtive Part* of the amplitude. It corresponds to the excitation of real particles as opposed to virtual intermediate states.

The most general formulation of unitarity in the diagram language involves the *Landau-Cutkoski cutting formulae*. These are elaborated in Eden, Landshoff, Olive, and Polkinghorne's *Analytic S-Matrix*.

5.2 The Analytic S Matrix

5.2.1 Origin of Analyticity

Physical systems are *causal*. A particle sitting happily in equilibrium will sit there forever until it is disturbed. Because it has no pre-knowledge of the imminent disturbance, it will not respond until *after* it has been nudged. This shows up in the analytic properties of the Green function describing the response. Suppose we wish to solve the equation for a damped oscillator

$$\ddot{x} + 2\gamma \dot{x} + (\Omega^2 + \gamma^2)x = F(t). \tag{5.13}$$

The solution will be of the form

$$x(t) = \int_{-\infty}^{\infty} G(t,t')F(t')dt' \tag{5.14}$$

where the Green function $G(t,t') = 0$ if $t < t'$. In fact, in this case,

$$x(t) = \frac{1}{\Omega} \int_{-\infty}^{t} e^{-\gamma(t-t')} \sin \Omega(t-t') F(t')dt' \tag{5.15}$$

and

$$\begin{aligned} G(t,t') &= \frac{1}{\Omega} e^{-\gamma(t-t')} \sin \Omega(t-t') \quad t > t' \\ &= 0 \quad t < t'. \end{aligned} \tag{5.16}$$

Because the integral extends only from 0 to $+\infty$, the Fourier transform of $G(t, 0)$

$$\tilde{G}(\omega) = \frac{1}{\Omega} \int_0^\infty e^{i\omega t} e^{-\gamma t} \sin \Omega t \, dt \tag{5.17}$$

is nicely convergent when $\text{Im } \omega > 0$, as evidenced by

$$\tilde{G}(\omega) = -\frac{1}{(\omega + i\gamma)^2 - \Omega^2} \tag{5.18}$$

having no singularities in the upper half-plane.

Similarly the Fourier transform of any causal response function will be an *analytic function* of ω in the entire upper half-plane. This analyticity should extend to the k-space matrix elements of the S matrix, which describes the response of the system to the collision. Analyticity is a very powerful notion. Precise knowledge of an analytic function in an arbitrarily small neighborhood of any point serves to determine the function at all points in its domain of definition.

5.2.2 Unitarity and Branch Cuts

As we have seen, causality suggests that the S matrix is an analytic function. Because it is Lorentz-invariant, it will be an analytic function of those Lorentz invariants that can be constructed from the external momenta. Unitarity imposes many constraints on this matrix-valued function, and it was at one time hoped that these were strong enough to determine it uniquely. As a route to evaluating the S matrix, this turns out largely impractical for realistic four-dimensional theories, but it has been successfully used to construct exact S matrices for many interesting two-dimensional models. In this section we will explore some of the immediate consequences of combining analyticity with unitarity.

Consider two-body to two-body scattering:

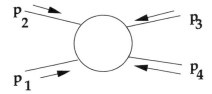

Fig 4. $s = (p_1 + p_2)^2$, $t = (p_1 + p_4)^2$, $u = (p_1 + p_3)^2$.

Suppose we gradually increase the variable s, which is the square of the center-of-mass energy of the incoming particles. Each time the inflowing energy has grown sufficiently to support a larger set of on-shell intermediate particles, a new contribution to the imaginary part of the amplitude appears. Such imaginary parts come from discontinuities, or *branch cuts*, in the analytic function representing the amplitude. The critical values of s at which more intermediate particles can occur are points at which a new branch cuts starts — i.e. *branch points*. For our two-particle collision in a theory of particles with mass m, the amplitude $T(s, t)$ will have a branch point singularity, called a *normal threshold*, at $s = 4m^2$, $s = 9m^2$ (assuming that an even number of particles can give rise to an odd number, as in $\lambda \varphi^3$), $s = 16m^2$ etc., corresponding to on-shell intermediate states in the s channel with two, three, and four particles, respectively.

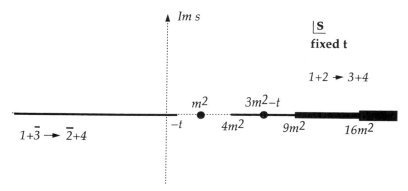

Fig 5. Normal threshold branch cuts.

At fixed negative t, the variable u varies as we vary s. The u-channel will contribute a single-particle pole at $u = m^2$ and can support multiparticle intermediate states if $u > 4m^2$. Since $s + t + u = 4m^2$, the pole will appear at $s = 3m^2 - t$ and the u channel cuts will contribute to Im T if $s < -t$. In the figure the amplitude is real between $s = -t$ and $s = 4m^2$, so by the Schwarz reflection principle it must satisfy $T(s^*, t) = T^*(s, t)$. The values of T above and below the cuts are therefore complex conjugates of each other.

For real scattering $1 + 2 \to 3 + 4$ we must have $s > 4m^2$. In extracting the scattering amplitude for this process we must be mindful of the $i\epsilon$'s in the propagators, and the physical value of T is obtained for s having an infinitesimal positive imaginary part — i.e we must evaluate the amplitude immediately above the cut on the positive real s axis. If we look at the region where $u > 4m^2$ we are dealing with the scattering process $1 + \bar{3} \to \bar{2} + 4$. Here the overbar indicates that the incoming particle is now the antiparticle. To get the physical T matrix we must now approach the axis from below. If all the particles 1, 2, 3, 4 and their antiparticles are identical then this process must be the same as the first[1] and we must have the *crossing symmetry* relation

$$T(s + i\epsilon, t) = T(4m^2 - s - t - i\epsilon, t). \tag{5.19}$$

If the particles are not all identical, then the u process may differ from the s process. For example, if we have $\pi^- + p \to \pi^- + p$ the crossed process will be $\pi^+ + p \to \pi^+ + p$, and then the crossing relation becomes

$$T_{(\pi^+, p)}(s + i\epsilon, t) = T_{(\pi^-, p)}(2m_\pi^2 + 2m_p^2 - s - t - i\epsilon, t). \tag{5.20}$$

---

[1]Quod est inferius est sicut quod est superius, et quod est superius est sicut quod est inferius, ad perpetranda miracula rei unius. (Hermes Trismegistus, *The Emerald Tablet*.)

A further crossing takes the process from $1+\bar{3} \to \bar{2}+4$ to $\bar{3}+\bar{4} \to \bar{1}+\bar{2}$. This takes us back to where we were in the s plane. We deduce that the amplitudes for $1+2 \to 3+4$ is the same as for $\bar{4}+\bar{3} \to \bar{1}+\bar{2}$. This is an example of the *CPT theorem* — a deep result that we will not have time to describe.

5.2.3 Resonances, Widths, and Lifetimes

When there is enough energy to create a real on-shell particle as an intermediate state, the particle can always decay back into the particles it was assembled from — and perhaps to other particles as well. The resultant finite lifetime for the particle serves to prevent there being an actual infinity arising from the on-shell particle's propagator, since this is no longer simply $1/(s - M^2)$ evaluated at $s = M^2$.

Suppose we have two particles. One, called χ, is heavy and has mass M. The other, φ, is light and of mass m. Suppose also that an interaction $\mathcal{L}_I = \frac{1}{2}\chi\varphi^2$ allows $\chi \leftrightarrow \varphi + \varphi$ processes. If $M^2 > 4m^2$, then we can create χ as an intermediate state. If the lifetime of the χ particle is long enough, the χ will signal itself as a sharp peak, or *resonance*, when we plot the $\varphi\varphi$ scattering cross-section.

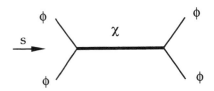

Fig 6. *The particle χ as an intermediate resonance state.*

The χ propagator has self-energy corrections with light φ particles in the loops.

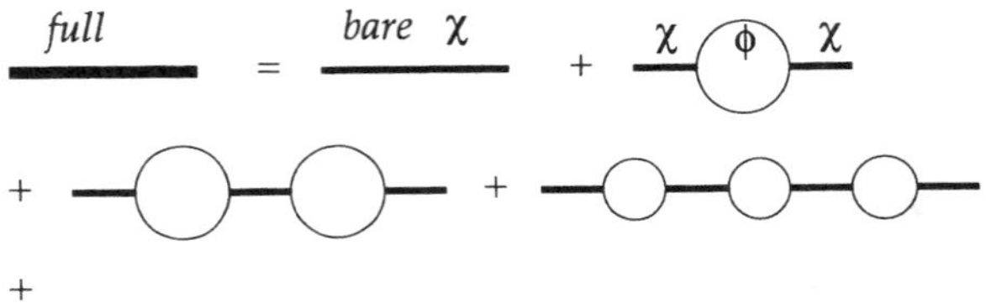

Fig 7. *Propagator sum.*

Summing these diagrams gives

$$\frac{i}{p^2 - M^2} + \frac{i}{p^2 - M^2}\{i\Pi(p^2)\}\frac{i}{p^2 - M^2} + \ldots \quad (5.21)$$

where $\Pi(p^2)$ is the φ-particle bubble.

The series is a geometric progression, and sums to

$$G(p^2) = \frac{i}{p^2 - M^2 + i\epsilon + \Pi(p^2)}. \quad (5.22)$$

We know that Π has a divergent real part. This we merely lump into M^2 to renormalize the mass of the χ particle, and then forget. It also has a finite imaginary part

$$2\,\text{Im}\,\Pi(p^2) = \theta(p^2 - 4m^2)\frac{1}{2}\int \frac{d^3k_1}{(2\pi)^3}\frac{1}{2E_{k_1}}\frac{d^3k_2}{(2\pi)^3}\frac{1}{2E_{k_2}}(2\pi)^4\delta^4(p-k_1-k_2)|\lambda|^2. \tag{5.23}$$

If λ is small, this imaginary part will also be small and so $G(p^2)$ will have a pole somewhere near M^2. To find its location we approximate

$$p^2 - M^2 + \Pi(p^2) \approx p^2 - \left(M - \frac{1}{2M}\Pi(p^2)\right)^2$$
$$\approx p^2 - \left(M - \frac{1}{2M}\Pi(M^2)\right)^2 \tag{5.24}$$

We see that the pole moves from M to $M - i\frac{1}{2}\Gamma$, where

$$\Gamma = \frac{1}{2M}\frac{1}{2}\int \frac{d^3k_1}{(2\pi)^3}\frac{1}{2E_{k_1}}\frac{d^3k_2}{(2\pi)^3}\frac{1}{2E_{k_2}}(2\pi)^4\delta^4(p-k_1-k_2)|\lambda|^2 \tag{5.25}$$

is recognized as the inverse lifetime of the χ particle. In fact the pole gets tucked under the normal threshold cut onto the second sheet in the T matrix.

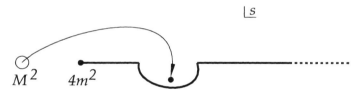

Fig 8. The pole, which is on the real axis when $M^2 < 4m^2$ and the particle is unable to decay, gets tucked under the cut (which has been displaced so the pole can be seen) when $M^2 > 4m^2$.

There is of course a pole at the complex conjugate point on the second sheet in the upper half plane, but this is a long way from where we are evaluating T (we have to go all the way round the branch cut at $s = 4m^2$ to get to it) so it has negligible effect.

If we regard the location of the pole as the eigen-energy of the χ particle, we see that the time evolution of the χ state should be

$$|\psi(t)|^2 \propto \left|e^{-i(M-i\frac{1}{2}\Gamma)t}\right|^2 = e^{-\Gamma t}, \tag{5.26}$$

so the mass shift corresponds to the decay lifetime as expected. Meanwhile, near resonance, the χ particle propagator is approximately

$$G(p^2) = \frac{1}{p^2 - (M - i\frac{1}{2}\Gamma)^2}, \tag{5.27}$$

so the total cross-section, which is proportional to the imaginary part of G, will look like

$$\sigma_{tot}(s) \propto \frac{1}{(\sqrt{s} - M)^2 + \frac{1}{4}\Gamma^2}. \quad (5.28)$$

This is a classic Breit-Wigner lorentzian peak with width Γ.

5.3 Some Loop Diagrams

5.3.1 Wick Rotation

In performing integrals such as

$$I = \int \frac{d^d p}{(2\pi)^4} \frac{1}{[p^2 - M^2 + i\epsilon]^m}, \quad (5.29)$$

we can use the routing of the contour past the poles, and the fact that (for large enough m) the integrand falls off rapidly at infinity, to perform a *Wick rotation* into the region of euclidean momenta and inner products. We simply set $p_0 = ip_0^E$ and integrate over p_0^E from $-\infty$ to $+\infty$.

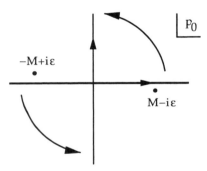

Fig 9. We can rotate the p_0 contour through 90° without changing the value of the integral.

Now we have

$$I = i(-1)^m \int \frac{d^d p^E}{(2\pi)^4} \frac{1}{[p^2 + M^2]^m} \quad (5.30)$$

where everything is straight forwardly euclidean.

Of course this manoeuver is only legal when $2m > d$, and usually we will require it when $2m < d$ and the integral is divergent. Later we will discuss the physics behind legitimating this kind of legerdemain.

Ouroboros Graphs

An amusing application of Wick rotation is to find the zero-point vacuum-energy density for the free theory. We do this by using perturbation theory to find the

5.3 Some Loop Diagrams

shift in energy when we change the mass^2 from m^2 to $m^2 + \delta m^2$. In this case the perturbation is quadratic, $\mathcal{H}_I = \frac{1}{2}\delta m^2$, and we can sum the vacuum graphs in closed form. For n insertions of $\mathcal{H}_I$ we have a contribution to $iW = -(Vol)(\mathcal{E}-\mathcal{E}_0)$ of

$$(Vol)\int \frac{d^4k}{(2\pi)^4}\frac{1}{2n}\left((-i\delta m^2)\frac{i}{k^2-m^2+i\epsilon}\right)^n. \tag{5.31}$$

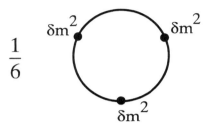

Fig 10. The third-order constribution to iW.

The symmetry factor $1/(2n)$ comes from the n-fold cyclic symmetry, and from the two directions we can traverse the circle. (The factor of $1/2$ is therefore absent when we consider charged particles that have directed lines.)

After Wick rotation the energy shift becomes

$$\mathcal{E} - \mathcal{E}_0 = -\sum_n (-1)^n \frac{1}{2n}\int \frac{d^4k}{(2\pi)^4}\left(\frac{\delta m^2}{k^2+m^2}\right)^n. \tag{5.32}$$

Since

$$\ln(1+x) = x - \frac{x^2}{2} + \frac{x^3}{3} - \cdots. \tag{5.33}$$

we can perform the sum on n to get

$$\mathcal{E} - \mathcal{E}_0 = \frac{1}{2}\int \frac{d^4k}{(2\pi)^4}\ln\left(1+\frac{\delta m^2}{k^2+m^2}\right)$$
$$= \frac{1}{2}\int \frac{d^4k}{(2\pi)^4}\left(\ln(k^2+(m^2+\delta m^2)) - \ln(k^2+m^2)\right). \tag{5.34}$$

We can interpret (5.34) as asserting that the ground-state energy density for a particular mass^2 is given by

$$\mathcal{E}(m^2) = \frac{1}{2}\int \frac{d^4k}{(2\pi)^4}\ln(k^2+m^2). \tag{5.35}$$

We may think of the logarithm of $k^2 + m^2$ as representing a propagator that has bitten its tail and does not know where it begins or ends. I like to call this diagram the *Ouroboros* diagram, after the Greek word for "tail eater." In alchemy the symbol of a snake eating its own tail is known as the Ouroboros and represents the eternal process of death and rebirth. It is the origin of the modern symbol "∞," and provides

58   5. Loops, Unitarity, and Analyticity

an apposite name for the contributions to the vacuum persistence amplitude from processes where a particle-antiparticle pair appears from nothing and then devours itself.

To show that the Ouroboros diagram gives the usual expression for the zero-point energy, we differentiate $\mathcal{E}(m^2)$ with respect to m^2 and find

$$\frac{\partial \mathcal{E}(m^2)}{\partial m^2} = \frac{1}{2} \int \frac{d^4k}{(2\pi)^4} \frac{1}{k^2 + m^2}. \tag{5.36}$$

Now perform the k_0 integral to get

$$\frac{\partial \mathcal{E}(m^2)}{\partial m^2} = \frac{1}{4} \int \frac{d^3k}{(2\pi)^3} \frac{1}{\sqrt{\mathbf{k}^2 + m^2}}. \tag{5.37}$$

Integrating up again, we find, up to an m-independent constant,

$$\mathcal{E}(m^2) = \frac{1}{2} \int \frac{d^3k}{(2\pi)^3} \sqrt{\mathbf{k}^2 + m^2}. \tag{5.38}$$

The vacuum energy is therefore the sum of the $\frac{1}{2}\hbar\omega$'s from each mode, as it should be. For charged particles we get twice this answer. This makes sense because there are now two types of particle.

5.3.2 Feynman Parameters

There are several useful identities for untangling Feynman integrals. The first of them is *Schwinger's trick*,[2] which in its general form reads

$$a_1^{-n_1} a_2^{-n_2} \ldots a_m^{-n_m} = \frac{1}{\prod_1^m \Gamma(n_i)} \int_0^\infty dt_1 \, dt_2 \ldots dt_m t_1^{n_1-1} \ldots t_m^{n_m-1} e^{-ta_1 \ldots -ta_m}. \tag{5.39}$$

By a clever change of variable we convert this to the general form of *Feynman's trick*. We introduce variables α_i by setting $t_i = \alpha_i t$, where $0 < \alpha_i < 1$ and $0 < t < \infty$, and use a delta function to impose the condition $\sum_i \alpha_i = 1$. (The α's are the squares of direction cosines.) Then (5.39) becomes

$$\frac{1}{\prod_1^m \Gamma(n_i)} \int_0^\infty dt \int d[\alpha] \delta(\sum \alpha_i - 1) t^{\sum n_i - 1} e^{-t(\sum \alpha_i a_i)} (\alpha_1^{n_1-1} \ldots \alpha_m^{n_m-1}) \tag{5.40}$$

$$= \frac{\Gamma(\sum n_i)}{\prod_1^m \Gamma(n_i)} \int d[\alpha] \delta(\sum \alpha_i - 1) \frac{\prod_1^m \alpha_i^{n_i-1}}{(\sum_1^m \alpha_i a_i)^{\sum n_i}}. \tag{5.41}$$

---

[2]Julian Seymour Schwinger. Born Feb. 12, 1918, New York, NY. Died July 16, 1994, Los Angeles, CA, Nobel Prize in Physics 1968.

5.3.3 Dimensional Regularization

As an example of (5.41) take $a_1 = a_2 = 1$, $n_1 = x$, $n_2 = y$ whence

$$1 = \frac{\Gamma(x+y)}{\Gamma(x)\Gamma(y)} \int d\alpha \, \alpha^{x-1}(1-\alpha)^{y-1}, \tag{5.42}$$

which should be familiar as the beta function identity of Euler.

Let us apply this to the integral

$$I = \int \frac{d^4p}{(2\pi)^4} \frac{1}{(p+k)^2 + m^2} \frac{1}{p^2 + m^2}. \tag{5.43}$$

There are no $i\epsilon$'s in this integral because I have performed the Wick rotation to the euclidean region. All the inner products are now ordinary (not lorentzian) ones. Since the integral is only a function of k^2, the physical Minkowski-space momenta may be recovered by analytically continuing k^2 from positive (euclidean) value to negative (Minkowski) values.

We apply Feynman in the form

$$\frac{1}{xy} = \int_0^1 d\alpha \frac{1}{(\alpha x + (1-\alpha)y)^2}, \tag{5.44}$$

to get

$$I = \int_0^1 d\alpha \int \frac{d^4p}{(2\pi)^4} \frac{1}{[\alpha(p^2 + 2pk + k^2) + m^2 + (1-\alpha)p^2]^2}. \tag{5.45}$$

The integral is divergent. To extract a finite expression we use t'Hooft and Veltman's *dimensional regularization* trick and evaluate the integral in d dimensions, where d is not necessarily an integer. The basic identity comes from looking at

$$\int \frac{d^d p}{(2\pi)^d} \frac{1}{[p^2 + 2pk + M^2]^m}. \tag{5.46}$$

At first we pretend that d is an integer and evaluate

$$= \frac{1}{\Gamma(m)} \int_0^\infty dt \int \frac{d^d p}{(2\pi)^d} t^{m-1} e^{-t(p^2 + 2pk + M^2)}$$

$$= \frac{1}{\Gamma(m) 2^d \pi^{d/2}} \int_0^\infty dt\, t^{m-d/2-1} e^{-t(M^2 - k^2)}$$

$$= \frac{\Gamma(m - d/2)}{\Gamma(m)} \frac{1}{2^d \pi^{d/2}} \left(M^2 - k^2\right)^{d/2 - m}. \tag{5.47}$$

The original expression (5.46) diverges when $d > 2m$. This divergence manifests itself in the factor $\Gamma(m - d/2)$ which has a pole when the argument of the gamma function is zero or a negative integer. We regard (5.47) as an analytic function in d and explicitly exhibit the pole together with a finite part.

In our case we want $d \to 4$, $m = 2$, "k"$= \alpha k$, $M^2 = m^2 + \alpha k^2$, so we find

$$I = \lim_{d \to 4} \frac{1}{16\pi^2} \frac{1}{\Gamma(2)} \Gamma(2 - \frac{d}{2}) \int_0^1 d\alpha (m^2 + \alpha k^2 - \alpha^2 k^2)^{d/2 - 2} \tag{5.48}$$

5. Loops, Unitarity, and Analyticity

$$= \frac{1}{16\pi^2} \left\{ \frac{1}{2 - \frac{d}{2}} - \int_0^1 d\alpha \ln(m^2 + \alpha(1-\alpha)k^2) + \text{finite} \right\}. \tag{5.49}$$

What are we to do with this expression? We will later find out how to absorb the pole part into a renormalization of the couplings. As to the integral, it is finite and elementary. Evaluating the integral is not very informative however. Instead, let us apply some general theory to extract its analytic behavior as a function of k^2.

The roots of the expression $m^2 + \alpha(1-\alpha)k^2$ lie at

$$\alpha_\pm = \frac{+k^2 \pm \sqrt{(k^2)^2 + 4m^2k^2}}{2k^2}$$

$$= \frac{1}{2}(1 \pm \sqrt{1 + 4m^2/k^2}), \tag{5.50}$$

and the logarithm is singular at these points.

How do these singularities affect the amplitude? Recall how singularities appear in functions defined as contour integrals. Suppose the function

$$F(a) = \int_\alpha^\beta dz f(z, a) \tag{5.51}$$

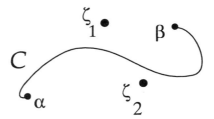

Fig 11. The contour defining $F(a)$. $\zeta_{1,2}$ are a pair of poles of $f(z, a)$ whose position depends on a.

is defined by a contour integral as shown in the figure. As we vary a the poles $\zeta_{1,2}$ will wander about in the complex z plane. If one of them strays onto the contour, then the integrand $f(z, a)$ will be become infinite at that point — but nothing serious will happen to $F(a)$ provided we can move the contour out of the way. By Cauchy's theorem moving the contour will not alter the value of $F(a)$, but it will restore the finiteness of the integrand. Problems, and resultant singularities in $F(a)$, will occur only if the contour cannot escape. This will occur when one of the poles hits the fixed endpoints of the contour, α or β, or when the contour becomes trapped between the pair of poles as they come together. The former situation gives rise to an *endpoint singularity*, and the latter to a *pinch* singularity. For example, the Legendre function

$$Q_n(z) = \frac{1}{2^{n+1}} \int_{-1}^{+1} \frac{(1-t^2)^n}{(z-t)^{n+1}} dt, \quad \text{Re}(n+1) > 0 \tag{5.52}$$

is singular when $z = \pm 1$ because the singularity at $z = t$ runs into the endpoints of the contour.

In the case of our Feynman integral the branch cuts of the logarithm at $\alpha_\pm$ pinch the α contour in the Minkowski region at $k^2 = -4m^2$. The bubble diagram therefore becomes singular at exactly the point that it should from unitarity.

Some further discussions on the nature of dimensional regularization as well as formulae for the real space propagator in d dimensions are to be found in Appendix C.

6

Formal Developments

This chapter will mostly be devoted to formalism. We will introduce generalized Green functions and show how to evaluate them perturbatively. We will then establish some of their properties that are independent of perturbation theory, and finally we will show how to extract the S matrix from them.

6.1 Gell-Mann Low Theorem

In addition to computing cross-section and decay rates, it is often desirable to be able to calculate matrix elements of products of operators. Such quantities are experimentally accessible when we regard some part of a scattering process as completely under control. For example, when we scatter electrons off hadrons, the quantum electrodynamic process by which the electron emits a virtual photon is throroughly understood, and we can regard the photon as an external probe of the hadron, measuring the quantity

$$\langle N|T\{J^\mu(x)J^\nu(y)\}|N\rangle, \tag{6.1}$$

where $|N\rangle$ is the hadronic ground state and $J^\mu(x)$ the electromagnetic current. The tool for evaluating these quantities is the *Gell-Mann Low* theorem.

Recall that the time evolution of a Heisenberg field is given by

$$\varphi_H(x,t) = e^{iHt}\varphi_S(x)e^{-iHt}, \tag{6.2}$$

while that of the interaction picture state is

$$\varphi_I(x,t) = e^{iH_0 t}\varphi_S(x)e^{-iH_0 t}. \tag{6.3}$$

6.1 Gell-Mann Low Theorem

Consequently

$$\varphi_H(x,t) = e^{iHt}e^{-iH_0t}\varphi_I(x,t)e^{iH_0t}e^{-iHt}$$
$$= U^\dagger(t,0)\varphi_I(x,t)U(t,0). \tag{6.4}$$

Using $U^\dagger(t,0) = U(0,t)$ we can rewrite this as

$$\varphi_H(x,t) = U(0,t)\varphi_I(x,t)U(t,0). \tag{6.5}$$

We will use this identity to compute generalized Green functions or *correlators*. These are matrix elements of the form

$$G(x_1,\ldots,x_n) = \langle 0|T\{\varphi_H(x_1)\ldots\varphi_H(x_n)\}|0\rangle, \tag{6.6}$$

where $|0\rangle$ is the true ground state of the interacting system. At the moment we only know how to compute

$$G_0(x_1,\ldots,x_n) = {}_0\langle 0|T\{\varphi_I(x_1)\ldots\varphi_I(x_n)\}|0\rangle_0, \tag{6.7}$$

where $|0\rangle_0$ is the ground state of H_0.

We can adiabatically evolve the non-interacting vacuum state into the true $|0\rangle$ by taking $H = H_0 + \epsilon(t)V$ with $\epsilon = 0$ at $t = -\infty$ and $\epsilon = 1$ at $t = 0$. Therefore,

$$|0\rangle = U(0,-\infty)|0\rangle_0 \ (\times phase). \tag{6.8}$$

Similarly, by switching ϵ off at $t = \infty$, we can put the vacuum back to $|0\rangle_0$:

$$|0\rangle_0 \propto U(+\infty,0)|0\rangle = U^\dagger(0,+\infty)|0\rangle. \tag{6.9}$$

In other words

$$|0\rangle = U(0,+\infty)|0\rangle_0 \ (\times phase'), \tag{6.10}$$

and so, taking the adjoint of this,

$$\langle 0| = {}_0\langle 0|U(+\infty,0) \ (\times phase'). \tag{6.11}$$

We could absorb the $(\times phase)$ in either (6.8) or (6.10) into the definition of $|0\rangle$, but not in both simultaneously. Therefore, we must keep this factor in mind.

We sometimes write

$$\Omega^{(\pm)} = U(0,\pm\infty). \tag{6.12}$$

These are called the *Møller wave operators*. Up to $(\times phase)$ we now have

$$\langle 0|T\{\varphi_H(x_1)\ldots\varphi_H(x_n)\}|0\rangle \ (\times phase) =$$
$${}_0\langle 0|U(\infty,0)U(0,t_{last})\varphi_I(t_{last})U(t_{last},0)U(0,t_{next})\varphi_I(t_{next})\ldots U(0,-\infty)|0\rangle_0$$
$$= {}_0\langle 0|U(\infty,t_{last})\varphi_I(t_{last})U(t_{last},t_{next})\varphi_I(t_{next})\ldots U(t_{first},-\infty)|0\rangle_0$$
$$= {}_0\langle 0|T\left\{\varphi_I(x_1)\ldots\varphi_I(x_n)e^{-i\int_{-\infty}^{\infty} H_I(t)dt}\right\}|0\rangle_0. \tag{6.13}$$

The subscripts $first, last$, and so on refer to the ordering of the times x_i^0 from the first (earliest) to the last (latest).

We get rid of the $phase$ by dividing by the same expression without the φ's.

We have thus proved:
Theorem:

$$\langle 0|T\{\varphi_H(x_1)\ldots\varphi_H(x_n)\}|0\rangle$$
$$= \frac{{}_0\langle 0|T\{\varphi_I(x_1)\ldots\varphi_I(x_n)e^{-i\int H_I(t)dt}\}|0\rangle_0}{{}_0\langle 0|T\{e^{-i\int H_I(t)dt}\}|0\rangle_0}. \quad (6.14)$$

COMMENT: The expansion of the numerator should be obvious: we simply expand out the exponential and time-order everything. The numerator is dealt with by omitting the vacuum bubbles.

As an example, the *dressed* φ propagator, $\langle 0|T\{\varphi(x)\varphi(y)\}|0\rangle$ in $\lambda\varphi^3$ theory is given in Fig. 1.

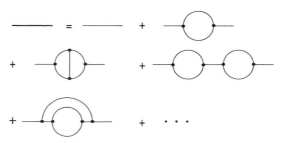

Fig 1. Some contributions to the φ propagator.

6.2 Lehmann-Källén Spectral Representation

Let $\langle 0|T\{\varphi(x)\varphi(y)\}|0\rangle = iG(x,y)$, so

$$iG(x,y) = \theta(x^0 - y^0)\sum_{p,\alpha}\langle 0|\varphi(x)|p,\alpha\rangle\langle p,\alpha|\varphi(y)|0\rangle + (x \leftrightarrow y). \quad (6.15)$$

Here p, α label possible intermediate states. In the interacting theory these will not only be single-particle states, but may be any multi particle state with quantum numbers α and 4-momentum p that can be created by φ. Recall that

$$\varphi(x) = e^{iPx}\varphi(0)e^{-iPx}, \quad (6.16)$$

so

$$\langle p,\alpha|\varphi(x)|0\rangle = \text{c-number}\, e^{ipx}. \quad (6.17)$$

Now define

$$2\pi\rho(p) = \sum_\alpha |\langle 0|\varphi(0)|p,\alpha\rangle|^2. \quad (6.18)$$

6.2 Lehmann-Källén Spectral Representation

By Lorentz invariance and positivity of the energy of the physical states, we have

$$2\pi\rho = 2\pi\theta(p^0)\rho(p^2). \tag{6.19}$$

Therefore

$$\begin{aligned}
iG(x, y) &= \theta(x^0 - y^0) \int \frac{d^4p}{(2\pi)^4} 2\pi\theta(p^0)\rho(p^2)e^{-ip(x-y)} + (x \leftrightarrow y) \\
&= \int_0^\infty dm^2\, \rho(m^2) \left\{ \int \frac{d^4p}{(2\pi)^4} 2\pi\delta(p^2 - m^2)\theta(p^0)e^{-ip(x-y)} + (x \leftrightarrow y) \right\} \\
&= i \int_0^\infty dm^2\, \rho(m^2) G_0(x, y; m^2),
\end{aligned} \tag{6.20}$$

where $G_0(x, y; m^2)$ is the *free* propagator for particles of mass m.

In exactly the same way we can prove

$$\langle 0|[\varphi(x), \varphi(y)]|0\rangle = \int_0^\infty dm^2\, \rho(m^2)\Delta_0(x, y; m^2), \tag{6.21}$$

where $\Delta_0(x, y; m^2)$ is the free field commutator for particles of mass m, and $\rho(m^2)$ is the same function as in the previous equation. This gives a valuable *sum rule* for the *spectral weight* $\rho(m^2)$. Comparing the initial data on the equal-time surface for both sides of this equation, and recalling that

$$\partial_{y^0}\Delta(x, y) = [\varphi(\mathbf{x}, t), \partial_{y^0}\varphi(\mathbf{y}, t)] = i\delta^3(\mathbf{x} - \mathbf{y}) \tag{6.22}$$

for both free *and* interacting fields, we find that we must have

$$1 = \int_0^\infty dm^2\, \rho(m^2). \tag{6.23}$$

Now $\rho(m^2)$ should contain contributions from the single-particle φ particle state, which has a definite mass, and also contributions from many-particle states, which may have a continuum of invariant masses. If we define the *wave-function renormalization constant* Z, the efficiency with which the φ field couples to the φ particle, by

$$\langle p|\varphi(x)|0\rangle = \sqrt{Z}e^{ipx}, \tag{6.24}$$

then we expect

$$\rho(m^2) = Z\delta(m^2 - m_\varphi^2) + \sigma(m^2). \tag{6.25}$$

Here $\sigma(m^2)$ is the contribution from the many-particle intermediate states, which (for $\lambda\varphi^3$ theory) start at $4m_\varphi^2$. Thus,

$$1 = Z + \int_{4m_\varphi^2}^\infty dm^2\, \sigma(m^2). \tag{6.26}$$

Since $\sigma(m^2) \geq 0$ (with equality only if we have a free theory), we have the important constraint $Z \leq 1$.

6. Formal Developments

We can therefore write

$$G(p) = \frac{Z}{p^2 - m_\varphi^2} + \int_{4m_\varphi^2}^\infty dm^2 \, \sigma(m^2) \frac{1}{p^2 - m^2}. \tag{6.27}$$

Equation (6.27) is in the form of a *dispersion relation* exhibiting both a cut and a pole.

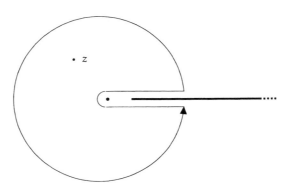

Fig 2. Contour for the dispersion relation.

Recall from complex analysis that an analytic function can be expressed by means of Cauchy's theorem in the form

$$f(z) = \frac{1}{2\pi i} \oint_C \frac{f(\zeta)}{\zeta - z} dz. \tag{6.28}$$

When the function $f(z)$ is real on the real axis except for a cut from a to ∞, and when the function falls off rapidly enough that the contribution from the large-radius circle can be neglected, then $f(z)$ can be written

$$f(z) = \frac{1}{\pi} \int_a^\infty \frac{\operatorname{Im} f(s)}{s - z} ds. \tag{6.29}$$

So we identify $\operatorname{Im} G/\pi$ with $\rho(m^2)$.

As an aside I should point out that the ρ sum rule is the generalization of a similar sum rule in quantum mechanics. For the unit-mass harmonic oscillator of frequency Ω, the analog of the propagator $iG_0(t) = \langle 0|T\{x(t)x(0)\}|0\rangle$ is easily calculated to be

$$\langle 0|T\{x(t)x(0)\}|0\rangle = \frac{e^{-i\Omega|t|}}{2\Omega}. \tag{6.30}$$

Consider the same quantity for a generic system with one degree of freedom:

$$\langle 0|T\{x(t)x(0)\}|0\rangle = \sum_n |\langle 0|x(0)|n\rangle|^2 \, e^{-i(E_n - E_0)|t|}$$

$$= \int dE \left\{ \sum_n \delta(E - (E_n - E_0)) 2E \, |\langle 0|x(0)|n\rangle|^2 \right\} \frac{e^{-iE|t|}}{2E}$$

$$= \int dE \rho(E) i G_0(t, E), \tag{6.31}$$

where

$$\rho(E) = \sum_n \delta(E - (E_n - E_0)) 2E |\langle 0|x(0)|n\rangle|^2. \tag{6.32}$$

Now

$$\int dE \, \rho(E) = \sum_n 2(E_n - E_0) |\langle 0|x(0)|n\rangle|^2$$
$$= \langle 0|[x, [H, x]]|0\rangle = \langle 0|[x, -ip]|0\rangle = 1. \tag{6.33}$$

This result for $\sum_n 2(E_n - E_0) |\langle 0|x(0)|n\rangle|^2$ is called the Thomas-Reiche-Kuhn sum rule in atomic physics, and the f-sum rule in solid state.

6.3 LSZ Reduction Formulae

6.3.1 Amputation of External Legs

The momentum-space Feynman diagram expansion for the correlation function

$$G(x_1, \ldots, x_n) = \langle 0|T\{\varphi_H(x_1) \ldots \varphi_H(x_n)\}|0\rangle \tag{6.34}$$

of Section 6.1 looks very much like the Feynman diagram expansion of the S matrix. The principal differences are that in correlation functions

- The momenta running into the external legs are not necessarily on the mass-shell.
- The external legs represent propagators, and these contain self-energy insertions of the sort we thought reasonable to omit when calculating S.

Given these these minor differences it is clear that we should be able to extract the S matrix from the correlation functions. To do this we must pull off their legs, and put the momenta on-shell.

Look at

$$\langle 0|T\{\varphi(x_1) \ldots \varphi(x_n)\varphi(y_1) \ldots \varphi(y_m)\}|0\rangle \tag{6.35}$$

and take the momentum running into each of the y variables to be nearly on the mass-shell for incoming particles while the momentum running into the x variables are nearly on-shell for outgoing particles. For nearly-on-shell momentum, the poles in the propagators on the external legs should dominate the expression and it should be true that

$$\int d^4[x] d^4[y] e^{-ip_i x_i + ik_j y_j} \langle 0|T\{\varphi(x_1) \ldots \varphi(x_n)\varphi(y_1) \ldots \varphi(y_m)\}|0\rangle$$
$$\approx \prod_i \frac{i\sqrt{Z}}{p_i^2 - m^2} \prod_j \frac{i\sqrt{Z}}{k_j^2 - m^2} \langle \mathbf{p}_1, \ldots, \mathbf{p}_n|S|\mathbf{k}_1, \ldots, \mathbf{k}_m\rangle. \tag{6.36}$$

Granted this, we can write

$$\langle \mathbf{p}_1, \ldots, \mathbf{p}_n | S | \mathbf{k}_1, \ldots, \mathbf{k}_m \rangle$$
$$= \frac{i^{(n+m)}}{(\sqrt{Z})^{(n+m)}} \int d^4[x] d^4[y] e^{-ip_i x_i + ik_j y_j} \prod_i (\partial_{x_i}^2 + m^2) \prod_j (\partial_{y_j}^2 + m^2)$$
$$\times \langle 0 | T \{ \varphi(x_1) \ldots \varphi(x_n) \varphi(y_1) \ldots \varphi(y_n) \} | 0 \rangle. \tag{6.37}$$

In writing down (6.37) I have tacitly assumed that the incoming and outgoing states were not just connected by a single straight-through propagator, so we are always considering S-matrix elements between states whose initial and final momenta are all different.

Equation (6.37) is called the LSZ (Lehman, Symanzik and Zimmerman) reduction formula. In "deriving" it I have elided over numerous subtleties to do with wavepackets and the necessity of being able to clearly separate the asymptotic particles. A good discussion is to be found in Gasiorowicz' *Elementary Particle Physics*. I will give a brief summary in the next section.

6.3.2 In and Out States and Fields

While I will not develop the LSZ formalism in any great detail, there are ingredients worth discussing. The first is the notion of *in* and *out* states and fields. These are introduced to escape from the necessity of adiabatic switching on and off of the interactions.

In the Heisenberg picture the states do not evolve with time. This does not mean that nothing happens. A state $|\mathbf{k}_1, \ldots \mathbf{k}_n; in\rangle$ that in the far distant past describes n well-separated particles with momenta[1] $\mathbf{k}_1, \ldots \mathbf{k}_n$ will not have this simple description when examined in the far future. There it will best be described as a linear combination of states $|\mathbf{p}_1, \ldots \mathbf{p}_m; out\rangle$ that contain well-separated particles of momentum $\mathbf{p}_1, \ldots \mathbf{p}_m$ at that time. In general a state with a simple incoming description (an *in* state) can be expanded in terms of those with simple late-time descriptions (*out* states) as

$$|\mathbf{k}_1, \ldots \mathbf{k}_n; in\rangle = \sum_{\mathbf{p}} |\mathbf{p}_1, \ldots \mathbf{p}_m; out\rangle S_{\mathbf{p}_1,\ldots \mathbf{p}_m; \mathbf{k}_1,\ldots \mathbf{k}_n}, \tag{6.38}$$

where $S_{\mathbf{p}_1,\ldots \mathbf{p}_m; \mathbf{k}_1,\ldots \mathbf{k}_n}$ are elements of the S matrix. These can also be written

$$S_{\mathbf{p}_1,\ldots \mathbf{p}_m; \mathbf{k}_1,\ldots \mathbf{k}_n} = \langle \mathbf{p}_1, \ldots \mathbf{p}_m; out | \mathbf{k}_1, \ldots \mathbf{k}_n; in \rangle. \tag{6.39}$$

Yet another equivalent way of saying this is that the operator whose matrix elements are $S_{\mathbf{k};\mathbf{p}}$ (I will use the symbols $\mathbf{k}$, $\mathbf{p}$ as shorthand for the collection of $\mathbf{k}$'s and $\mathbf{p}$'s) acts as

$$S | \mathbf{k}_1, \ldots \mathbf{k}_n; out \rangle = | \mathbf{k}_1, \ldots \mathbf{k}_n; in \rangle. \tag{6.40}$$

---

[1] Strictly speaking I must use wavepackets made of momenta close to these. True plane waves are never well separated

6.3 LSZ Reduction Formulae

Since $|\mathbf{k}; in\rangle = S|\mathbf{k}; out\rangle$ is equivalent to

$$\langle \mathbf{k}; out| S^\dagger = \langle \mathbf{k}; in|, \quad (6.41)$$

we find

$$S_{\mathbf{k};\mathbf{p}} = \langle \mathbf{k}; out|S|\mathbf{p}; out\rangle = \langle \mathbf{k}; in|S|\mathbf{p}; in\rangle. \quad (6.42)$$

In the absence of bound states, either of the set of states $|\mathbf{k}; in\rangle$ or $|\mathbf{p}; out\rangle$ will form a basis of the Hilbert space. Only the vacuum state and the single-particle states are common to both sets. (A single stable particle with momentum $\mathbf{k}$ in the far distant past will be a single particle with momentum $\mathbf{k}$ forever.) Since the *in* and *out* states have the same labels and inner products as a free-particle Fock space, we can define operators $\hat{a}_{in}(\mathbf{k})$, $\hat{a}_{out}(\mathbf{k})$ and their adjoints and construct the *in/out* bases by applying creation operators to the vacuum $|0\rangle$:

$$|\mathbf{k}_1, \ldots \mathbf{k}_n; out\rangle = \hat{a}^\dagger_{out}(\mathbf{k}_1) \ldots \hat{a}^\dagger_{out}(\mathbf{k}_n)|0\rangle. \quad (6.43)$$

The *in* and *out* creation and annihilation operators are connected via the S matrix so that, for example, $\hat{a}^\dagger_{in}(\mathbf{k}) = S\hat{a}^\dagger_{out}(\mathbf{k})S^{-1}$.

From $\hat{a}_{in}(\mathbf{k})$, $\hat{a}_{out}(\mathbf{k})$, and their *out* equivalents we can construct *in* and *out* fields $\varphi_{in}(x)$ and $\varphi_{out}(x)$. These fields satisfy the canonical commutation relations

$$\delta(x_0 - x'_0)[\varphi_{in}(x), \partial_0 \varphi_{in}(x')] = i\delta^4(x - x'). \quad (6.44)$$

To derive the LSZ formalism we assume that there is a Heisenberg field $\tilde{\varphi}(x)$, which "interpolates" between the *in* and *out* fields in that, in some sense,

$$\lim_{t \to \infty} \tilde{\varphi}(x) = \varphi_{out}(x), \quad \lim_{t \to -\infty} \tilde{\varphi}(x) = \varphi_{in}(x). \quad (6.45)$$

This limit has to be interpreted with care. It can only be correct as a limit of matrix elements

$$\lim_{t \to \infty} \langle \Phi|\tilde{\varphi}(x)|\Psi\rangle = \langle \Phi|\varphi_{out}(x)|\Psi\rangle \quad (6.46)$$

between normalizable states. This is because only normalizable states correspond to wave packets that will eventually become well-separated. Convergence only in this *weak operator topology* sense is not strong enough to ensure that the limit of a product is the product of the limits. Consequently, it is generally not true that the limit of a commutator is the commutator of the limits.

$$\lim_{t \to \infty} \langle \Phi|[\tilde{\varphi}(x), \tilde{\varphi}(x')]|\Psi\rangle \neq \langle \Phi|[\varphi_{out}(x), \varphi_{out}(x')]|\Psi\rangle. \quad (6.47)$$

Because of this we cannot deduce that $\tilde{\varphi}(x)$ has the scale set by the canonical commutation relations.

This is fortunate because we would like to identify $\varphi(x)$ with some scalar multiple of $\tilde{\varphi}(x)$. We know that both the Heisenberg field $\varphi(x)$ that we used in deriving the Lehman-Källén sum rule and the $\varphi_{in/out}$ fields obey canonical commutation relations. On the other hand, $\varphi(x)$ also satisfies

$$\langle p|\varphi(x)|0\rangle = \sqrt{Z}e^{ipx} \quad (6.48)$$

while $\varphi_{in}(x)$ has

$$\langle p|\varphi_{in}(x)|0\rangle = e^{ipx}. \tag{6.49}$$

This requires that we should identify

$$\varphi(x) = \sqrt{Z}\tilde{\varphi}(x). \tag{6.50}$$

If we *were* allowed to interchange the limiting process $\sqrt{Z}\varphi(x) \to \varphi_{in}(x)$ and operator multiplication, we would have a conflict beween the left and right hand sides of (6.47).

We usually call $\tilde{\varphi}(x)$ the *renormalized* field and $\varphi(x)$ the *bare* field. Since $\sqrt{Z}$ is usually given by divergent expressions and is cut-off dependent it is necessary to use $\tilde{\varphi}(x)$ when computing physical quantities. The $\sqrt{Z}$'s in (6.37) serve to make this conversion.

With this preamble I can sketch how the LSZ formalism works for the hermitian KG field. Recall from equation (2.30) that

$$\hat{a}_\mathbf{k} = (e^{-ikx}, \varphi), \quad \hat{a}^\dagger_\mathbf{k} = (\varphi, e^{-ikx}) \tag{6.51}$$

where the parentheses denote the KG equation inner-product. We may replace the plane-wave solution e^{-ikx} of the KG equation by a normalizable wavepacket solution $f_\mathbf{k}(x)$ that contains a range of momenta centered around $\mathbf{k}$. We will take this wavepacket to be sufficiently spread-out in space that we may regard $f_\mathbf{k}(x)$ as almost being the plane-wave state $\mathbf{k}$ while yet remaining normalizable.

With this in mind, consider the S matrix element

$$\begin{aligned} S_{\mathbf{p}_1,\mathbf{p}_2;\mathbf{k}_1,\mathbf{k}_2} &= \langle \mathbf{p}_1, \mathbf{p}_2; out|\mathbf{k}_1, \mathbf{k}_2; in\rangle \\ &= \langle \mathbf{p}_1, \mathbf{p}_2; out|\hat{a}^\dagger_{in}(\mathbf{k}_1)|\mathbf{k}_2; in\rangle. \end{aligned} \tag{6.52}$$

Using the weak limit

$$\lim_{t\to-\infty} \varphi(x) = \sqrt{Z}\varphi_{in}(x) \tag{6.53}$$

we can replace $\hat{a}^\dagger_{in}(\mathbf{k}_1)$ by $\lim_{t\to-\infty} \hat{a}^\dagger(\mathbf{k}_1, t)$ where

$$\hat{a}^\dagger(\mathbf{k}_1, t) = \\ = \frac{i}{\sqrt{Z}} \int_{x_0=t} \left(\varphi(x)\partial_0 f_{\mathbf{k}_1}(x) - (\partial_0\varphi(x))f_{\mathbf{k}_1}(x)\right) d^3x. \tag{6.54}$$

Now consider the four-dimensional Green's identity

$$\begin{aligned} i\int d^4x &\left(\varphi(\partial^2 + m^2)f_{\mathbf{k}_1} - ((\partial^2 + m^2)\varphi)f_{\mathbf{k}_1}\right) \\ &= i\int_{x_0=+\infty} \left(\varphi\partial_0 f_{\mathbf{k}_1} - (\partial_0\varphi)f_{\mathbf{k}_1}\right) d^3x \\ &\quad - i\int_{x_0=-\infty} \left(\varphi\partial_0 f_{\mathbf{k}_1} - (\partial_0\varphi)f_{\mathbf{k}_1}\right) d^3x \end{aligned} \tag{6.55}$$

[the integration by parts in the spatial directions is allowed because $f_{\mathbf{k}_1}(x)$ is normalizable]. Provided that $\mathbf{k}_1$ is distinct from both $\mathbf{p}_1$ and $\mathbf{p}_2$, we may use (6.55) and the fact that $f_{\mathbf{k}_1}(x)$ obeys the KG equation to see that

$$\langle \mathbf{p}_1, \mathbf{p}_2; out|\hat{a}_{in}^\dagger(\mathbf{k}_1)|\mathbf{k}_2; in\rangle$$
$$= \frac{i}{\sqrt{Z}} \int d^4x f_{\mathbf{k}_1}(\partial^2 + m^2)\langle \mathbf{p}_1, \mathbf{p}_2; out|\varphi(x)|\mathbf{k}_2; in\rangle. \tag{6.56}$$

The condition on $\mathbf{k}_1$, $\mathbf{p}_1$, and $\mathbf{p}_2$ ensures that the $\hat{a}_{out}^\dagger(\mathbf{k}_1)$, which is generated by the $t = +\infty$ term, annihilates $\langle \mathbf{p}_1, \mathbf{p}_2; out|$.

Proceeding in this manner we can systematicaly trade-off the annihilation and creation operators for fields, and so establish (6.37).

6.3.3 Borcher's Classes

Again I can only sketch this topic. In deriving the scattering matrix from the correlation functions of field operators nothing requires that the interpolating field φ should be related to any field appearing in the lagrangian. Indeed, the LSZ formalism makes no direct reference to any lagrangian or hamiltonian. All we need is an interpolating field operator and its correlation functions, and we can compute an S-matrix. The previous section suggests that a candidate interpolating field needs only the correct Lorentz transformation properties, a non-vanishing matrix element between the vacuum and the single-particle states

$$\langle p|\varphi(x)|0\rangle = \sqrt{Z}e^{ipx}, \tag{6.57}$$

and then the S matrix for the particles it couples to can be computed from its time-ordered vacuum expectation values (TVEV's) according to (6.37). If you think about it for a moment, you will realize that this assertion cannot be quite corect. The field $\varphi_{in}(x)$ has the required properties, but has free-field TVEV's and so gives a free-theory S matrix. The subtle error is that, when proving that any two fields with the stated properties give the same S matrix, we tacitly assume the Lorentz covariance of TVEV's simultaneously involving both sets of fields. This is not the case unless the two sets of fields *commute at space-like separation*. Otherwise a Lorentz transformation may alter the order of the field opertors and so alter the value of the TVEV. The set of fields which connect the vacuum to the required states and are *mutually local* in that they commute at space-like separation is called a *Borcher's class*. All members of this class give the same S matrix. For example the correlation functions of $\varphi(x)$ and of $\sin \varphi(x)$ give the same S-matrix.

7
Fermions

We now begin to study quantum fields that have no classical equvalent.

7.1 Dirac Equation

In 1928, realizing that the nonpositive-definite inner product that comes along with the Klein-Gordon equation had its origin in the second-order time derivative, Dirac[1] decided to try to find an alternative equation having only a single time derivative. He tried an equation of the form

$$\left(-i\gamma^\mu \partial_\mu + m\right)\psi = 0, \tag{7.1}$$

where the γ^μ were some undefined symbols that he assumed to commute with p_μ, although not necessarily among themselves. Seeking a plane-wave solution of the form $\psi \propto e^{ipx}$, he found

$$\gamma^\mu p_\mu + m = 0. \tag{7.2}$$

No doubt inspired by the notion of a complex conjugate, he then observed that, if we multiply this by the still undefined quantity $\gamma^\mu p_\mu - m$, we find

$$\gamma^\mu \gamma^\nu p_\mu p_\nu - m^2 = 0. \tag{7.3}$$

---

[1] Paul Adrien Maurice Dirac, Born: August 8, 1902, Bristol, Gloucestershire, UK. Died: October 20, 1984, Tallahassee, FL., Nobel Prize for Physics 1933.

By demanding that

$$\{\gamma^\mu, \gamma^\nu\} \equiv \gamma^\mu\gamma^\nu + \gamma^\nu\gamma^\mu = 2g^{\mu\nu}, \quad (7.4)$$

he made (7.3) equivalent to

$$p^2 = m^2. \quad (7.5)$$

This is the mass-shell condition. Dirac had therefore produced an equation compatible with the relativistic energy-momentum relation. Of course he still needed a set of objects with anti-commutation relations

$$\{\gamma^\mu, \gamma^\nu\} = 2g^{\mu\nu}. \quad (7.6)$$

Such a set is said to form a *Clifford* algebra.[2] Dirac, probably not knowing of Clifford's work, was able to find a representation of the algebra by 4×4 matrices. These are now known as the Dirac gamma matrices. In the appendix we show that we can find suitable $2^N \times 2^N$ matrices for any euclidean space of dimension $2N$.

The Lorentz covariance of the Dirac equation is guaranteed if there exists a matrix representation $S(L)$ of the Lorentz group so that for any Lorentz transformation $L^\mu{}_\nu$ there exists a matrix $S(L)$ such that

$$S(L)\gamma^\mu S^{-1}(L) = (L^{-1})^\mu{}_\nu \gamma^\nu. \quad (7.7)$$

[The reader should make sure that she understands why the representation property requires L^{-1}, and not L, on the right-hand side of (7.7).] If $z^\mu = L^\mu{}_\nu x^\nu$, then

$$\frac{\partial}{\partial x^\nu} = \frac{\partial z^\mu}{\partial x^\nu}\frac{\partial}{\partial z^\mu} = L^\mu{}_\nu \frac{\partial}{\partial z^\mu}. \quad (7.8)$$

From this we have that

$$\gamma^\nu \frac{\partial}{\partial x^\nu} = \gamma^\nu L^\mu{}_\nu \frac{\partial}{\partial z^\mu} = S^{-1}(L)\gamma^\mu S(L)\frac{\partial}{\partial z^\mu}. \quad (7.9)$$

This shows that if

$$\left(-i\gamma^\mu \frac{\partial}{\partial x^\mu} + m\right)\psi = 0, \quad (7.10)$$

then

$$\left(-i\gamma^\mu \frac{\partial}{\partial z^\mu} + m\right)S\psi = 0. \quad (7.11)$$

The existence of such an $S(L)$ is demonstrated in Appendix D.

---

[2] Clifford, William Kingdon Clifford, Born May 4, 1845, Exeter, Devon, UK. Died March 3, 1879, Madeira Islands, Portugal.

7.2 Spinors, Tensors, and Currents

7.2.1 Field Bilinears

In the appendix we show how to construct Dirac matrices for an even-dimensional space with a euclidean metric $g_{\mu\nu} = \delta_{\mu\nu}$. These matrices are hermitian. To obtain matrices for a Minkowski signature metric where the 1, 2, 3 diagonal elements are negative, it suffices to set $\gamma_a^{Minkowski} = i\gamma_a^{Euclid}$, $a = 1, 2, 3$ and $\gamma_0^{Minkowski} = \gamma_0^{Euclid}$. The resulting matrices are no longer all hermitian, but instead obey

$$(\gamma^\mu)^\dagger = \gamma^0 \gamma^\mu \gamma^0. \tag{7.12}$$

Because of this modified hermiticity condition, it is convenient to define a different form of conjugate for the field ψ. We define

$$\bar\psi = \psi^\dagger \gamma^0. \tag{7.13}$$

This has the valuable property that

$$\psi \to S\psi \Rightarrow \bar\psi \to (\psi^\dagger S^\dagger)\gamma^0 = \psi^\dagger \gamma^0 S^{-1} = \bar\psi S^{-1}. \tag{7.14}$$

We see that $\bar\psi\psi$ is a Lorentz invariant. We also see that

$$\bar\psi \gamma^\mu \psi \to \bar\psi S^{-1} \gamma^\mu S \psi = L^\mu{}_\nu \bar\psi \gamma^\nu \psi, \tag{7.15}$$

so this combination is a Lorentz vector.

Define[3] $\gamma^5 = i\gamma^0 \gamma^1 \gamma^2 \gamma^3$. We find that

$$\gamma_5^2 = 1, \qquad (\gamma_5)^\dagger = \gamma_5. \tag{7.16}$$

Look at the transformation properties of the current $\bar\psi \gamma^5 \gamma^\mu \psi$. We find

$$\bar\psi \gamma^5 \gamma^\mu \psi \to i\bar\psi(S^{-1}\gamma^0 S S^{-1}\gamma^1 S S^{-1}\gamma^2 S S^{-1}\gamma^3 S S^{-1}\gamma^\mu S)\psi$$
$$= iL^0{}_\alpha L^1{}_\beta L^2{}_\gamma L^3{}_\delta L^\mu{}_\nu \bar\psi(\gamma^\alpha \gamma^\beta \gamma^\gamma \gamma^\delta \gamma^\nu)\psi. \tag{7.17}$$

If $\alpha, \beta, \gamma, \delta$ are all different, then this is equal to

$$\epsilon_{\alpha\beta\gamma\delta} L^0{}_\alpha L^1{}_\beta L^2{}_\gamma L^3{}_\delta L^\mu{}_\nu \bar\psi \gamma^5 \gamma^\nu \psi$$
$$= (\det L) L^\mu{}_\nu \bar\psi \gamma^5 \gamma^\nu \psi. \tag{7.18}$$

On the other hand, if any pair of indices coincide, we use $(\gamma^\alpha)^2 = 2g^{\alpha\alpha}$ and $L^i{}_\alpha L^j{}_\alpha = \delta^{ij}$ to deduce that the expression is zero. The quantity $\bar\psi \gamma^5 \gamma^\mu \psi$ is therefore a pseudo-vector.

Similarly $\bar\psi \gamma^5 \psi$ is a pseudo-scalar and $\bar\psi \sigma_{\mu\nu} \psi$ is a symmetric tensor.

If we count components, we find that there are six $\sigma_{\mu\nu}$'s, four γ^μ's, four $\gamma^5 \gamma^\mu$'s, one γ^5, and finally one identity matrix 1. The total number of Lorentz components is therefore 16. This is also the number of entries in any 4×4 matrix. Indeed it is easy to show that the set $\{\gamma^\mu, \gamma^5 \gamma^\mu, \sigma_{\mu\nu}, \gamma^5, 1\}$ is linearly independent and spans the space of 4×4 matrices.

---

[3] This is the Minkowski signature, γ^5, and differs by a factor of i from the euclidean signature version in the appendix.

7.2.2 Conservation Laws

From
$$(-i\gamma^\mu \partial_\mu + m)\psi = 0 \tag{7.19}$$
we find
$$+i\partial_\mu \psi^\dagger (\gamma^\mu)^\dagger + m\psi^\dagger = 0 \tag{7.20}$$
or
$$i\partial_\mu \bar{\psi}\gamma^\mu + m\bar{\psi} = 0. \tag{7.21}$$
Therefore
$$\partial_\mu \left(\bar{\psi}\gamma^\mu \psi\right) = \bar{\psi}\gamma^\mu \partial_\mu \psi + \left(\partial_\mu \bar{\psi}\right)\gamma^\mu \psi = 0. \tag{7.22}$$

Thus the current $J_\mu = \bar{\psi}\gamma_\mu \psi$ is conserved. Note that $J_0 = \psi^\dagger \psi$. This is clearly positive definite and could, if we wished, be regarded as a conserved probability density. With this interpretation Dirac had achieved his goal of a paradox-free relativistic quantum mechanics. Unfortunately, as we will see in the next section, he had paid the rather high price of having a quantum-mechanical system with no lowest energy state. Today we do not interpret $\psi^\dagger \psi$ as a probability density. Just as with the KG equation, we regard it as the operator representing the *charge* density i.e. the difference between the density of particles and antiparticles.

We also find that
$$\partial_\mu \left(\bar{\psi}\gamma^5 \gamma^\mu \psi\right) = -2im\bar{\psi}\psi. \tag{7.23}$$
Thus the pseudo-vector current, $J_\mu^5 = \bar{\psi}\gamma^5 \gamma_\mu \psi$, is conserved if the particle is massless.

7.3 Holes and the Dirac Sea

7.3.1 Positive and Negative Energies

Let us seek a hamiltonian for the Dirac particles. From
$$(-i\gamma^\mu \partial_\mu + m)\psi = 0 \tag{7.24}$$
we get
$$i\partial_t \psi = (+m\gamma^0 - i\gamma^0 \gamma^a \partial_a)\psi$$
$$= (\beta m + \alpha \cdot p)\psi \tag{7.25}$$
where
$$\alpha^a = \gamma^0 \gamma^a \quad \beta = \gamma^0 \tag{7.26}$$
is the traditional notation for these matrices. Comparing with
$$i\partial_t \psi = \hat{H}\psi \tag{7.27}$$

7. Fermions

gives us the Dirac hamiltonian $\hat{H} = (\beta m + \alpha \cdot p)$.

In analyzing the eigenstates of $\mathcal{H}$ it is convenient to consider separately the cases $m = 0$ and $m \neq 0$.

We adopt a *chiral*, or *Weyl*[4], matrix representation for the γ's. Here we take

$$\gamma^5 = \begin{pmatrix} 1 & 0 \\ 0 & -1 \end{pmatrix}, \quad \gamma^0 = \begin{pmatrix} 0 & 1 \\ 1 & 0 \end{pmatrix}, \quad \gamma^a = \begin{pmatrix} 0 & -\sigma^a \\ \sigma^a & 0 \end{pmatrix}. \tag{7.28}$$

We find

$$\hat{H} = \begin{pmatrix} \sigma \cdot \mathbf{p} & m \\ m & -\sigma \cdot \mathbf{p} \end{pmatrix}. \tag{7.29}$$

The spin operator associated with rotations turns out to be

$$\Sigma = \begin{pmatrix} \sigma & 0 \\ 0 & \sigma \end{pmatrix}, \tag{7.30}$$

so the eigenvalues of the σ's that appear in $\hat{H}$ are the physical spin of the particle.

For massless particles (it is for this case that the Weyl basis is most useful) we have two classes of solutions distinguished by their *chirality*, i.e. the eigenvalue of γ^5. (Note that γ^5 commutes with $\hat{H}$ for massless particles. This is why J^5_μ is conserved: it counts the density of $\gamma^5 = +1$ particles minus the number of $\gamma^5 = -1$ particles.) First chirality $+1$ states are of the form

$$\psi = \begin{pmatrix} \psi_R \\ 0 \end{pmatrix}. \tag{7.31}$$

They have energy given by the eigenvalues $\pm |\mathbf{p}|$ of $\sigma \cdot \mathbf{p}$, and thus have positive energy when the spin and momentum point in the same direction, as with a right-handed screw. We say that these right-handed states have positive *helicity*. The negative chirality states are of the form

$$\psi = \begin{pmatrix} 0 \\ \psi_l \end{pmatrix}. \tag{7.32}$$

Here the positive energy solutions have the spin and momentum pointing in opposite directions i.e. they have negative helicity. Since the right and left chiralities are not coupled, we could take either one on its own and still have a Lorentz invariant theory. The resulting two-component spinors are called *Weyl fermions*. Right-handed Weyl fermions satisfy the *Weyl equation*

$$i(\partial + \sigma \cdot \nabla)\psi_R = 0. \tag{7.33}$$

Neutrinos are candidate Weyl fermions. They only interact via gravity or weak interactions. In weak processes only the left-handed chiral component is involved. There is some evidence for a small neutrino mass, but if they were strictly massless any neutrinos created by weak interactions would remain left-handed for ever, and

---

[4]Hermann Weyl, Born November 9, 1885, Elmshorn, near Hamburg. Died December 8, 1955, Zürich, Switzerland.

7.3 Holes and the Dirac Sea

the corresponding field will obey the left-handed Weyl equation. Any right-handed counterparts, perhaps left over from the big bang, will decouple and be indetectable except for effects on a cosmological scale (hot dark matter).

There are also negative energy eigensolutions. Here for chirality $+1$ particles, $\sigma \cdot \mathbf{p} < 0$. We will soon learn to associate these with antiparticles. The antiparticle of a right-handed Weyl particle has its spin and momentum 3-vector pointing in opposite directions. Thus a chirality $+1$ *particle* has positive helicity, while its *antiparticle* has negative helicity.

To find the solutions when $m \neq 0$, we can use some linear algebra projection operator tricks. We first note that

$$\begin{aligned} \hat{H}^2 &= (\gamma^0(\gamma^i p_i + m))^2 \\ &= -(\gamma^i p_i - m)(\gamma^0)^2(\gamma^j p_j + m) \\ &= \mathbf{p}^2 + m^2. \end{aligned} \quad (7.34)$$

The right-hand side is proportional to the identity in spinor space. This means that $\hat{H}^2$ has eigenfunctions $\psi = \chi e^{ipx}$, where χ is any constant 4-spinor, and the 4-momentum is on-shell, $p^2 = m^2$, so that the energy takes the values $\pm\sqrt{\mathbf{p}^2 + m^2}$.

Now we use Dirac's spectral trick from his book on quantum mechanics to observe that for any hamiltonian obeying $\hat{H}^2 = E^2$, where $E = +\sqrt{\mathbf{p}^2 + m^2}$, we can decompose the identity operator as

$$\begin{aligned} 1 &= \frac{1}{2E}(E + \hat{H}) + \frac{1}{2E}(E - \hat{H}) \\ &= P_+ + P_-. \end{aligned} \quad (7.35)$$

Here $P_\pm$ obey

$$P_\pm^2 = \frac{1}{4E^2}(E^2 + \hat{H}^2 \pm 2E\hat{H}) = \frac{1}{4E^2}(2E^2 \pm 2E\hat{H}) = P_\pm \quad (7.36)$$

and

$$P_+ P_- = \frac{1}{4E^2}(E^2 - \hat{H}^2) = 0. \quad (7.37)$$

We also have that

$$\hat{H} P_\pm = \frac{1}{2E}(\pm \hat{H}^2 + E\hat{H}) = \pm E P_\pm. \quad (7.38)$$

The $P_\pm$ are therefore projection operators onto the eigenspaces of eigenvalue $\pm E$. The space of constant spinors therefore decomposes $\mathcal{H} = \mathcal{H}_+ \oplus \mathcal{H}_-$, where $\mathcal{H}_\pm$ are the spaces of positive and negative energy solutions, and $\mathcal{H}_\pm = P_\pm \mathcal{H}$.

By applying $P_\pm$ to any constant spinor χ, we produce eigenstates of $\hat{H}$. If we start with a χ which in an eigenstate of spin, we end up with an energy eigenstate with the same spin.

7.3.2 Holes

Originally it seemed that the negative energy states $\in \mathcal{H}_-$ were an undesirable feature of the theory, since they suggested that the energy of the system is not bounded below. Later, in 1930, Dirac came up with the notion of what is now called the *Dirac sea*. He realized that his equation could only apply to *fermions*, i.e, particles which obey the Pauli exclusion principle. Given this, we simply assume that all the negative energy states are already occupied by particles so there is no room for any more. Then, although the single-particle energies can be arbitrarily negative, the energy of the many-body states is bounded below: the addition or removal of a particle always increases the energy.

To grasp this, think of massless chirality $+1$ particles where $E = \sigma \cdot \mathbf{p}$. If we delete a sea-particle from a state with momentum $-\mathbf{k}$, spin $\sigma \parallel \mathbf{k}$, and energy $-|E| = +\sigma \cdot (-\mathbf{k})$, we have increased the energy of the system by $+|E|$, the momentum by $\mathbf{k}$, and the spin by $-\sigma$. The resultant *hole* appears as a particle with positive energy but its spin and momentum are antiparallel. This is an *antiparticle*.

It is worth pointing out that all particles, positive and negative chirality, particles and antiparticles, all have velocity parallel to their momentum. This is not quite obvious. There are many minus signs to confuse us. Remember that the velocity of a packet of waves with a dispersion relation $\omega(\mathbf{p})$ is determined by the group velocity

$$\mathbf{v}_g = \frac{\partial \omega}{\partial \mathbf{p}}. \tag{7.39}$$

For example, consider negative energy states where

$$E_\mathbf{p} = -\sqrt{\mathbf{p}^2 + m^2}. \tag{7.40}$$

The velocity is

$$\mathbf{v}_g = \frac{-\mathbf{p}}{\sqrt{\mathbf{p}^2 + m^2}}. \tag{7.41}$$

This is in the direction $-\mathbf{p}$ — but the momentum of the hole *is* in the direction $-\mathbf{p}$!

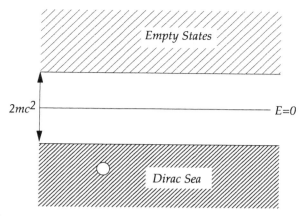

Fig 1. A hole in the Dirac sea.

7.4 Quantization

The Dirac equation can be obtained by variation of the action

$$S = \int d^4x \left[\bar{\psi} \left(i\gamma^\mu \partial_\mu - m \right) \psi \right]. \tag{7.42}$$

If we vary $\bar{\psi}$, we find

$$\left(-i\gamma^\mu \partial_\mu + m \right) \psi = 0, \tag{7.43}$$

and if we vary ψ, we find

$$i\partial_\mu \bar{\psi} \gamma^\mu + m\bar{\psi} = 0. \tag{7.44}$$

The momentum conjugate to ψ is $i\psi^\dagger$, suggesting that we should quantize by requiring

$$[\psi_\alpha(\mathbf{x}), \psi_\beta^\dagger(\mathbf{x}')] \stackrel{?}{=} \delta_{\alpha\beta} \delta^3(\mathbf{x} - \mathbf{x}'). \tag{7.45}$$

This does not work. We run into the unbounded spectrum problems that were solved by the hole theory and the assumption that the Dirac particles obeyed the exclusion principle. We must quantize the ψ field as a *fermion*, and this requires taking

$$\{\psi_\alpha(\mathbf{x}), \psi_\beta^\dagger(\mathbf{x}')\} = \delta_{\alpha\beta} \delta^3(\mathbf{x} - \mathbf{x}'). \tag{7.46}$$

For obtaining the mode expansion it is preferable to work directly with the Dirac equation rather than with the hamiltonian. The information we obtain is the exactly the same, but it is packaged rather more conveniently.

In this section I will use the standard notation

$$\not{p} = \gamma^\mu p_\mu. \tag{7.47}$$

7. Fermions

We look for solutions of the form

$$\psi = u(p)e^{-ipx} \quad \text{or} \quad v(p)e^{ipx}, \tag{7.48}$$

and find that we need

$$(\not{p} - m)u(p) = 0 = \bar{u}(p)(\not{p} - m),$$
$$(\not{p} + m)v(p) = 0 = \bar{v}(p)(\not{p} + m). \tag{7.49}$$

As in the previous section we exploit the mass-shell condition

$$(\not{p} - m)(\not{p} + m) = p^2 - m^2 = 0, \tag{7.50}$$

to write

$$1 = \frac{1}{2m}(m + \not{p}) + \frac{1}{2m}(m - \not{p})$$
$$= \Lambda_+(p) + \Lambda_-(p), \tag{7.51}$$

where $\Lambda_\pm(p)$ are projection operators

$$\Lambda_\pm^2(p) = \Lambda_\pm(p), \quad \Lambda_+(p)\Lambda_-(p) = 0. \tag{7.52}$$

We use these to find suitable u's and v's, two of each corresponding to the two components of spin. We will not need their particular forms.

Normalize the resulting $u^{(r)}(p)$, $v^{(r)}(p)$ ($r = 1, 2$) by requiring

$$\bar{u}^{(r)}(p)u^{(s)}(p) = -\bar{v}^{(r)}(p)v^{(s)}(p) = \delta^{rs}. \tag{7.53}$$

Since $0 = \bar{u}^{(r)}(\{\not{p} - m, \gamma^\mu\})u^{(s)}$, we have

$$2p^\mu \bar{u}^{(r)}(p)u^{(s)}(p) = 2m\bar{u}^{(r)}(p)\gamma^\mu u^{(s)}(p), \tag{7.54}$$

so this normalization is equivalent to the statement

$$u^{\dagger(r)}(p)u^{(s)}(p) = \frac{E_\mathbf{p}}{m}\delta^{rs}. \tag{7.55}$$

Similarly

$$v^{\dagger(r)}(p)v^{(s)}(p) = \frac{E_\mathbf{p}}{m}\delta^{rs}. \tag{7.56}$$

We have the useful completeness relations

$$\Lambda_+(p) = \sum_r u^{(r)}(p)\bar{u}^{(r)}(p) = \frac{1}{2m}(m + \not{p}),$$
$$\Lambda_-(p) = -\sum_r v^{(r)}(p)\bar{v}^{(r)}(p) = \frac{1}{2m}(m - \not{p}). \tag{7.57}$$

Many books use this $E_\mathbf{p}/m$ normalization for the spinors.

We wish to quantize the ψ field so that

$$\{\psi_\alpha(\mathbf{x}), \psi_\beta^\dagger(\mathbf{x}')\} = \delta_{\alpha\beta}\delta^3(\mathbf{x} - \mathbf{x}'). \tag{7.58}$$

If we write

$$\psi_\alpha(\mathbf{x}) = \sum_r \int \frac{d^3\mathbf{k}}{(2\pi)^3} \frac{m}{E_\mathbf{k}} \left\{ \hat{a}_{\mathbf{k}r} u_\alpha^{(r)}(p) e^{-ipx} + \hat{b}_{\mathbf{k}r}^\dagger v_\alpha^{(r)}(p) e^{+ipx} \right\}, \quad (7.59)$$

where

$$\{\hat{a}_{\mathbf{k}r}, \hat{a}_{\mathbf{k}'s}^\dagger\} = (2\pi)^3 \frac{E_\mathbf{p}}{m} \delta^{rs} \delta^3(\mathbf{k}-\mathbf{k}'),$$

$$\{\hat{b}_{\mathbf{k}r}, \hat{b}_{\mathbf{k}'s}^\dagger\} = (2\pi)^3 \frac{E_\mathbf{p}}{m} \delta^{rs} \delta^3(\mathbf{k}-\mathbf{k}'). \quad (7.60)$$

we find that it works.

We also find that

$$\hat{H} = \sum_r \int \frac{d^3\mathbf{k}}{(2\pi)^3} \frac{m}{E_\mathbf{k}} E_\mathbf{k} \left(\hat{a}_{\mathbf{k}'r}^\dagger \hat{a}_{\mathbf{k}r} - \hat{b}_{\mathbf{k}r} \hat{b}_{\mathbf{k}'s}^\dagger \right). \quad (7.61)$$

Dropping an infinte constant (the energy of the filled Dirac sea), we can write this as

$$\hat{H} = \sum_r \int \frac{d^3\mathbf{k}}{(2\pi)^3} \frac{m}{E_\mathbf{k}} E_\mathbf{k} \left(\hat{a}_{\mathbf{k}'r}^\dagger \hat{a}_{\mathbf{k}r} + \hat{b}_{\mathbf{k}r}^\dagger \hat{b}_{\mathbf{k}'s} \right). \quad (7.62)$$

In a similar vein we find

$$\hat{P}^\mu = \sum_r \int \frac{d^3\mathbf{k}}{(2\pi)^3} \frac{m}{E_\mathbf{k}} P^\mu \left(\hat{a}_{\mathbf{k}'r}^\dagger \hat{a}_{\mathbf{k}r} + \hat{b}_{\mathbf{k}r}^\dagger \hat{b}_{\mathbf{k}'s} \right) \quad (7.63)$$

for the momentum operator.

7.4.1 *Normal and Time-Ordered Products*

When we defined the time-ordered and normal-ordered products in Chapter 3, we moved the various fields into the desired order while deliberately ignoring any c-numbers that would come from commuting them past one another. In other words we treated the fields as of they were c-numbers themselves. When we define these products for Fermi fields, we continue to suppress the c-numbers, but we must *not* ignore the minus signs that come from interchanging two Fermi fields. In other words we treat the fields as *anticommuting c-numbers*, or *Grassmann*[5] variables. This means, for example, that the time ordered product

$$T\{\psi_\alpha(x)\psi_\beta(x')\} \quad (7.64)$$

is defined as

$$\theta(x_0 - x_0')\psi_\alpha(x)\psi_\beta(x') - \theta(x_0' - x_0)\psi_\beta(x')\psi_\alpha(x), \quad (7.65)$$

and is *antisymmetric* under the interchange $x \leftrightarrow y$.

---

[5]Hermann Günther Grassmann. Born April 15, 1809, Stettin, Prussia [now Szczecin, Poland.] Died. Sept. 26, 1877, Stettin, Germany.

7. Fermions

The propagator $iS_{\alpha\beta}(x, y)$ is defined by

$$iS_{\alpha\beta}(x, y) = \langle 0|T\{\psi_\alpha(x)\bar\psi_\beta(x')\}|0\rangle. \tag{7.66}$$

Now

$$\begin{aligned}
(-i\partial_x + m)_{\alpha'\alpha}&\langle 0|T\{\psi_\alpha(x)\bar\psi_\beta(x')\}|0\rangle \\
&= -i\gamma^0_{\alpha'\alpha}\delta(x_0 - y_0)\{\psi_\alpha(x), \bar\psi_\beta(y)\} \\
&= -i\delta^4(x-y)(\gamma^0)^2_{\alpha'\beta} = -i\delta_{\alpha'\beta}\delta^4(x-y).
\end{aligned} \tag{7.67}$$

Thus we expect that

$$iS_{\alpha\beta}(x, y) = i\int \frac{d^4k}{(2\pi)^3} e^{-ik(x-x')} \frac{(\slashed{k} + m)_{\alpha\beta}}{k^2 - m^2 + i\epsilon}. \tag{7.68}$$

A computation putting in a complete set of states confirms this, including the $i\epsilon$, which does not follow merely from the requirement that the propagator be a Green function for the Dirac equation.

Sometimes we will use the notation

$$iS(k) = \frac{i}{\slashed{k} - m} = i\frac{\slashed{k} + m}{k^2 - m^2} \tag{7.69}$$

for the momentum-space fermion propagator.

When we use Wick's theorem to evaluate diagrams, there will be various minus signs arising from grouping the operators to form propagators. In order to contract a pair of fields into a propagator, they need to be placed adjacent to one another, and in the right order to make $G(x, y) = \langle\psi(x)\bar\psi(y)\rangle$. For example, if we wish to take the product

$$\psi_1\bar\psi_2\psi_3\bar\psi_4 \tag{7.70}$$

and pair ψ_1 with $\bar\psi_2$ and ψ_3 with $\bar\psi_4$, then we are already in the right order and get $G(1, 2)G(3, 4)$. If we wish to pair $\bar\psi_2$ with ψ_3 and ψ_1 with $\bar\psi_4$, we get two minus signs from jumping ψ_1 over the two intermediate fields, and another from reordering $\bar\psi_2\psi_3 = -\psi_3\bar\psi_2$. The result is $-G(1, 4)G(3, 2)$.

8
QED

In this chapter we will obtain the Feynman rules for quantum electrodynamics. The success of Feynman's economical approach to the interaction of photons and electrons took field theory from being a fine art that could only be practiced by the most able and mathematically gifted and made it into a craft that could be practiced by the working physicist.

8.1 Quantizing Maxwell's Equations

First we must learn how to apply quantum mechanics to electromagnetic waves. This is a subject whose complications stem from the *gauge invariance* of the electromagnetic field equations. The final diagrammatic rules are far simpler than the formalism needed to derive them.

8.1.1 *Hamiltonian Formalism*

To set up a hamiltonian formalism for electromagnetism it is necessary to use the potentials A_μ as the dynamical fields. Because Maxwell's equations only determine the field strengths $F_{\mu\nu} = \partial_\mu A_\nu - \partial_\nu A_\mu$, and because a gauge transformation $A_\mu \to A_\mu + \partial_\mu \Lambda(x)$ does not affect $F_{\mu\nu}$, these potentials A_μ are not uniquely determined by the equations of motion.

This gauge invariance is a concomitant of the masslessness of the photon. A Lorentz-vector field naturally comes with four components. If it is to describe a spin-one particle, we expect only three of them to be independent. When the

Lagragian for a vector field contains a mass term, it automatically forces A_μ to obey $\partial_\mu A^\mu = 0$. This condition eliminates one degree of freedom leaving the expected three. A *massless* spin-one particle, however, should have only *two* modes — right and left circularly polarized photons. (This follows from the representation theory of the Poincaré group for a massless particle.) Something has to eliminate the one more degree of freedom, and this something is the gauge invariance.

8.1.2 Axial Gauge

If we demand a unique solution for the potentials, it necessary to impose some gauge condition on the A_μ. Perhaps the simplest condition is the *axial gauge*, where we set $A_0 \equiv -\phi = 0$. Then we can take as our lagrangian

$$L = \int d^3x \left[\frac{1}{2} \{ \mathbf{E}^2 - \mathbf{B}^2 \} + \mathbf{J} \cdot \mathbf{A} \right]. \tag{8.1}$$

Here

$$\mathbf{E} = -\dot{\mathbf{A}}, \quad \mathbf{B} = \nabla \wedge \mathbf{A}. \tag{8.2}$$

From its definition, $\nabla \cdot \mathbf{B} = 0$. We observe that $\mathbf{A}$ is canonically conjugate to $-\mathbf{E}$. We can now construct the hamiltonian in the usual manner and find

$$H = \int d^3x \left[\frac{1}{2} \{ \mathbf{E}^2 + \mathbf{B}^2 \} - \mathbf{J} \cdot \mathbf{A} \right]. \tag{8.3}$$

The six Hamilton equations are

$$\nabla \wedge \mathbf{E} = -\frac{\partial \mathbf{B}}{\partial t}, \quad \nabla \wedge \mathbf{B} = \mathbf{J} + \frac{\partial \mathbf{E}}{\partial t}. \tag{8.4}$$

These equations still do not yet uniquely determine $\mathbf{A}$. We are free to make any *time-independent* gauge transformation $\mathbf{A} \to \mathbf{A} + \nabla \Lambda(\mathbf{x})$. Furthermore, the fourth Maxwell equation, Gauss' law

$$\nabla \cdot \mathbf{E} = \rho, \tag{8.5}$$

does *not* appear as a consequence of either Hamilton's or Lagrange's equations. This is not surprising since, having set $A_0 = 0$, ρ does not appear in L. The three Maxwell equations we *do* have, however, tell us that

$$\frac{\partial}{\partial t} (\nabla \cdot \mathbf{E} - \rho) = \nabla \cdot (\nabla \wedge \mathbf{B} - \mathbf{J}) - \frac{\partial \rho}{\partial t}, \tag{8.6}$$

and, since $\nabla \cdot (\nabla \wedge \mathbf{B}) = 0$, the left-hand side of (8.6) is zero provided charge is conserved, i.e. provided $\dot{\rho} + \nabla \cdot \mathbf{J} = 0$. Consequently, if Gauss' law holds at $t = 0$, it holds eternally.

To quantize in this gauge we set

$$[E_i(\mathbf{x}, t), A_j(\mathbf{x}', t)] = i \delta_{ij} \delta^3(\mathbf{x} - \mathbf{x}'). \tag{8.7}$$

We might be tempted to impose $\nabla \cdot \mathbf{E} - \rho = 0$ as an operator-valued initial condition. This will not work though. Equation (8.7) implies that

$$U_\Lambda \mathbf{A} U_\Lambda^{-1} = \mathbf{A} + \nabla \Lambda(\mathbf{x}), \tag{8.8}$$

where

$$U_\Lambda = \exp\left(i \int d^3 x \mathbf{E} \cdot \nabla \Lambda(\mathbf{x})\right) = \exp\left(-i \int d^3 x \Lambda(\mathbf{x}) \cdot \nabla \mathbf{E}\right). \tag{8.9}$$

Indeed on including matter fields with charge operator $\hat{\rho}$, we find that $\nabla \cdot \mathbf{E} - \hat{\rho}$ is the generator of gauge transformations. It cannot be zero therefore.

If we restrict our interactions and observables to the class of gauge-invariant expressions, these will commute with $\nabla \cdot \mathbf{E} - \hat{\rho}$, and we can impose the condition

$$(\nabla \cdot \mathbf{E} - \hat{\rho}) |physical\rangle = 0 \tag{8.10}$$

as a restriction on the set of states we will call *physical*, confident that if the system begins in this subspace, it will remain in it.

Although often described as one, gauge invariance is not a symmetry in the conventional sense. It is more a statement that some of the variables we are using to describe the system are redundant. The physical state condition (8.10) asserts that a physically meaningful wave function should not depend on the redundant gauge degrees of freedom.

8.1.3 Lorentz Gauge

The formalism in the previous section is intuitive and popular with lattice gauge theorists. It is not manifestly Lorentz invariant however. For practical calculations it is preferable to use a formalism that explicitly preserves the Lorentz symmetry. Unfortunately none of the traditional approaches to the covariant gauges has a simple Hilbert space interpretation. The modern approach is to use BRST cohomology, but we will not explore this. For now I will just develop enough of the old fashioned Gupta-Bleuler formalism to derive the QED Feynman rules.

The covariant action is

$$S = \int d^4 x \left\{ -\frac{1}{4} F_{\mu\nu} F^{\mu\nu} + j^\mu A_\mu \right\}. \tag{8.11}$$

This is consistent with (8.1) after we set $A_\mu = (-\phi, \mathbf{A})$, so that $E^a = F_{a0}$, $B^a = \frac{1}{2} \epsilon^{abc} F_{bc}$, and $j^\mu = (\rho, \mathbf{J})$. (No particular meaning should be attached to the up or down position of the indices on E or B, since they are not parts of a four-vector.) Varying A_μ gives the equation of motion

$$-\partial^2 A^\mu + \partial^\mu \partial_\nu A^\nu = j^\mu. \tag{8.12}$$

Equation (8.12) implies that the current must be conserved, $\partial_\mu j^\mu = 0$. We will always assume that this is so even when we modify the equations of motion in such a way that it is no longer an automatic consequence of them.

To quantize in the *Lorentz gauge* we add to the lagrangian density a term

$$\mathcal{L}_{Lorentz} = -\frac{1}{2}\lambda \left(\partial^\mu A_\mu\right)^2. \tag{8.13}$$

This is called a "gauge fixing term," and by an abuse of terminology different choices of λ are referred to as different "gauges." In particular the choice $\lambda = 1$ is called the *Feynman gauge*. Physics should be (and is) independent of our choice of λ.

The equations of motion now become

$$-\partial^2 A^\mu + (1-\lambda)\partial^\mu \partial_\nu A^\nu = j^\mu. \tag{8.14}$$

Taking the divergence of this and assuming $\partial_\mu j^\mu = 0$, we find

$$\lambda \partial^2 (\partial^\mu A_\mu) = 0, \tag{8.15}$$

so the quantity $(\partial^\mu A_\mu)$ is *free-field*. It does not couple to the current or anything else. Classically we would just set $\partial^\mu A_\mu = 0$ and forget about it. In quantum mechanics this is not an option because $\partial^\mu A_\mu$ is required to have nonzero commutators with other fields. In particular, $\lambda \partial^\mu A_\mu$ is now the momentum variable canonically conjugate to A_0.

For general values of λ the canonical structure is rather complicated. For the Feynman gauge, where $\lambda = 1$, things are easier because, after some integration by parts, the action becomes

$$S = \int d^4x \left\{ -\frac{1}{2}(\partial^\mu A_\nu)^2 + j^\mu A_\mu \right\}, \tag{8.16}$$

or

$$S = \int d^4x \left\{ +\frac{1}{2}\sum_{a=1}^{3}(\partial^\mu A_a)(\partial_\mu A_a) - \frac{1}{2}(\partial^\mu A_0)(\partial_\mu A_0) + j^\mu A_\mu \right\}. \tag{8.17}$$

This is just the action for four "scalar" fields labeled by the index $\nu = 0, 1, 2, 3$ — although A_0 has the wrong sign for its kinetic energy. If we ignore this minor irritation, we would expect to quantize by requiring

$$[A_\mu(\mathbf{x}, t), \dot{A}_\nu(\mathbf{x}', t)] = -i g_{\mu\nu} \delta^3(\mathbf{x} - \mathbf{x}'). \tag{8.18}$$

If we set

$$A_\mu = \int \frac{d^3k}{(2\pi)^3} \frac{1}{2E_\mathbf{k}} \left\{ \hat{a}_\mu(\mathbf{k}) e^{-ikx} + \hat{a}_\mu^\dagger(\mathbf{k}) e^{+ikx} \right\}, \tag{8.19}$$

we find that

$$[\hat{a}_\mu(\mathbf{k}), \hat{a}_\nu^\dagger(\mathbf{k}')] = -g_{\mu\nu} 2E_\mathbf{k}(2\pi)^3 \delta^3(\mathbf{k} - \mathbf{k}'). \tag{8.20}$$

Equation (8.20) looks nice and covariant — but we quickly see that $\hat{a}_0^\dagger(\mathbf{k})$ creates negative norm-squared states, or *ghosts*. This is catastrophic. We interpret the square

8.1 Quantizing Maxwell's Equations

of the norm as a probability, and negative probabilties make no sense.[1] We will be saved from paradox if we again impose a condition defining a subspace of physical states. Since we are trying to implement the gauge condition $\partial^\mu A_\mu = 0$, one might expect that we should require

$$(\partial^\mu A_\mu)|physical\rangle \stackrel{wrong!}{=} 0. \tag{8.21}$$

This turns out to be too strong a condition. Instead we require

$$(\partial^\mu A_\mu)^+|physical\rangle = 0, \tag{8.22}$$

where the $(\ldots)^+$ means the positive frequency or annihilation operator part. (This decomposition makes sense only because $\partial^\mu A_\mu$ is free field.)

If, for example, **k** is in the z direction, the physical subspace condition means that

$$(\hat{a}_0 + \hat{a}_3)|physical\rangle = 0. \tag{8.23}$$

When we look at the expectation of $\hat{a}_3^\dagger \hat{a}_3 - \hat{a}_0^\dagger \hat{a}_0$ between physical states (where $\hat{a}_0|physical\rangle = -\hat{a}_3|physical\rangle$), we find that it vanishes. This means that when computing the expectation value of

$$\hat{H} = \int \frac{d^3k}{(2\pi)^3} \frac{1}{2E_k} E_k (\hat{a}_1^\dagger \hat{a}_1 + \hat{a}_2^\dagger \hat{a}_2 + \hat{a}_3^\dagger \hat{a}_3 - \hat{a}_0^\dagger \hat{a}_0) \tag{8.24}$$

only the physically meaningful transverse 1, 2 modes contribute to the energy. For a further discussion of the Hilbert space structure see the appendix on indefinite metric spaces.

What about the propagator $iG_{\mu\nu}(x, x') = \langle 0|T\{A_\mu(x)A_\nu(x')\}|0\rangle$? This, as always, is a Green function for the equation obeyed by the field. In the present case the equation is

$$-\partial^2 A^\mu + (1-\lambda)\partial^\mu \partial_\nu A^\nu = j^\mu. \tag{8.25}$$

In Fourier space this reads

$$k^2 A^\mu - (1-\lambda)k^\mu k_\nu A^\nu = j^\mu,$$

$$(k^2)(\delta^\mu_\nu - (1-\lambda)\frac{k^\mu k_\nu}{k^2})A^\nu = j^\mu. \tag{8.26}$$

The inverse of the $\mu\nu$ matrix is easily found and we find for the propagator

$$iG_{\mu\nu}(x, x') = -i \int \frac{d^4k}{(2\pi)^3} \frac{e^{-ik(x-x')}}{k^2 + i\epsilon} \left(g_{\mu\nu} + \frac{1-\lambda}{\lambda} \frac{k_\mu k_\nu}{k^2} \right). \tag{8.27}$$

We will represent this propagator by a wavy line.

---

[1] We might be tempted to regard $\hat{a}_0^\dagger$ as an *annihilation* operator, and require the vacuum to obey $\hat{a}_0^\dagger|0\rangle = 0$, $\hat{a}_i|0\rangle = 0$, $i = 1, 2, 3$. Unfortunately, because space and time are treated differently, this $|0\rangle$ state would not be a Lorentz singlet.

88   8. QED

Fig 1. The photon propagator.

The physics should be independent of our choice of λ — so if we are confident that we have all the diagrams, we can simply set $\lambda = 1$ and forget about the projection term. Note the minus sign in front of the right-hand side of (8.27). This is consonant with (8.20) and ensures that, when treated as we did the Yukawa potential in Chapter 3, like charges repel and unlike charges attract.

It is convenient to define polarization vectors $\epsilon_\mu^{(r)}(\mathbf{k})$, so that we can write

$$A_\mu = \int \frac{d^3k}{(2\pi)^3} \frac{1}{2E_{\mathbf{k}}} \left\{ \hat{a}_{(r)} \epsilon_\mu^{(r)}(\mathbf{k}) e^{-ikx} + \hat{a}_{(r)}^\dagger \epsilon_\mu^{(r)*}(\mathbf{k}) e^{+ikx} \right\}. \tag{8.28}$$

Here $r = 0, 1, 2, 3$ labels a basis for the polarization directions. We must make sure that any initial and final-state polarizations are *transverse*. That is, $\epsilon_\mu^{(r)} = (0, \mathbf{e})$, where $\mathbf{k} \cdot \mathbf{e} = 0$. For $\mathbf{k}$ directed along the positive z axis, right- and left-handed circularly polarized light have $\epsilon^\mu = (0, 1, \pm i, 0)/\sqrt{2}$. Because of the peculiar historical conventions of optics, right-circularly polarized light has angular momentum component $-\hbar$ along the direction of propagation, while left-circularly polarized light has component $+\hbar$.

8.2   Feynman Rules for QED

The action for QED combines the action of the Dirac field with that of the Maxwell field.

$$S = \int d^4x \left\{ \bar{\psi} \left(-i\gamma^\mu (\partial_\mu - ieA_\mu) + m \right) \psi - \frac{1}{4} F^{\mu\nu} F_{\mu\nu} \right\}. \tag{8.29}$$

The interaction term $\mathcal{L}_I = \bar{\psi} \gamma^\mu \psi A_\mu$ is Lorentz invariant, local and hermitian. In addition the whole action is gauge-invariant under the substitution

$$\psi \to e^{i\Lambda(x)} \psi, \quad A_\mu \to A_\mu + \partial_\mu \Lambda. \tag{8.30}$$

In working with this action bear in mind that $\psi(x)$ annihilates electrons at the point x with amplitude $u(k)e^{-ikx}$, or creates positrons at the same point with amplitude $v(k)e^{+ikx}$. Similarly $\bar{\psi}(x)$ creates electron with amplitude $\bar{u}(k)e^{+ikx}$, and annihilates positrons with amplitude $\bar{v}(k)e^{-ikx}$.

The Feynman rules are:

1. At each order draw all distinct graphs with n vertices contributing to the process of interest.
2. For each photon line include a factor $-ig_{\mu\nu}/(p^2 + i\epsilon)$.
3. For each vertex a factor of $ie\gamma^\mu (2\pi)^4 \delta(\sum p)$.
4. For each fermion line a factor $i(\slashed{p} + m)/(p^2 - m^2 + i\epsilon)$.
5. Integrate $\int d^4p/(2\pi)^4$ over the 4-momentum.
6. For each external line a factor of:

- $u(p)$ for each incoming electron.
- $\bar{v}(p)$ for each incoming positron.
- $\bar{u}(p)$ for each outgoing electron.
- $v(p)$ for each outgoing positron.
- $\epsilon_\mu^{(\lambda)}$ polarization vector for each photon enering a vertex.

7. A ± 1 for any relative interchanges of fermions in the final state (include particle/antiparticle permutations).
8. A $(-1)^L$ for L closed fermion loops.

The only part of this that seems to merit comment is the factor of (-1) for each closed fermion loop. This originates in the rearrangement necessary to perform the contractions. It plays a vital role in enforcing the exclusion principle. To see this consider the diagrams in Fig. 2. These represent specific terms in the sum over 3-momenta in some of the Rayleigh-Schrödinger perturbation series diagrams that result from a selecting a time-ordering of the vertices in the corresponding Feynman diagrams. Time is running upwards.

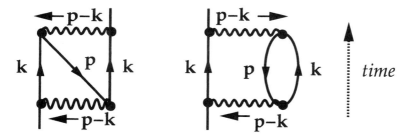

Fig 2. The term corresponding to left-hand diagram cancels against the term represented by the right-hand diagram. This enforces the exclusion principle.

In the left-hand diagram we see that there are two fermions with the same momentum **k** in the intermediate state. This is not allowed by Fermi statistics. The diagram on the right, however, has exactly the same intermediate state, energy denominators, and matrix elements as the other. It differs only in its topology — i.e. it contains a closed fermion loop. The resultant factor of (-1) ensures that the right-hand term cancels the illegal left-hand term.

When Feynman first wrote down his rules for QED, he was wedded to the notion that all the electrons and positrons in the universe were just one single particle busily bouncing backwards and forwards in time. (As in the left-hand diagram of Fig.2.) He therefore argued that closed fermion loop diagrams were not needed. It was only after the problem with the exclusion principle was pointed out to him that he reluctantly conceded the necessity of closed loop worldlines.

Note that one *can* forget about fermion loops with an odd number of photons attached. They are zero by charge conjugation invariance (Furrey's theorem).

8.2.1 Møller Scattering

As an example of the Feynman rules we will consider Møller scattering. This is the scattering of an electron by another electron. At lowest order there are two contributing graphs:

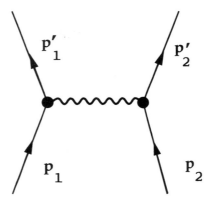

Fig 3. Direct Møller scattering.

and its exchange partner:

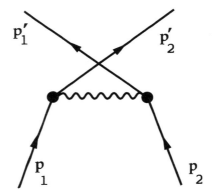

Fig 4. Exchange Møller scattering.

These diagrams contribute to

$$S = (2\pi)^4 i\delta(p_1 + p_2 - p'_1 - p'_2)T \tag{8.31}$$

the expression

$$T = e^2 \left[\frac{[\bar{u}(p'_1)\gamma^\mu u(p_1)][\bar{u}(p'_2)\gamma_\mu u(p_2)]}{(p_1 - p'_1)^2 + i\epsilon} - \frac{[\bar{u}(p'_2)\gamma^\mu u(p_1)][\bar{u}(p'_1)\gamma_\mu u(p_2)]}{(p_1 - p'_2)^2 + i\epsilon} \right]. \tag{8.32}$$

Note the minus sign in the second term which comes from the relative interchange of the final states.

In writing down the expressions corresponding to the graphs, I have set $\lambda = 1$ in the photon propagator so that the numerator only contains a factor of $g_{\mu\nu}$. Let us check gauge invariance by verifying that it makes no difference if we replace $g_{\mu\nu}$ by $g_{\mu\nu} + \zeta k_\mu k_\nu$. The addition to the denominator in the first term is proprtional to

$$\bar{u}(p_1')\gamma^\mu u(p_1)(p_1' - p_1)_\mu \bar{u}(p_2')\gamma^\nu u(p_2)(p_1' - p_1)_\nu \tag{8.33}$$

but $(\not{p} - m)u(p) = \bar{u}(p)(\not{p} - m) = 0$, so this vanishes.

If the electrons were distinguishable only the first term would exist. Let us suppose this to be the case. For example we can consider $e\mu$ scattering.

Most scattering experiments are conducted with unpolarized beams and with detectors that are insensitive to the polarization state of the outgoing particles. When we do not specify the initial and final spin states, then we must *average* over the initial state spins and *sum* over the final state spins. Thus the effective cross section is derived from

$$\frac{1}{2}\sum_{s_1}\frac{1}{2}\sum_{s_2}\sum_{s_1'}\sum_{s_2'}|T(s_1, s_2 \to s_1', s_2')|^2 \tag{8.34}$$

as

$$|T|^2 = \frac{1}{4}\sum_{s_1,s_2,s_1',s_2'}\left|\frac{(\bar{u}\gamma_\mu u)(\bar{u}\gamma^\mu u)}{q^2 + i\epsilon}\right|^2. \tag{8.35}$$

Now

$$|\bar{u}(f)\Gamma u(i)|^2 = [\bar{u}(f)\Gamma u(i)][\bar{u}(i)\bar{\Gamma}u(f)], \tag{8.36}$$

where $\bar{\Gamma} = \gamma^0\Gamma^\dagger\gamma_0$, so $\bar{\gamma}^\mu = \gamma^\mu$. Recall

$$\sum_s u_\beta(p,s)\bar{u}_{\beta'}(p,s) = \frac{1}{2m}(\not{p} + m)_{\beta\beta'}, \tag{8.37}$$

so

$$|\bar{u}(f)\Gamma u(i)|^2 = \text{Tr}\left\{\frac{1}{2m}(\not{p}_f + m)\Gamma\frac{1}{2m}(\not{p}_i + m)\bar{\Gamma}\right\}. \tag{8.38}$$

Therefore,

$$|T|^2 = \frac{e^4}{4q^4}\text{Tr}\left\{\frac{1}{2m}(\not{p}_1' + m)\gamma^\mu\frac{1}{2m}(\not{p}_1 + m)\gamma^\nu\right\}$$
$$\times \text{Tr}\left\{\frac{1}{2m}(\not{p}_2' + m)\gamma_\mu\frac{1}{2m}(\not{p}_2 + m)\gamma_\nu\right\}. \tag{8.39}$$

We evaluate the traces using

$$\text{Tr}(\not{a}\not{b}\not{c}\not{d}) = 4[(a \cdot b)(c \cdot d) + (a \cdot d)(c \cdot c) - (a \cdot c)(b \cdot d)], \tag{8.40}$$

$$Tr(\not{a}\not{b}) = 4(a \cdot b) \quad \text{Tr}(\not{a}\not{b}\not{c}) = 0. \tag{8.41}$$

These are easily proved from the properties of the gamma matrices. We find that

$$\text{Tr}\left\{\frac{1}{2m}(\not{p}'_1+m)\gamma^\mu\frac{1}{2m}(\not{p}_1+m)\bar{\gamma}^\nu\right\} = \frac{1}{m^2}((p'_1)^\mu p_1^\nu + p_1^\mu(p'_1)^\nu - g^{\mu\nu}(p'_1\cdot p_1 - m^2)), \quad (8.42)$$

and finally

$$|T|^2 = \frac{e^4}{4q^4}[(p_1\cdot p'_2)(p_1\cdot p_2) + (p'_2\cdot p_1)(p_2\cdot p'_1) - m^2(p'_2\cdot p_2 + p_1\cdot p_1) + 2m^4]. \quad (8.43)$$

8.3 Ward Identity and Gauge Invariance

8.3.1 The Ward Identity

The Ward identity is a route to establishing gauge invariance. To derive it we look at

$$\frac{\partial}{\partial x^\mu}\langle 0|T\{j^\mu(x)\bar{\psi}(y)\psi(z)\}|0\rangle, \quad (8.44)$$

where $j^\mu(x) = \bar{\psi}(x)\gamma^\mu\psi(x)$ is the fermion number current. Now current conservation, $\partial_\mu\bar{\psi}\gamma^\mu\psi(x) = 0$, remains an *operator* identity even in an interacting system, so, taking into acount the discontinuities caused by the time-ordering, we find

$$\partial_\mu\langle 0|T\{j^\mu(x)\bar{\psi}(y)\psi(z)\}|0\rangle = \delta(x_0 - y_0)\langle 0|T\{[j^0(x), \bar{\psi}(y)]\psi(z)\}|0\rangle \\ + \delta(x_0 - z_0)\langle 0|T\{\bar{\psi}(y)[j^0(x), \psi(z)]\}|0\rangle. \quad (8.45)$$

The commutators of the fields with the charge densities are

$$\delta(x_0 - y_0)[j^0(x), \bar{\psi}(y)] = \bar{\psi}(y)\delta^4(x - y), \\ \delta(x_0 - z_0)[j^0(x), \psi(z)] = -\psi(z)\delta^4(x - z). \quad (8.46)$$

Therefore

$$\partial_\mu\langle 0|T\{j^\mu(x_0\bar{\psi}(y)\psi(z)\}|0\rangle = \delta^4(x - y)\langle 0|T\{\bar{\psi}(y)\psi(z)\}|0\rangle \\ - \delta^4(x - z)\langle 0|T\{\bar{\psi}(y)\psi(z)\}|0\rangle. \quad (8.47)$$

This identity is true for both free and interacting fields. It is easily verified for the free theory. We use the trivial identity $(\not{k}+\not{q}-m)-(\not{q}-m) = \not{k}$ to find

$$\frac{1}{\not{q}-m}k_\mu\gamma^\mu\frac{1}{\not{k}+\not{q}-m} = \frac{1}{\not{q}-m} - \frac{1}{\not{q}+\not{k}-m}, \quad (8.48)$$

which is momentum space form of (8.47).

8.3 Ward Identity and Gauge Invariance

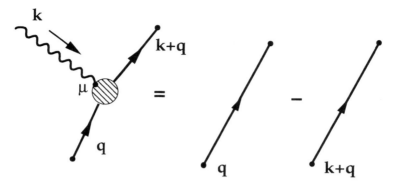

Fig 5. *The Ward identity*

We can express the Ward identity for an interacting system in terms of the full momentum space propagator $S(k)$ and the *irreducible vertex* $\Gamma_\mu(q, k+q)$ (shown as a hatched blob in Fig. 5):

$$k_\mu S(q)\Gamma^\mu(q, k+q)S(k+q) = S(q) - S(k+q) \tag{8.49}$$

or as

$$S^{-1}(k+q) - S^{-1}(q) = k_\mu \Gamma^\mu(q, k+q). \tag{8.50}$$

Here $S^{-1}(q)$ is the matrix in spinor space inverse to $S(q)$.

8.3.2 Applications

The Vacuum Polarization Bubble

As an application of the Ward identity consider the vacuum polarization bubble that renormalizes the photon propagator at order e^2.

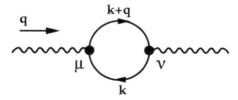

Fig 6. *The photon polarization bubble.*

This is given by

$$\Pi^{\mu\nu}(q) = \int \frac{d^4k}{(2\pi)^4} \text{tr}\,(\gamma^\mu S(k+q)\gamma^\nu S(k)). \tag{8.51}$$

From the ward identity we have

$$q_\mu \Pi^{\mu\nu}(q) = \int \frac{d^4k}{(2\pi)^4} (\text{tr}\,(S(k)\gamma^\nu) - \text{tr}\,(S(k+q)\gamma^\nu)). \tag{8.52}$$

8. QED

If we can shift the variable of integration $k \to k + q$ in the first term in (8.52), then the integral is zero. (This change of integration variable is *not* always safe, but it is here.). Accepting this we can gain some information about the structure of $\Pi^{\mu\nu}(q)$. By Lorentz invariance we know it has to be of the form

$$\Pi^{\mu\nu}(q) = g^{\mu\nu} F_1(q^2) + q^\mu q^\nu F_2(q^2), \tag{8.53}$$

but $q_\mu \Pi^{\mu\nu}(q) = 0$ implies

$$q^\nu F_1 + q^2 q^\nu F_2 = 0, \tag{8.54}$$

or

$$\Pi^{\mu\nu}(q) = F_1(q^2) \left(g^{\mu\nu} - \frac{q^\mu q^\nu}{q^2} \right). \tag{8.55}$$

The bubble is therefore *transverse* as is the original un-gauge-fixed inverse photon propagator.

External Fields

Another application of the Ward identity is to demonstrate the gauge invariance of the process where an electron scatters off an external field. Here we assume that the interaction is of the form $H_I = \int d^3x \, e A_\mu^{ext}(x) j^\mu(x)$, where $A_\mu^{ext}(x)$ is a prescribed field, i.e. one not influenced by back-reaction from the electron.

The electron propagator in this field is given by the sum

$$i S_{A^{ext}}(y, x) = \langle 0 | T\{\psi(y) \bar{\psi}(x)\} | 0 \rangle_{A^{ext}}$$

$$= \sum_n \frac{(-ie)^n}{n!} \int d^4z_1 \ldots d^4z_n A_{\mu_1}^{ext}(z_1) \ldots A_{\mu_n}^{ext}(z_n) \times$$

$$\langle 0 | T\{\psi(y) j^{\mu_1}(z_1) \ldots j^{\mu_n}(z_1) \bar{\psi}(x)\} | 0 \rangle$$

$$= \sum_n \frac{(-ie)^n}{n!} \int d^4z_1 \ldots d^4z_n A_{\mu_1}^{ext}(z_1) \ldots A_{\mu_n}^{ext}(z_n) \times$$

$$i S(y, z_n) \gamma^{\mu_n} i S(z_n, z_{n-1}) \gamma^{\mu_{n-1}} \ldots \gamma^{\mu_1} i S(z_1, x). \tag{8.56}$$

(The $n!$ cancels because of there are $n!$ possible orders for the vertices.)

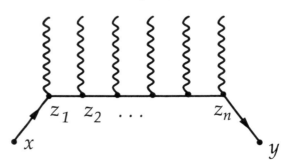

Fig 7. *The photon propagator in an external field.*

8.3 Ward Identity and Gauge Invariance

If we replace $A_\mu^{ext}(x)$ by $A_\mu^{ext}(x)+\partial_\mu\delta\Lambda(x)$, where $\delta\Lambda(x)$ is small, then repeated uses of (8.47) shows that the change in the propagator is

$$\delta\langle 0|T\{\psi(y)\bar\psi(x)\}|0\rangle_{A^{ext}} = ie(\delta\Lambda(x)-\delta\Lambda(y))\langle 0|T\{\psi(y)\bar\psi(x)\}|0\rangle_{A^{ext}} \quad (8.57)$$

If we integrate up to a finite Λ, this becomes

$$\langle 0|T\{\psi(y)\bar\psi(x)\}|0\rangle_{A^{ext}} \to e^{ie(\Lambda(x)-\Lambda(y))}\langle 0|T\{\bar\psi(x)\psi(y)\}|0\rangle_{A^{ext}}, \quad (8.58)$$

which is what we should perhaps have expected from the ways things work in quantum mechanics.

In momentum space these diagrams become the sum

$$S(p,p+k)$$
$$= \sum_n (-ie)^n \left(\frac{1}{\slashed{p}+\sum\slashed{k}-m}\gamma^{\mu_n}\ldots\gamma^{\mu_1}\frac{1}{\slashed{p}-m}\right)\tilde{A}_{\mu_n}(k_n)\ldots\tilde{A}_{\mu_1}(k_1),$$
$$(8.59)$$

and one can use (8.48) to establish that the change in the momentum space propagator is

$$-\left(\frac{1}{\slashed{p}+\sum\slashed{k}-\slashed{k}_n-m}\gamma^{\mu_{n-1}}\ldots\gamma^{\mu_1}\frac{1}{\slashed{p}-m}\right)\delta\Lambda(k_n)\tilde{A}_{\mu_{n-1}}(k_{n-1})\ldots\tilde{A}_{\mu_1}(k_1)$$
$$+\left(\frac{1}{\slashed{p}+\sum\slashed{k}--m}\gamma^{\mu_n}\ldots\gamma^{\mu_2}\frac{1}{\slashed{p}+\slashed{k}_1-m}\right)\tilde{A}_{\mu_n}(k_n)\ldots\tilde{A}_{\mu_2}(k_2)\delta\Lambda(k_1).$$
$$(8.60)$$

Now suppose instead of the propagators on the external legs we put $u(p)$ and $\bar{u}(p+\sum k)$ so as to get the scattering amplitude.

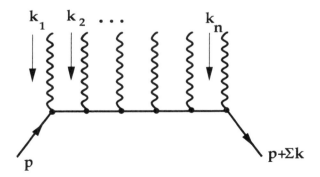

Fig 8. Electron scattering off an external field.

We can do this by putting a factor of $S^{-1}(p)u(p)$ on the right-hand end and $\bar{u}(p+\sum k)S^{-1}(p+\sum k)$ on the extreme left. These factors are nonzero in the intitial amplitude because the $S(p)$ eats the $S^{-1}(p)$. In the second term in the expression for the change (8.60) we are left with an uncancelled $S^{-1}(p)$ in $\ldots S^{-1}(p)u(p)$ but $S^{-1}(p)u(p) = (\slashed{p}-m)u(p) = 0$. Similarly the other term is zero because

$\bar{u}(p + \sum k) S^{-1}(p + \sum k) = 0$ We see therefore that the scattering amplitude, unlike the propagator, is unaffected by the gauge transformation.

9
Electrons in Solids

The next two chapters are devoted nonrelativistic many-body physics. Most of the fundamental ideas of the preceding chapters, such as Wick's theorem and Feynman diagrams, can be carried over to these systems with very little modification. The principal complications stem from the absence of Lorentz invariance and its convenient linking of momentum and energy into a single four-vector.

Many-body systems have at least one preferred reference frame: that where the center of mass of the particles is at rest. Typically the behavior of quantities in terms of the energy in this frame, $\hbar\omega$, and the momentum, $\hbar\mathbf{k}$, is very different depending on whether ω is large and $\mathbf{k}$ small, or vice versa. Often no single approximation will be able to capture the entire range of ω, $\mathbf{k}$, and so computations are less structured than in relativistic field theory. On the other hand, it is usually easier to tap one's physical intuition for a guide to how to proceed.

9.1 Second Quantization

We begin by reviewing many-fermion Hilbert space. Much of this material has been implicit in the earlier chapters, but we reconsider it here because we wish to focus more on the "particle" than on the "field" interpretation of the physics.

The Hilbert space for a many-electron system is constructed by taking antisymmetric (exterior) tensor powers of a single-electron Hilbert space. This purely formal construction is rather misleadingly called "second quantization."

9. Electrons in Solids

Suppose we find the normalized energy eigenfunctions $\psi_n(x)$ for the hamiltonian

$$H = -\frac{1}{2m}\nabla^2 + V(x) \tag{9.1}$$

for a single electron. These obey

$$H\psi_n = E_n\psi_n \tag{9.2}$$

and form a complete orthonormal basis for the single-electron Hilbert space $\mathcal{H}^{(1)}$. Let the operators $\hat{a}_n$, $\hat{a}_n^\dagger$ obey the anti-commutation relations

$$\{\hat{a}_n, \hat{a}_m^\dagger\} = \delta_{mn}, \tag{9.3}$$

and define the Heisenberg field

$$\hat{\psi}(x,t) = \sum_n \hat{a}_n \psi_n(x) e^{-iE_n t} \tag{9.4}$$

to be the "second-quantized" electron field. We find it obeys

$$\{\hat{\psi}(x,t), \hat{\psi}^\dagger(x',t)\} = \delta^3(x-x'). \tag{9.5}$$

This is the anticommutator we would get were we to take as our lagrangian density the expression

$$\mathcal{L} = i\hat{\psi}^\dagger \partial_t \hat{\psi} - \frac{1}{2m}|\nabla\hat{\psi}|^2 - \hat{\psi}^\dagger V(x)\hat{\psi}. \tag{9.6}$$

From this we read off that $\hat{\pi}(x) \equiv i\hat{\psi}^\dagger(x)$ is the momentum conjugate to $\hat{\psi}$. Quantizing by imposing anticommutation relations so as to get Fermi statistics leads us to (9.5).

The operators $\hat{\psi}(x,t)$, $\hat{\psi}^\dagger(x,t)$ act on the "big" many-particle space

$$\mathcal{H} = \bigoplus_{N=0}^{\infty} \left\{ \bigwedge^N \mathcal{H}^{(1)} \right\}. \tag{9.7}$$

This is the sum of the N-th exterior powers of the single-particle Hilbert-space $\mathcal{H}^{(1)}$. Each N-particle space $\bigwedge^N \mathcal{H}^{(1)}$ is spanned by states of the form

$$|m_1, m_2, \ldots\rangle = (\hat{a}_1^\dagger)^{m_1}(\hat{a}_2^\dagger)^{m_2}\ldots|0\rangle, \tag{9.8}$$

where $|0\rangle$ is the no-particle state annihilated by all the $\hat{a}_i$, and $\sum m_i = N$. Because $(\hat{a}_i^\dagger)^2 = 0$, the m_i can only take the values 1 or 0.

We write the "second-quantized" hamiltonian as

$$\hat{H} = \int d^3x\, \hat{\psi}^\dagger(x) H \hat{\psi}(x) = \sum_n E_n \hat{a}_n^\dagger \hat{a}_n. \tag{9.9}$$

Its action on $\bigwedge^1 \mathcal{H}^{(1)}$ coincides with the action of H on $\mathcal{H}^{(1)}$.

The particle number operator is

$$\hat{Q} = \int d^3x\, \hat{\rho}(x) = \int d^3x\, \hat{\psi}^\dagger(x)\hat{\psi}(x) = \sum_n \hat{a}_n^\dagger \hat{a}_n, \tag{9.10}$$

9.1 Second Quantization

and the number current is

$$\hat{\mathbf{J}} = \frac{1}{2mi}\left(\hat{\psi}^\dagger \nabla \hat{\psi} - (\nabla \hat{\psi}^\dagger)\hat{\psi}\right). \tag{9.11}$$

Using $\hat{H}$ to determine the dynamics we easily find that

$$\partial_t \hat{\rho} + \nabla \cdot \hat{\mathbf{J}} = 0 \tag{9.12}$$

is obeyed as an equation between operators.

To see where the usual wavefunction expressions have gone, consider a single-particle state $|m\rangle = \hat{a}_m^\dagger |0\rangle$. We find that the wavefunction can be recovered by evaluating

$$\langle 0|\hat{\psi}(x,t)|m\rangle = \langle 0|\sum_n \hat{a}_n \psi_n(x)e^{-iE_n t}\hat{a}_m^\dagger|0\rangle = \psi_m(x)e^{-iE_m t}. \tag{9.13}$$

Furthermore, if $|\Psi_0\rangle$ is the ground state made by filling the N lowest energy levels, we have

$$\langle 0|\prod_{i=1}^{N}\hat{\psi}(x_i)|\Psi_0\rangle = \begin{vmatrix} \psi_1(x_1) & \psi_2(x_1) & \cdots & \psi_N(x_1) \\ \psi_1(x_2) & \psi_2(x_2) & \cdots & \psi_N(x_2) \\ \vdots & \vdots & \ddots & \vdots \\ \psi_1(x_N) & \psi_2(x_N) & \cdots & \psi_N(x_N) \end{vmatrix}. \tag{9.14}$$

This is the familiar Slater-determinant expression for the ground state in the wavefunction picture.

N-particle states that can be expressed by a single $N \times N$ Slater determinant are said to be *decomposable*. Such states form a very small subset of the totality of N-particle states. The Hartree-Fock approximation to the quantum N-body problem consists of seeking the best trial wavefunction in this small subset.

The idea of interpreting the Schrödinger wavefunction $\psi(x)$ as an operator-valued field $\hat{\psi}(x)$ is essentially due to Jordan.[1]

As we should by now expect, time and normal-ordering is defined with minus signs

$$\langle 0|T\{\hat{\psi}^\dagger(x)\hat{\psi}(x')\}|0\rangle \stackrel{\text{def}}{=} \theta(x_0 - x_0')\langle 0|\hat{\psi}^\dagger(x)\hat{\psi}(x')|0\rangle \\ - \theta(x_0' - x_0)\langle 0|\hat{\psi}(x')\hat{\psi}^\dagger(x)|0\rangle. \tag{9.15}$$

---

[1] Ernst Pascual Jordan. Born October 18, 1902 Hannover, Germany. Died July 31, 1980, Hannover.

9.2 Fermi Gas and Fermi Liquid

9.2.1 One-Particle Density Matrix

When $V=0$ and the system is large, the sum (9.4) goes over to an integral

$$\psi(\mathbf{x},t) = \int \frac{d^3k}{(2\pi)^3} a_\mathbf{k} e^{i\mathbf{k}\cdot\mathbf{x}-i\varepsilon(\mathbf{k})t}. \tag{9.16}$$

We have now dropped the hats on the $\psi(x)$'s as there is no longer any chance of confusion with the wavefunctions $\psi_n(x)$. Note that there is no need for relativistic normalization here. We therefore have no $\frac{1}{2E}$ factors in either this expression or in the anticommutation relations obeyed by the $\hat{a}, \hat{a}^\dagger$'s. The latter are

$$\{a_\mathbf{k}, \hat{a}^\dagger_{\mathbf{k}'}\} = (2\pi)^3 \delta^3(\mathbf{k}-\mathbf{k}'). \tag{9.17}$$

The single-particle energy $\varepsilon(\mathbf{k})$ is given by

$$\varepsilon(\mathbf{k}) = \frac{1}{2m}\mathbf{k}^2. \tag{9.18}$$

Some authors replace the single-particle energy by the quantity $\xi_\mathbf{k} = \varepsilon(\mathbf{k}) - \varepsilon_f$. Here $\varepsilon_f = \varepsilon(\mathbf{k}_f)$ is the Fermi energy. The ground state $|\Psi_0\rangle$ is the Fermi-sea made by filling all single-particle states up to the Fermi energy. This, of course, corresponds to filling all the negative $\xi_\mathbf{k}$ levels — so we have an analogy between the Fermi sea and the infinitely deep Dirac sea of relativistic electrons. Beware, though, that *holes* in the Fermi sea have *negative* mass: the state made removing an electron with momentum $-\mathbf{k}$ and negative $\xi_\mathbf{k}$, has positive energy, $|\xi_\mathbf{k}|$, momentum $\mathbf{k}$, and velocity $\mathbf{v} = -\mathbf{k}/m$. Its velocity and momentum therefore point in *opposite* directions[2].

The total number of particles is

$$\begin{aligned} N &= 2\times(Vol)\int \frac{d^3k}{(2\pi)^3}\theta(\varepsilon_f - \varepsilon(\mathbf{k})) \\ &= 2\times(Vol)\times \frac{4}{3}\pi(k_f)^3 \frac{1}{(2\pi)^3} = 2\times(Vol)\times\frac{1}{6\pi^2}(k_f)^3. \end{aligned} \tag{9.19}$$

(The ×2 is for the two posssible spin values.)

The density is therefore

$$\rho = 2\times \frac{k_f^3}{6\pi^2}. \tag{9.20}$$

Let us now define the propagator to be

$$iG_{\alpha\beta}(x,x') = \langle\Psi_0|T\{\psi_\alpha(x)\psi^\dagger_\beta(x')\}|\Psi_0\rangle, \tag{9.21}$$

---

[2] Holes in the valance band of a semiconductor have positive mass, and are a much better analog of holes in the Dirac sea.

9.2 Fermi Gas and Fermi Liquid

where α, β label the spin components. As usual, the factor of i is inserted to make this a Green function for the Schrödinger equation

$$(i\partial_t + \frac{1}{2m}\partial^2)G_{\alpha\beta}(x, x') = \delta^4(x - x')\delta_{\alpha\beta}. \tag{9.22}$$

If

$$G_{\alpha\beta}(x, 0) = \int \frac{d^3k}{(2\pi)^3} \frac{d\omega}{2\pi} G(\mathbf{k}, \omega) e^{i\mathbf{k}\cdot\mathbf{x} - i\omega t}, \tag{9.23}$$

being a Green function requires $(\omega - \varepsilon(\mathbf{k}))G(\mathbf{k}, \omega) = \delta_{\alpha\beta}$. When we use the definition (9.21), we get

$$iG_{\alpha\beta}(x, x') = \int \frac{d^3k}{(2\pi)^3} \left\{ \theta(t - t')\langle\Psi_0|\hat{a}_\mathbf{k}\hat{a}_\mathbf{k}^\dagger|\Psi_0\rangle \right.$$

$$\left. - \theta(t' - t)\langle\Psi_0|\hat{a}_\mathbf{k}^\dagger\hat{a}_\mathbf{k}|\Psi_0\rangle \right\} e^{i\mathbf{k}\cdot(\mathbf{x}-\mathbf{x}') - i\varepsilon(\mathbf{k})(t-t')}$$

$$= \int \frac{d^3k}{(2\pi)^3} \left\{ (\theta(t - t')\theta(|\mathbf{k}| - k_f) - \theta(t' - t)\theta(k_f - |\mathbf{k}|)) \right.$$

$$\left. \times e^{i\mathbf{k}\cdot(\mathbf{x}-\mathbf{x}') - i\varepsilon(\mathbf{k})(t-t')} \right\}. \tag{9.24}$$

From this we find

$$G(\mathbf{k}, \omega)_{\alpha\beta} = \frac{1}{\omega - \varepsilon(\mathbf{k}) + i\epsilon\,\text{sgn}(|\mathbf{k}| - k_f)}\delta_{\alpha\beta}, \tag{9.25}$$

which does indeed satisfy $(\omega - \varepsilon(\mathbf{k}))G(\mathbf{k}, \omega) = \delta_{\alpha\beta}$.

Let us check the new "$i\epsilon$" prescription appearing in (9.25) by using it to compute the particle density. Forget spin for now. We want to evaluate

$$\langle\Psi_0|\psi^\dagger(\mathbf{x}, t)\psi(\mathbf{x}, t)|\Psi_0\rangle = -\lim_{\tau\to 0^+} \langle\Psi_0|T\{\psi(\mathbf{x}, t)\psi^\dagger(\mathbf{x}, t + \tau)\}|\Psi_0\rangle. \tag{9.26}$$

This equals

$$-\lim_{\tau\to 0^+} \left\{ \int \frac{d^3k}{(2\pi)^3} \frac{d\omega}{2\pi} \frac{i}{\omega - \varepsilon(\mathbf{k}) + i\epsilon\,\text{sgn}(|\mathbf{k}| - k_f)} e^{i\omega\tau} \right\}. \tag{9.27}$$

Since $\tau > 0$, we can use Jordan's lemma to close the ω contour in the upper half-plane. The pole at $\omega = \varepsilon(\mathbf{k}) - i\epsilon\,\text{sgn}(|\mathbf{k}| - k_f)$ lies within the contour and contributes a factor of $2\pi i$ only if $|\mathbf{k}| < k_f$. We end up with

$$\rho = -i\frac{2\pi i}{2\pi} \int \frac{d^3k}{(2\pi)^3}\theta(k_f - |\mathbf{k}|), \tag{9.28}$$

which is correct. From now on I will abbreviate $\epsilon\,\text{sgn}(|\mathbf{k}| - k_f)$ as $\epsilon(|\mathbf{k}| - k_f)$.

If we keep $\mathbf{x}$, $\mathbf{x}'$ separate, then

$$\langle\Psi_0|\psi^\dagger(\mathbf{x}, t)\psi(\mathbf{x}', t)|\Psi_0\rangle = -\lim_{\tau\to 0^+} \langle\Psi_0|T\{\psi(\mathbf{x}, t)\psi^\dagger(\mathbf{x}', t + \tau)\}|\Psi_0\rangle$$

$$= \int \frac{d^3k}{(2\pi)^3} e^{i\mathbf{k}\cdot(\mathbf{x}-\mathbf{x}')} n(\mathbf{k}), \tag{9.29}$$

where

$$n(\mathbf{k}) = \langle \Psi_0 | \hat{a}_\mathbf{k}^\dagger \hat{a}_\mathbf{k} | \Psi_0 \rangle = \theta(k_f - |\mathbf{k}|). \tag{9.30}$$

We see therefore that Fourier transforming the *one-particle density matrix*, $\langle \Psi_0 | \psi^\dagger(\mathbf{x}, t) \psi(\mathbf{x}', t) | \Psi_0 \rangle$, with respect to $(\mathbf{x} - \mathbf{x}')$ yields the momentum-space particle distribution. This statement remains true in the interacting theory.

The momentum-space particle distribution of real electron systems can be directly measured by *angle-resolved photoemission*. Here one shines synchroton light on the sample and analyses the momentum of the electrons knocked out of the solid by the energetic photons. The momentum $\mathbf{k}$ is conserved in this process but even photons with enough energy to knock an electron out of the sample have wavelengths of several hundred angstroms, so their $\mathbf{k}$ is negligable compared to that of the electrons whose typical wavelength is of the order of the lattice spacing (a few angstroms).

A collection of noninteracting fermions is usually called a *Fermi gas*. Landau[3] showed that many even strongly interacting electron systems are what he called *Fermi liquids*. In these systems we can treat the system as if it were composed of almost free *quasiparticles*, so many of their properties are Fermi-gas like. This is the explanation for the remarkable success of the single-electron picture in solid-state physics. Landau's original argument was couched in terms of the lack of available phase space for scattering processes near the Fermi surface. A modern version rephrases this to assert that all but forward scattering processes are *irrelevant* in the sense of the renormalization group.

In a Fermi liquid it remains true that all states with energy $\varepsilon < \varepsilon_f$ are occupied. It is *not* true, however, that the corresponding momentum distribution has all momentum states with $|\mathbf{k}| < k_f$ occupied. When we look at $n(\mathbf{k})$, we find a *discontinuity* at k_f whose magnitude is Z, which corresponds to a pole term in the Lehmann expansion of $G(\mathbf{k}, \omega)$:

$$G(\mathbf{k}, \omega) = \frac{Z}{\omega - \varepsilon(\mathbf{k}) + i\epsilon(|\mathbf{k}| - k_f)} + \text{other excitations.} \tag{9.31}$$

As usual $Z < 1$.

---

[3]Lev Davidovich Landau, Born January 22, 1908, Baku, Azerbaijan, Russian Empire. Died April 1, 1968, Moscow USSR. Nobel Prize for Physics 1962.

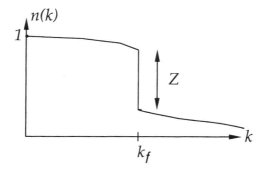

Fig 1. Momentum distribution in a Fermi liquid.

9.2.2 Linear Response

Semiclassical Boltzmann Equation

If any inhomogeneities are at long wavelength compared to p_f, we can extend the $n(\mathbf{p})$ of the previous section to a distribution $n(\mathbf{p}, \mathbf{x}, t)$ giving the particle density in $\mathbf{p}$ space for each $\mathbf{x}$. Of course this is not possible in general because of the uncertainty principle. Because of this restriction to slow variations, $n(\mathbf{p}, \mathbf{x}, t)$ is only a semiclassical concept.

Bboltzmann's equation is a continuity equation for $n(\mathbf{p}, \mathbf{x}, t)$. It asserts that

$$\frac{\partial n}{\partial t} + \mathbf{v}_\mathbf{p} \cdot \nabla_\mathbf{p} n + \mathbf{v}_\mathbf{x} \cdot \nabla_\mathbf{x} n = I(n). \tag{9.32}$$

Here

$$\mathbf{v}_\mathbf{p} = -\frac{\partial \varepsilon(\mathbf{p}, \mathbf{x})}{\partial \mathbf{x}}, \quad \mathbf{v}_\mathbf{x} = \frac{\partial \varepsilon(\mathbf{p}, \mathbf{x})}{\partial \mathbf{p}} \tag{9.33}$$

are the rate of change of $\mathbf{p}$ and the group velocity $\mathbf{v}_\mathbf{x}$ given by Hamilton's theory of rays. The energy $\varepsilon(\mathbf{p}, \mathbf{x})$ is simply

$$\varepsilon(\mathbf{p}, \mathbf{x}) = \frac{1}{2m}|\mathbf{p}|^2 + V(x). \tag{9.34}$$

Liouville's theorem tells us that

$$\nabla_\mathbf{p} \cdot \mathbf{v}_\mathbf{p} + \nabla_\mathbf{x} \cdot \mathbf{v}_\mathbf{x} = 0 \tag{9.35}$$

so (9.33) might be written

$$\frac{\partial n}{\partial t} + \nabla_\mathbf{p} \cdot n\mathbf{v}_\mathbf{p} + \nabla_\mathbf{x} \cdot n\mathbf{v}_\mathbf{x} = I(n), \tag{9.36}$$

which makes it a bit more obvious that the equation is related to a $\partial_t \rho + \nabla \cdot \rho \mathbf{v} = 0$ conservation equation. The only difference from a conventional continuity equation is the quantity $I(n)$, which is a collision term representing the rate at which particles are scattered from one $\mathbf{p}$ to another. These collisions generate entropy.

104   9. Electrons in Solids

We will ignore them for now and just concentrate on coherent collective effects. The collisionless Boltzmann equation is often called the *Vlasov* equation.

We want to consider deviations from equilibrium $n = n_0 + \delta n$, where $n_0 = \theta(k_f - |\mathbf{p}|)$. Note that $\partial_t n_0 = 0$ and $\nabla_{\mathbf{x}} n_0 = 0$, while $\nabla_{\mathbf{p}} n_0$ contains a delta function at $|\mathbf{p}| = k_f$

$$\nabla_{\mathbf{p}} n_0 = \frac{\partial n_0}{\partial \varepsilon} \frac{\partial \varepsilon}{\partial \mathbf{p}} = -\delta(|\mathbf{p}| - k_f)\frac{\mathbf{v}_x}{v_f}. \tag{9.37}$$

To solve (9.33) perturbatively we need only δn in the $\partial_t n$ and $\nabla_{\mathbf{x}} n$ terms, and should keep only n_0 in the $\nabla_{\mathbf{p}} n$ term. Writing

$$\delta n(\mathbf{p}, \mathbf{x}, t) = \int \frac{d^3 k}{(2\pi)^3} \frac{d\omega}{2\pi} e^{i\mathbf{k}\cdot\mathbf{x} - i\omega t} \delta\tilde{n}(\mathbf{p}, \mathbf{k}, \omega), \tag{9.38}$$

we find that

$$-i\omega\delta\tilde{n} + i\tilde{V}(\mathbf{k}, \omega)\frac{\mathbf{k}\cdot\mathbf{v}_x}{v_f}\delta(|\mathbf{p}| - k_f) + i\mathbf{k}\cdot\mathbf{v}_x\delta\tilde{n} = 0. \tag{9.39}$$

Here $\tilde{V}(\mathbf{k}, \omega)$ is the Fourier transform of $V(x)$. Thus,

$$(\omega - \mathbf{k}\cdot\mathbf{v}_x)\delta\tilde{n}(\mathbf{p}, \mathbf{k}, \omega) = \frac{\mathbf{k}\cdot\mathbf{v}_x}{v_f}\delta(|\mathbf{p}| - k_f)\tilde{V}(\mathbf{k}, \omega). \tag{9.40}$$

We see that δn is nonzero only on the Fermi surface. Remember in interpreting this formula that $\mathbf{v}_x = \mathbf{p}/m$.

9.2.3 Diagram Approach

We can obtain (9.40) from field theory.

Let us write

$$\begin{aligned}
n_0(\mathbf{x}, \mathbf{p}) &= \int d^3 y \langle \Psi_0 | T\{\psi^\dagger(\mathbf{x} + \tfrac{1}{2}\mathbf{y}, t + \epsilon)\psi(\mathbf{x} - \tfrac{1}{2}\mathbf{y}, t)\}|\Psi_0\rangle e^{-i\mathbf{p}\cdot\mathbf{y}} \\
&= -i\int d^3 y \int \frac{d^3 q}{(2\pi)^3}\frac{d\omega}{2\pi}\frac{e^{i(\mathbf{q}-\mathbf{p})\cdot\mathbf{y}}e^{i\epsilon\omega}}{\omega - \varepsilon(\mathbf{q}) + i\epsilon(|\mathbf{q}| - q_f)} \\
&= \theta(|\mathbf{p}| - p_f)
\end{aligned} \tag{9.41}$$

and consider the effect on this of the perturbation $V(x, t)\psi^\dagger(\mathbf{x}, t)\psi(\mathbf{x}, t)$, where $V(x, t)$ is slowly varying in space and time.

9.2 Fermi Gas and Fermi Liquid

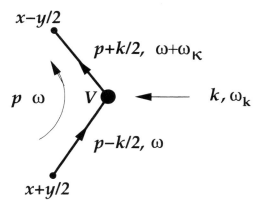

Fig 2. The momentum distribution function n(x, t) to first-order in the perturbation V. The variable $\mathbf{k}$ and ω_k are the Fourier conjugates of $\mathbf{x}$ and t.

If we Fourier transform to replace $\mathbf{x}$ and t by $\mathbf{k}$ and ω_k, we have

$$\delta \tilde{n}(p, k, \omega_k) =$$
$$\int \frac{d\omega}{2\pi} \frac{-i}{\omega + \omega_k - \varepsilon(\mathbf{p} + \tfrac{1}{2}\mathbf{k}) + i\epsilon(|\mathbf{p} + \tfrac{1}{2}\mathbf{k}| - p_f)}$$
$$\times (-i\tilde{V}(\mathbf{k}, \omega_k)) \frac{i}{\omega - \varepsilon(\mathbf{p} - \tfrac{1}{2}\mathbf{k}) + i\epsilon(|\mathbf{p} - \tfrac{1}{2}\mathbf{k}| - p_f)} e^{i\epsilon\omega}$$
$$= -i \int \frac{d\omega}{2\pi} \frac{\tilde{V}(\mathbf{k}, \omega_k)}{\omega_k - (\varepsilon(\mathbf{p} + \tfrac{1}{2}\mathbf{k}) - \varepsilon(\mathbf{p} - \tfrac{1}{2}\mathbf{k}))} \left[\frac{1}{\omega - \varepsilon(\mathbf{p} - \tfrac{1}{2}\mathbf{k}) + i\epsilon(|\mathbf{p} - \tfrac{1}{2}\mathbf{k}| - p_f)} \right.$$
$$\left. - \frac{1}{\omega + \omega_k - \varepsilon(\mathbf{p} + \tfrac{1}{2}\mathbf{k}) + i\epsilon(|\mathbf{p} + \tfrac{1}{2}\mathbf{k}| - p_f)} \right] e^{i\epsilon\omega}$$
$$= \frac{-1}{\omega_k - (\varepsilon(\mathbf{p} + \tfrac{1}{2}\mathbf{k}) - \varepsilon(\mathbf{p} - \tfrac{1}{2}\mathbf{k}))} \left[n_0(\mathbf{p} + \tfrac{1}{2}\mathbf{k}) - n_0(\mathbf{p} - \tfrac{1}{2}\mathbf{k}) \right] \tilde{V}(\mathbf{k}, \omega_k) \quad (9.42)$$

(I have omitted some $i\epsilon$'s in the last two equations. They are not important at the moment, though we will need to worry about them later.) So far we have made no approximations beyond working to first order in V. We now use the condition that V is slowly vaying in space, so that $\mathbf{k}$ is never large. We can therefore approximate (9.42) by

$$\delta n(\mathbf{p}, \mathbf{k}, \omega_k) = -\frac{1}{\omega_k - \mathbf{v_x} \cdot \mathbf{k}} \mathbf{v_x} \cdot \mathbf{k} \frac{\partial n_0}{\partial \varepsilon} \tilde{V}(\mathbf{k}, \omega_k), \quad (9.43)$$

which with

$$\frac{\partial n_0}{\partial \varepsilon} = -\delta(\varepsilon - \varepsilon_f) = -\frac{1}{v_f} \delta(|\mathbf{p}| - p_f) \quad (9.44)$$

is recognized to be the same expression we obtained from the semiclassical Boltzmann equation.

9.2.4 Applications

In this section we will discuss some applications of (9.43). My intention is not so much to calculate things that can only be obtained from field theory, but to show how the field theory reproduces physics that we could describe by less fancy methods. The reason for doing this is so that when you see a diagram calculation in a research paper, you will have a good idea of what sort of physical phenomena are being included in the calculation.

Most of the applications I will discuss will use the particle-hole bubble

$$\Pi(\mathbf{k}, \omega_\mathbf{k}) = -\left(\int \frac{d^3 p}{(2\pi)^3} \frac{1}{\omega_\mathbf{k} - (\varepsilon(\mathbf{p} + \mathbf{k}) - \varepsilon(\mathbf{p}))} \left[n_0(\mathbf{p} + \mathbf{k}) - n_0(\mathbf{p}) \right] \right) \tag{9.45}$$

or its small $\mathbf{k}$ limit where the integrand is replaced by

$$-\frac{1}{\omega_\mathbf{k} - \mathbf{v_x} \cdot \mathbf{k}} \mathbf{v_x} \cdot \mathbf{k} \frac{1}{v_f} \delta(|\mathbf{p}| - p_f). \tag{9.46}$$

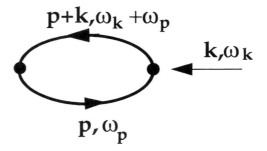

Fig 3. The electron-hole bubble. $\Pi(\mathbf{k}, \omega_\mathbf{k})$.

In particular we will make much use of the equation

$$\delta\tilde{\rho}(\mathbf{k}, \omega_\mathbf{k}) = \Pi(\mathbf{k}, \omega_\mathbf{k})\tilde{\phi}(\mathbf{k}, \omega_\mathbf{k}) \tag{9.47}$$

For $\omega_\mathbf{k} = 0$, the expression (9.45) is called the *Lindhard function*. We will denote it by $\chi(\mathbf{k})$. The integral is elementary and can be done exactly to give

$$\chi(\mathbf{k}) = \frac{mk_f}{2\pi^2} \left[\frac{1}{2} + \frac{1 - x^2}{4x} \ln\left| \frac{1 + x}{1 - x} \right| \right], \tag{9.48}$$

where $x = k/(2k_f)$. Note that the expression in square brackets is unity at $x = 0$.

Debye Screening

The first problem we will discuss is the Debye-Hükel screening of a static ($\omega_\mathbf{k} = 0$) charge distribution by the electron gas. The unscreened Coulomb potential of a unit point charge is

$$\phi_0(r) = \frac{1}{4\pi r} \tag{9.49}$$

9.2 Fermi Gas and Fermi Liquid

and its Fourier transform is therefore

$$\tilde{\phi}_0(\mathbf{k}) = \int d^3x \frac{1}{4\pi r} e^{i\mathbf{k}\cdot\mathbf{r}} = \frac{1}{\mathbf{k}^2}. \tag{9.50}$$

For slowly varying potentials we expect the electron gas to to respond to an external potential by pulling in or expelling electrons to give a density change. This we estimate by using the Thomas-Fermi approximation

$$\delta\rho(x) = \left.\frac{\partial\rho}{\partial\mu}\right|_{\varepsilon_f} e\phi(x) = e\frac{mk_f}{\pi^2}\phi(x). \tag{9.51}$$

Here we have used (9.20), $\mu = \varepsilon_f = \frac{1}{2}mk_f^2$ and $\rho_0 = k_f^3/3\pi^2$.

It is only this density *change* that produces any effect because the charge-density of the uniform electron gas is cancelled by that of the background ions. As we are interested only in long-wavelength effects, we may imagine the ion charge to be uniformly spread throughout space. (The electron gas with a uniform background charge is called the *Jellium* model.)

When we insert the density variation into Poisson's equation $-\nabla^2\phi = (-e)\rho$ we find

$$-\nabla^2\phi = -\left.\frac{\partial\rho}{\partial\mu}\right|_{\varepsilon_f} e^2\phi. \tag{9.52}$$

This means that the Coulomb potential is screened and takes the form

$$\phi(\mathbf{r}) = \int g(\mathbf{r}, \mathbf{r}')q_{ext}(\mathbf{r}')d^3r', \tag{9.53}$$

where

$$g(\mathbf{r}, 0) = \frac{1}{4\pi r}e^{-r\sqrt{e^2\frac{\partial\rho}{\partial\mu}}} = \frac{1}{4\pi r}e^{-k_0 r} \tag{9.54}$$

in real space and

$$\tilde{g}(\mathbf{k}) = \frac{1}{\mathbf{k}^2 + e^2\frac{\partial\rho}{\partial\mu}} = \frac{1}{\mathbf{k}^2 + k_0^2} \tag{9.55}$$

in Fourier space. Here

$$k_0^2 = e^2\frac{\partial\rho}{\partial\mu} = e^2\frac{\partial\rho}{\partial k_f}\left(\frac{\partial\varepsilon_f}{\partial k_f}\right)^{-1} = \frac{e^2}{\pi^2}mk_f. \tag{9.56}$$

We can recover this result diagramatically by summing the following diagrams with $\omega_\mathbf{k}$ set equal to zero since we are interested only in static effects.

108   9. Electrons in Solids

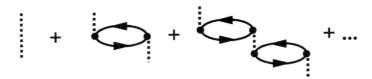

Fig 4. The diagrams summed to get the Debye-Hükel screening. The verical line is the instantaneous Coulomb interaction.

These diagrams give

$$\tilde{g}(\mathbf{k}) = \frac{1}{\mathbf{k}^2}\left[1 - \frac{e^2}{\mathbf{k}^2}\int \frac{d^3p}{(2\pi)^3}\frac{1}{v_f}\delta(|\mathbf{p}| - p_f) + \ldots\right], \quad (9.57)$$

the dots refering to a continuing geometric progression. Now

$$\frac{1}{v_f}\delta(|\mathbf{p}| - p_f) = \delta(\varepsilon - \varepsilon_f) \quad (9.58)$$

and

$$\int \frac{d^3p}{(2\pi)^3}\delta(\varepsilon - \varepsilon_f) = \frac{\partial}{\partial \varepsilon_f}\int \frac{d^3p}{(2\pi)^3}\theta(\varepsilon_f - \varepsilon) = \frac{\partial \rho}{\partial \varepsilon_f}. \quad (9.59)$$

Therefore

$$\tilde{g}(\mathbf{k}) = \frac{1}{\mathbf{k}^2}\frac{1}{1 + \frac{e^2}{\mathbf{k}^2}\frac{\partial \rho}{\partial \varepsilon_f}}$$

$$= \frac{1}{\mathbf{k}^2 + e^2\frac{\partial \rho}{\partial \varepsilon_f}} = \frac{1}{\mathbf{k}^2 + k_0^2} \quad (9.60)$$

as we expect. Our long-wavelength approximation is therefore equivalent to the Thomas-Fermi theory.

When we consider a delta-function point source, we cannot trust the smoothed-out Thomas-Fermi response function and must use instead the full Lindhard function. Then the singularity from the logarithm at $|\mathbf{k}| = 2k_f$ gives rise to an asymptotic oscillatory term

$$\phi(r) \approx \frac{1}{r^3}\cos 2k_f r. \quad (9.61)$$

These are the *Friedel oscillations*.

Plasma Oscillations

Next we study *plasma oscillations*. These are oscillations in the density of electrons. The restoring force is the electric field that results whenever there is any deviation from charge neutrality. Again we first study the classical theory. We consider small density fluctuations in a gas of charged particles. The particles obey Newton's law

$$m\dot{\mathbf{v}} = \frac{(-e)}{\mathbf{E}} \quad (9.62)$$

9.2 Fermi Gas and Fermi Liquid   109

and Gauss' law
$$\nabla \cdot \mathbf{E} = (-e)(\rho - \rho_0). \tag{9.63}$$

If the amplitude of the fluctuations is small, we may make the approximation $\mathbf{j} \equiv \rho \mathbf{v}$ by $\rho_0 \mathbf{v}$ in the continuity equation
$$\partial_t \rho + \nabla \cdot \mathbf{j} = 0. \tag{9.64}$$

With this small-amplitude approximation we find
$$\partial_t (e\mathbf{j}) = \frac{e^2}{m} \rho_0 \mathbf{E}. \tag{9.65}$$

On taking the divergence this becomes
$$\ddot{\rho} + \frac{e^2 \rho_0}{m}(\rho - \rho_0) = 0. \tag{9.66}$$

The solutions are density oscillations at the *plasma frequency* $\omega_p^2 = e^2 \rho_0/m$. In this purely classical model the oscillation frequency is independent of the wavelength so the *plasmon* modes have vanishing group velocity. We will see that this lack of propagation is modified when we take into account the existence of the Fermi sea.

Let us see how plasmons arise from diagrams. Because plasma oscillations have non-zero frequency, even at $\mathbf{k} = 0$, it makes sense to consider the region where $kv_f \ll \omega$. Given this we can expand the electron-hole bubble in powers of $v_f k/\omega$:

$$\int \frac{d^3p}{(2\pi)^3} \frac{1}{v_f} \delta(|\mathbf{p}| - p_f) \frac{\mathbf{v_x} \cdot \mathbf{k}}{\omega_k - \mathbf{v_x} \cdot \mathbf{k}}$$
$$= \frac{\partial \rho}{\partial \mu} \int \frac{d\Omega_p}{4\pi} \frac{\mathbf{v_x} \cdot \mathbf{k}}{\omega_k} \left(1 + \frac{\mathbf{v_x} \cdot \mathbf{k}}{\omega_k} + \left(\frac{\mathbf{v_x} \cdot \mathbf{k}}{\omega_k}\right)^2 + \left(\frac{\mathbf{v_x} \cdot \mathbf{k}}{\omega_k}\right)^3 + \cdots\right)$$
$$= \frac{\partial \rho}{\partial \mu} \left(\frac{v_f^2}{\omega_k^2} \frac{\mathbf{k}^2}{3} + \left(\frac{v_f^2}{\omega_k^2}\right)^2 \frac{(\mathbf{k}^2)^2}{5} + \cdots\right). \tag{9.67}$$

In obtaining this expression I have used the identities
$$\frac{1}{v_f^2} \int \frac{d\Omega_p}{4\pi} v_\alpha v_\beta = \frac{1}{3} \delta_{\alpha\beta} \tag{9.68}$$

and
$$\frac{1}{v_f^4} \int \frac{d\Omega_p}{4\pi} v_\alpha v_\beta v_\gamma v_\delta = \frac{1}{15} \left(\delta_{\alpha\beta}\delta_{\gamma\delta} + \delta_{\alpha\gamma}\delta_{\beta\delta} + \delta_{\alpha\delta}\delta_{\beta\gamma}\right) \tag{9.69}$$

[remember that $\mathbf{v} = \mathbf{v}(p) = \mathbf{p}/m$.] together with the observation that all odd powers of $\mathbf{v}$ must average to zero.

Keeping only the first term in (9.67) the bubble sum of Fig. 4 becomes
$$\frac{e^2}{\mathbf{k}^2} \frac{1}{1 - \frac{e^2}{\mathbf{k}^2} \frac{\partial \rho}{\partial \mu} \frac{v_f^2}{\omega_k^2} \frac{\mathbf{k}^2}{3}}. \tag{9.70}$$

Now $\rho \propto \varepsilon_f^{3/2}$ so

$$\left.\frac{\partial \rho}{\partial \varepsilon}\right|_{\varepsilon_f} = \frac{3}{2}\frac{\rho_0}{\varepsilon_f} = \frac{3\rho_0}{mv_f^2}. \tag{9.71}$$

Thus the pole in (9.70) occurs when

$$1 = e^2 \frac{3\rho_0}{mv_f^2} \frac{v_f^2}{\omega_{\mathbf{k}}^2} \frac{1}{3} = \frac{\rho_0 e^2}{m\omega_{\mathbf{k}}^2}, \tag{9.72}$$

which is precisely at the classical plasmon frequency $\omega_{\mathbf{k}}^2 = \omega_p^2 = \rho_0 e^2/m$.

Retaining the next term in (9.67), but approximating[4] the factor of $1/\omega_{\mathbf{k}}^4$ as $1/(\omega_{\mathbf{k}}^2 \omega_p^2)$ gives the next estimate of the pole position as

$$\omega^2(\mathbf{k}) = \omega_p^2 + \frac{3}{5}v_f^2 k^2 + \ldots \tag{9.73}$$

There is little purpose in expanding much further in **k**, because once there is enough energy for the electron and hole in the intermediate state to become real particles, the plasmon excitation becomes unstable to decay to this state, and so becomes a resonance in the particle-hole continuum. This is sketched in Fig. 5.

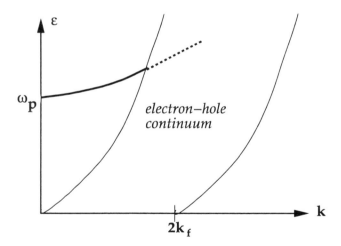

Fig 5. *The plasmon eventually becomes a resonance in the particle-hole continuum.*

Fig. 5 also shows the boundaries of the region in the $\omega, \mathbf{k}$ plane where we can have particle-hole excitations. These boundaries are given by the equations $\omega = \varepsilon(k) \pm kv_f$ and correspond to the two extreme cases in Fig.6.

---

[4] If we do not do this, we will need to solve a quadratic equation to find $\omega_{\mathbf{k}}^2$. The approximation finds the root that lies near ω_p^2 to within $O(k^2)$ precision.

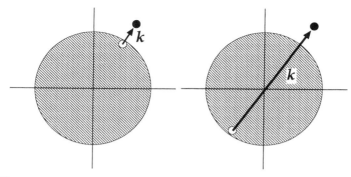

*Fig 6. The minimum and maximum values of **k** for a particle-hole excitation of given energy.*

Landau Damping

Once the $\omega(\mathbf{k})$ dispersion curve for the plasmon enters the particle-hole continuum, the plasmon becomes unstable to decay to an electron-hole pair. Just as in Chapter 5, this is signalled by the particle-hole bubble, $\Pi(\mathbf{k}, \omega_k)$ gaining an imaginary part. We can calculate the magnitude of the imaginary part by utilizing (9.42). Once we are in the electron-hole continuum there will be a set of momenta $\mathbf{p}$ lying below the Fermi surface such that $\mathbf{p} + \mathbf{k}$ lies above the Fermi surface, while $\omega(\mathbf{k}) = \varepsilon(\mathbf{p}+\mathbf{k}) - \varepsilon(\mathbf{p})$. The denominator of (9.42) vanishes for these $\mathbf{p}$ — so now is the time to reinstate the $i\epsilon$ terms that were omitted in our earlier discussion of the Lindhard function.

With these terms restored, $\Pi(\mathbf{k}, \omega)$ becomes

$$\int \frac{d^3p}{(2\pi)^3} \frac{[n(\mathbf{p}) - n(\mathbf{p}+\mathbf{k})]}{\omega_k - (\varepsilon(\mathbf{p}+\mathbf{k}) - \varepsilon(\mathbf{p})) + i\epsilon(|\mathbf{p}+\mathbf{k}| - p_f) - i\epsilon(|\mathbf{p}| - p_f)}$$

$$= \int \frac{d^3p}{(2\pi)^3} \frac{n(\mathbf{p})(1 - n(\mathbf{p}+\mathbf{k}))}{\omega_k - (\varepsilon(\mathbf{p}+\mathbf{k}) - \varepsilon(\mathbf{p})) + i\epsilon}$$

$$- \int \frac{d^3p}{(2\pi)^3} \frac{n(\mathbf{p}+\mathbf{k})(1 - n(\mathbf{p}))}{\omega_k - (\varepsilon(\mathbf{p}+\mathbf{k}) - \varepsilon(\mathbf{p})) - i\epsilon}. \quad (9.74)$$

For positive ω_k, the imaginary part comes from the first term and is

$$\text{Im } \Pi(\mathbf{k}, \omega) = \pi \int \frac{d^3p}{(2\pi)^3} \delta\left(\omega_k - (\varepsilon(\mathbf{p}+\mathbf{k}) - \varepsilon(\mathbf{p}))\right)[n(\mathbf{p})(1 - n(\mathbf{p}+\mathbf{k}))]. \quad (9.75)$$

This is the sum over the allowed particle-hole final states expected from unitarity.

We can gain insight into this expression from classical arguments. The quantity ω_k plays a dual role as the energy of the plasmon and as the frequency of the classical wave. When the frequency gains an imaginary part, this means that the energy in the plasma oscillation wave is being dissipated. The resultant decay of the wave is known as *Landau damping* after Landau's 1947 analysis of the analogous process is classical plasmas. To repeat this analysis we note that our Boltzmann

9. Electrons in Solids

equation remains valid at finite temperature provided we replace the step-function form of $n(\mathbf{p})$ by the Fermi function

$$n(\mathbf{p}) \to n_\beta(\mathbf{p}) = \frac{1}{e^{\beta(\varepsilon(\mathbf{p})-\mu)}+1}. \tag{9.76}$$

At high enough temperatures the Fermi function reduces to the Maxwell-Boltzmann distribution, and this was the case considered by Landau. Whatever distribution we use, once $\mathbf{k}$ is small compared to the range of variation of $n(\mathbf{p})$, we can approximate $\Pi(\mathbf{k}, \omega_\mathbf{k})$ by

$$\int \frac{d^3 p}{(2\pi)^3} \frac{\mathbf{k} \cdot \nabla n_\beta(\mathbf{p})}{\omega_\mathbf{k} - \mathbf{v} \cdot \mathbf{k} + i\epsilon}, \tag{9.77}$$

where $\mathbf{v} = \mathbf{p}/m$. Although we have not derived the $i\epsilon$ rule for the classical case, the form chosen here is very plausible in that it is consistent with causality. The imaginary part of $\Pi(\mathbf{k}, \omega_\mathbf{k})$ is therefore

$$\pi \int \frac{d^3 p}{(2\pi)^3} \left(\mathbf{k} \cdot \nabla n_\beta(\mathbf{p}) \right) \delta(\omega_\mathbf{k} - \mathbf{v} \cdot \mathbf{k}). \tag{9.78}$$

The delta function ensures that the only relevant electrons are those whose $\mathbf{k}$ component of their velocity is close to the phase velocity ω/k of the plasmon density wave. These electrons are able to "surf" the wave and exchange energy with it.

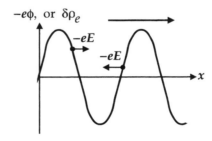

Fig 7. *The electron density for a plasmon wave moving to the right. Electrons surfing the leading edge of the wave crest are gaining energy. Those on the trailing edge are loosing energy.*

There is a small range of velocities for which the electrons become entrained by the wave. Within this range electrons initially moving slightly slower than the wave drift back with respect to it until they reach the leading edge of the wave crest where they are accelerated up to the wave speed, thus robbing energy from the wave. Those initially moving faster find themselves on the trailing edge where they are decelerated and so end up donating energy to the wave. Because the velocity distribution contains more slow particles than fast, the net effect is to damp the wave. The need for a variation of the distribution with energy is the origin of the $\mathbf{k} \cdot \nabla n_\beta(\mathbf{p})$ factor in (9.78).

Zero Sound

This is not really an "electrons in solids" topic but it fits so naturally here that I include it.

Zero sound replaces ordinary sound in liquid ^{3}He at low temperatures. Each ^{3}He atom has two electrons, two protons, and a neutron. This is an odd number of fermions, so when we are at low enough energies it will behave like a simple electrically neutral particle with Fermi statistics. Below about 1 K the helium atoms form a degenerate Fermi gas, but one not yet superfluid. (The superfluid critical temperature at about 1 or 2 mK depending on the pressure.)

The interaction between real helium atoms is quite complicated, but as first approximation they may be regarded as impenetrable spheres. We describe their mutual repulsion by a two-body delta-function interaction. whose second-quantized form is

$$H_{int} = \frac{1}{2}\lambda \int d^3x (\psi^\dagger \psi)^2, \qquad (9.79)$$

with $\lambda > 0$

Zero sound is a density-wave excitation similar to the plasma oscillation of the previous sections. The principal difference is that, the particles being electrically neutral, there are no long-range electrostatic forces to push the minimum frequency away from zero.

We look for excitations in the particle-hole (density fluctuation) channel by summing diagrams of the sort

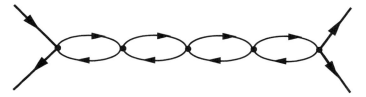

Fig 8. *The channel in which zero sound occurs.*

where the Feynman rule for the interaction is to include a factor of $-i\lambda$ for each vertex.

Zero-sound waves appear as a pole in this sum. We can also find them by seeking a solution to the equation

$$(\omega_k - \mathbf{v}_x \cdot \mathbf{k})\delta\tilde{n}(\mathbf{p}, \mathbf{k}, \omega_k) = \mathbf{v}_x \cdot \mathbf{k} \frac{1}{v_f}\delta(|\mathbf{p}| - p_f) \int \lambda \frac{d^3 p'}{(2\pi)^3} \lambda \delta\tilde{n}(\mathbf{p}', \mathbf{k}, \omega_k). \quad (9.80)$$

Consistency of the two sides of (9.80) requires that

$$\delta\tilde{n}(\mathbf{p}, \mathbf{k}, \omega_k) = \nu(\theta)\frac{1}{v_f}\delta(|\mathbf{p}| - p_f). \qquad (9.81)$$

Here ν is a function of the angle θ between $\mathbf{p}$ and $\mathbf{k}$ (and hence between $\mathbf{v}$ and $\mathbf{k}$). We are therefore allowing for the Fermi surface to change shape, as well as breathe in and out. With this form the delta functions cancel and we are left with

$$(\omega_{\mathbf{k}} - \mathbf{v}_{\mathbf{x}} \cdot \mathbf{k})v = \lambda \mathbf{v}_{\mathbf{x}} \cdot \mathbf{k} \left(\frac{\partial \rho}{\partial \varepsilon_f}\right) \int \frac{d\Omega}{4\pi} v. \qquad (9.82)$$

Equation (9.82) suggests a solution of the form

$$v = A \frac{\cos\theta}{s - \cos\theta} e^{i\mathbf{k}\cdot\mathbf{x} - i\omega t}, \qquad (9.83)$$

where $s = \omega_{\mathbf{k}}/|\mathbf{k}|v_f$ is the ratio of the zero-sound phase velocity to the Fermi velocity. Substituting this ansatz into (9.82), we find that the dispersion relation is determined by

$$\frac{s}{2}\ln\frac{s+1}{s-1} - 1 = \left(\frac{\partial\rho}{\partial\varepsilon_f}\lambda\right)^{-1}. \qquad (9.84)$$

As λ decreases to zero, we see that $s \to 1$ so that these waves propagate with velocity $c = \omega/k \to v_f$. They are therefore not related to ordinary (first) sound waves whose velocity, $c_{sound} = v_f/\sqrt{3}$, is determined by the compressibility. Also in the weak coupling limit the function (9.83) becomes more and more sharply peaked in the direction $\mathbf{k}$. This again is unlike ordinary sound waves where the Fermi surface remains spherical. Ordinary sound requires collisions that allow the particle distribution to relax to a local equilibrium. This relaxation takes some characteristic time τ. Zero sound appears at frequencies $\omega\tau \gg 1$, while conventional sound occurs at frequencies $\omega\tau < 1$.

9.3 Electrons and Phonons

In the previous sections we have considered the electrons as forming a uniform gas of negatively charged particles whose Coulomb field is almost entirely cancelled by that of the background ions. The background positive charge we imagined to be uniformly distributed throughout space.

Now we partly reverse the roles of the electrons and ions. We construct a simple model for the phonon modes in a metal by considering the ions as point charges of mass M immersed in an initially uniform electron gas. Small, long-wavelength displacements, $\delta\mathbf{r}$, of the ions give rise to a excess charge

$$e\delta\rho_{ion} = -NeZ\nabla \cdot \delta\mathbf{r} \qquad (9.85)$$

where N is the mean ion density and Ze their charge.[5] (The condition for equilibrium charge neutrality is $NZ = \rho_0$ where ρ_0 is the electron density) In the absence of any electron motion this would give rise to oscillations at the ion plasma frequency $\Omega_p^2 = N(eZ)^2/M$. This is much lower than the electron plasma frequency

---

[5]This is common, but somewhat misleading, notation. Z is *not* meant to be the atomic number, but is the number of electrons the atom contributes to the valance band.

9.3 Electrons and Phonons

allowing the electrons to passively follow the motion of the ions, screen their charge, and permit ion motion at even lower frequencies. The equation of motion for the ions is

$$M\delta\ddot{\mathbf{r}} = Ze\mathbf{E} = -Ze\nabla\phi, \qquad (9.86)$$

while

$$-\nabla^2\phi + k_0^2\phi = e\delta\rho_{ion} + eq_{ext}(\mathbf{r}). \qquad (9.87)$$

Here we have included an external charge source so we can examine the combined effect of the electrons and ions in screening it. In Fourier space we find that (9.85), (9.86), and (9.86), combine to give

$$\phi(k,\omega) = \frac{1}{k^2}\left\{1 - \frac{\Omega_p^2}{\omega^2} + \frac{k^2}{k_0^2}\right\}^{-1} q_{ext}(k,\omega). \qquad (9.88)$$

It is traditional to express this as

$$\phi(k,\omega) = \frac{1}{\epsilon(k,\omega)}\frac{1}{k^2}q_{ext}(k,\omega), \qquad (9.89)$$

where $\epsilon(k,\omega)$ is a frequency- and wavelength-dependent *dielectric function*. We have

$$\frac{1}{\epsilon(k,\omega)} = \left\{1 - \frac{\Omega_p^2}{\omega^2} + \frac{k^2}{k_0^2}\right\}^{-1} = \frac{1}{1+\frac{k_0^2}{k^2}}\left(\frac{\omega^2}{\omega^2 - \omega^2(k^2)}\right), \qquad (9.90)$$

where

$$\omega^2(k^2) = \frac{\Omega_p^2}{1+\frac{k_0^2}{k^2}}. \qquad (9.91)$$

Equation (9.90) has several consequences. First we see that the response to an external charge has a pole at $\omega^2 = \omega^2(k^2)$ indicating that a longitudinal density fluctuation can propagate at this frequency. For small k this frequency is $\omega = ck$ where the sound velocity is

$$c^2 = \frac{\Omega_p^2}{k_0^2}. \qquad (9.92)$$

Substituting the free-electron values (we are including spin)

$$\rho_0 = \frac{k_f^3}{3\pi^2}, \qquad \frac{1}{k_0^2} = \frac{1}{e^2}\frac{\pi^2}{mk_f}, \qquad (9.93)$$

we find

$$c^2 = \frac{1}{3}Z\frac{m}{M}v_f^2. \qquad (9.94)$$

This approximate relation between the sound speed and the Fermi velocity called the *Bohm-Staver* relation. Given the crudity of the model, it is tolerably well obeyed

— especially in the alkali metals (Na, K, etc.) where the bulk modulus of the metal is almost equal to that of a gas of free electrons.

Second we see that the effective Coulomb propagator mediating the interaction between the electrons becomes

$$g(k, \omega) = \frac{1}{k^2} \cdot \frac{1}{1 + \frac{k_0^2}{k^2}} \cdot \left(\frac{\omega^2}{\omega^2 - \omega^2(k^2)} \right). \tag{9.95}$$

This is frequency dependent — meaning that the force is no longer instantaneous. More important, for frequencies $\omega^2 < \omega^2(k^2)$, it corresponds to an *attractive* force. This arises because the passage of an electron draws the ions toward it and they remain gathered together long enough, even after the electron has departed, that they are able to attract another electron to this wake of enhanced positive charge. The maximum phonon frequency (here Ω_p, but more generally the Debye frequency) is always small compared to the Fermi energy, ε_f, so this region of attraction applies only to electrons with energies near to the Fermi surface. It is, nonetheless, the force that gives rise to to superconductivity in conventional low-temperature superconducting materials such as lead, niobium, and aluminum.

10
Nonrelativistic Bosons

The most commonly studied nonrelativistic boson system is liquid ^{4}He, and in particular its superfluid phase. This, therefore, will be the principal focus of this chapter.

10.1 The Boson Field

The many-body formalism for bosons parallels that developed in the previous chapter for fermions, only the statistics differs.

Again we can suppose we find the complete orthonormal set of energy eigenfunctions of the one-particle hamiltonian

$$H = -\frac{1}{2m}\nabla^2 + V(x), \qquad (10.1)$$

so

$$H\varphi_n = E_n\varphi_n. \qquad (10.2)$$

These form a basis for the single-boson space $\mathcal{H}^{(1)}$. Now we assume the operators $\hat{a}_n$, $\hat{a}_n^\dagger$ obey *commutation* relations

$$[\hat{a}_n, \hat{a}_m^\dagger] = \delta_{mn}. \qquad (10.3)$$

We define the second quantized-Heisenberg boson field

$$\hat{\varphi}(x, t) = \sum_n \hat{a}_n \varphi_n(x) e^{-iE_n t}, \qquad (10.4)$$

and find it obeys

$$[\hat{\varphi}(x,t), \hat{\varphi}^\dagger(x',t)] = \delta^3(x-x'). \tag{10.5}$$

Again this is the commutator we would get if we take as our lagrangian density the expression

$$\mathcal{L} = i\varphi^\dagger \partial_t \varphi - \frac{1}{2m}|\nabla \varphi|^2 - \varphi^\dagger V(x)\varphi, \tag{10.6}$$

with the resultant identification $\pi(x) \equiv i\varphi^\dagger(x)$ for the momentum conjugate to φ.

The operators $\hat{\varphi}(x,t)$, $\hat{\varphi}^\dagger(x,t)$ act on the "big" many-particle space

$$\mathcal{H} = \bigoplus_N \left\{ \bigotimes_{symm}^N \mathcal{H}^{(1)} \right\}. \tag{10.7}$$

This is the sum of the N-th *symmetric* powers of the single-particle Hilbert-space $\mathcal{H}^{(1)}$, and describes states containing N particles. Each $\bigotimes_{symm}^N \mathcal{H}^{(1)}$ is spanned by states of the form

$$|m_1, m_2, \ldots\rangle = \prod_{i=1}^{\infty} \frac{1}{\sqrt{m_i!}} (\hat{a}_i^\dagger)^{m_i} |0\rangle, \tag{10.8}$$

where $\sum m_i = N$, and $|0\rangle$ is the no-particle state annihilated by all the $\hat{a}_i$. Now, of course, Bose statistics means we can have many particles in the same state, so the m_i can take any integer value.

Once again the second-quantized hamiltonian is

$$\hat{H} = \int d^3x\, \hat{\varphi}^\dagger(x) H \hat{\varphi}(x) = \sum_n E_n \hat{a}_n^\dagger \hat{a}_n. \tag{10.9}$$

The particle number is

$$\hat{Q} = \int d^3x\, \hat{\rho}(x) = \int d^3x\, \hat{\varphi}^\dagger(x)\hat{\varphi}(x) = \sum_n \hat{a}_n^\dagger \hat{a}_n, \tag{10.10}$$

and the number-current density is

$$\hat{\mathbf{J}} = \frac{1}{2mi}\left(\hat{\varphi}^\dagger \nabla \hat{\varphi} - (\nabla \hat{\varphi}^\dagger)\hat{\varphi}\right). \tag{10.11}$$

We again find that

$$\partial_t \hat{\rho} + \nabla \cdot \hat{\mathbf{J}} = 0 \tag{10.12}$$

is obeyed as an operator equation.

From now on we will drop the hats on $\hat{\varphi}$ and simply write φ.

10.2 Spontaneous Symmetry Breaking

The Bose and Fermi formalisms are so similar. Where lies the difference? With Bose statistics there is no exclusion principle, and therefore no direct analog of the

10.2 Spontaneous Symmetry Breaking

Fermi sea. There is however the possibility of *Bose* (or Bose-Einstein) *Condensation* (BEC), where many particles occupy the lowest possible energy state. Actually multiple occupancy is probably not the best characterization of this phenomenon when interactions are present. It is preferable to think in terms of *spontaneously broken symmetry*. In the present case the symmetry that is broken is that of the $U(1)$ associated with number conservation.

If the ground state, $|\Phi_0\rangle$, is an eigenstate of charge, $\hat{Q}|\Phi_0\rangle = Q_0|\Phi_0\rangle$, then $e^{i\alpha\hat{Q}}|\Phi_0\rangle = e^{i\alpha Q_0}|\Phi_0\rangle \propto |\Phi_0\rangle$, and the ground state transforms as a singlet under the $U(1)$ symmetry. This is always the situation in the quantum mechanics of a system with a finite number of degrees of freedom. The Perron-Frobenius theorem tells us that ground state of such a system is nondegenerate, and any symmetry operation that commutes with the hamiltonian must leave the state in the same energy eigenspace.

Now our Bose field φ obeys

$$e^{i\alpha\hat{Q}}\varphi(x)e^{-i\alpha\hat{Q}} = e^{-i\alpha}\varphi(x). \tag{10.13}$$

Using this property and assuming that $|\Phi_0\rangle$ is indeed a singlet requires both

$$\langle\Phi_0|e^{i\alpha\hat{Q}}\varphi(x)e^{-i\alpha\hat{Q}}|\Phi_0\rangle = e^{i\alpha Q_0}e^{-i\alpha Q_0}\langle\Phi_0|\varphi(x)|\Phi_0\rangle = \langle\Phi_0|\varphi(x)|\Phi_0\rangle \quad (10.14)$$

and

$$\langle\Phi_0|e^{i\alpha\hat{Q}}\varphi(x)e^{-i\alpha\hat{Q}}|\Phi_0\rangle = e^{-i\alpha}\langle\Phi_0|\varphi(x)|\Phi_0\rangle. \tag{10.15}$$

This is only possible if $\langle\Phi_0|\varphi(x)|\Phi_0\rangle = 0$.

Suppose now that we have noninteracting particles and adjust the origin of the energy scale so that the normalized single-particle wavefunction, $\varphi_0(x) = 1/\sqrt{V}$, corresponds to energy $E_0 = 0$. (Shifting the energy to arrange this is an innocent manoeuver in the single-particle picture, but in the many-body language it is equivalent to adjusting the chemical potential, μ, of the system.) We may now put an arbitrary number of particles in the φ_0 state without changing the energy.

The state

$$|\phi\rangle = e^{\sqrt{V}(\phi\hat{a}_0^\dagger - \phi^*\hat{a}_0)}|0\rangle = e^{-\frac{1}{2}V|\phi|^2}\sum_{N_0=1}^{\infty}(\sqrt{V}\phi)^{N_0}\frac{1}{\sqrt{N_0!}}|N_0\rangle \tag{10.16}$$

is a sum of zero-energy eigenstates and is therefore itself a (normalized) zero-energy eigenstate. Now, however,

$$\varphi(x)|\phi\rangle = \phi|\phi\rangle \tag{10.17}$$

so

$$\langle\phi|\varphi(x)|\phi\rangle = \phi \neq 0. \tag{10.18}$$

Of course the state $|\phi\rangle$ is *not* an eigenstate of $\hat{Q}$. Instead

$$e^{-i\alpha\hat{Q}}|\phi\rangle = |e^{-i\alpha}\phi\rangle, \tag{10.19}$$

and the right-hand side is no longer the same state as $|\phi\rangle$. Indeed, since

$$\langle\phi'|\phi\rangle = e^{-\frac{1}{2}V|\phi-\phi'|^2}, \tag{10.20}$$

in the infinite volume limit the states $|e^{-i\alpha}\phi\rangle$ and $|\phi\rangle$ are orthogonal. In this case the $U(1)$ symmetry operation generated by $\hat{Q}$ takes us from one possible zero-energy state to another. There are very many such states since ϕ can be any complex number.

The particle density in the state $|\phi\rangle$ is

$$\langle\phi|\varphi^\dagger\varphi(x)|\phi\rangle = |\phi|^2. \tag{10.21}$$

In an interacting system the physical properties will be determined either by fixing the total number of particles present (a canonical ensemble) or by tuning the density using external pressure or a chemical potential (a grand canonical ensemble). In the thermodynamic limit there is no practical difference between these two ensembles, and in the latter we are free to consider a coherent superposition of particle-number states such as (10.16). Because the interaction does not change particle number we know that $\hat{Q}$ commutes with the hamiltonian. Consequently, when a state $|\phi\rangle$ giving a nonvanishing $\langle\varphi\rangle = \phi$ is an energy eigenstate, then $e^{-i\alpha\hat{Q}}|\phi\rangle = |e^{-i\alpha}\phi\rangle$ will again be an orthogonal state with the same energy. Even with interactions present, therefore, the ground state remains highly degenerate, but now the possible values of ϕ are restricted to lie on a circle with a fixed value of $|\phi|$. The $U(1)$ symmetry group acts transitively on this manifold of ground states.

When the ground state is no longer a singlet under the action of a symmetry group, many of the usual consequences of the symmetry — such as selection rules or particles falling into equal mass multiplets that form representations of the group — no longer hold in their simple form. We say that the symmetry is *spontaneously broken*. A quantity like $\langle\Phi_0|\varphi(x)|\Phi_0\rangle$, which must be zero when the ground state is a singlet under some symmetry transformation, but becomes nonzero in a phase with spontaneously broken symetry, is called an *order parameter*. It serves as a diagnostic for the symmetry breaking.

For those who are uncomfortable with the notion of states with no definite particle number, there is an alternative characterization of BEC or spontaneous symmetry breaking. This is the notion of *Off-diagonal Long-range Order*, or ODLRO, due to Oliver Penrose (the brother of Sir Roger, of twistor and Penrose tile fame). We look at the one-particle density matrix

$$R(\mathbf{x}, \mathbf{x}') = \langle\Phi|\varphi^\dagger(\mathbf{x})\varphi(\mathbf{x}')|\Phi\rangle$$
$$= \int \frac{d^3k}{(2\pi)^3}\frac{d^3k'}{(2\pi)^3} e^{-i\mathbf{k}\cdot\mathbf{x}+i\mathbf{k}'\cdot\mathbf{x}'} \langle\Phi|\hat{a}^\dagger_{\mathbf{k}'}\hat{a}_{\mathbf{k}}|\Phi\rangle. \tag{10.22}$$

In a homogeneous system this depends only on $(\mathbf{x} - \mathbf{x}')$, so we can average over the system to get

$$\langle\Phi|\varphi^\dagger(\mathbf{x})\varphi(\mathbf{x}')|\Phi\rangle = \frac{1}{Vol}\int d^3y \int \frac{d^3k}{(2\pi)^3}\frac{d^3k'}{(2\pi)^3} e^{-i\mathbf{k}\cdot(\mathbf{x}+\mathbf{y})+i\mathbf{k}'\cdot(\mathbf{x}'+\mathbf{y})} \langle\Phi|\hat{a}^\dagger_{\mathbf{k}'}\hat{a}_{\mathbf{k}}|\Phi\rangle$$

10.2 Spontaneous Symmetry Breaking

$$= \int \frac{d^3k}{(2\pi)^3} \frac{d^3k'}{(2\pi)^3} (2\pi)^3 \delta^3(\mathbf{k} - \mathbf{k}') e^{-i\mathbf{k}\cdot\mathbf{x}+i\mathbf{k}'\cdot\mathbf{x}'} \langle \Phi | \hat{a}_{\mathbf{k}'}^\dagger \hat{a}_{\mathbf{k}} | \Phi \rangle$$

$$= \frac{1}{Vol} \int \frac{d^3k}{(2\pi)^3} e^{-i\mathbf{k}\cdot(\mathbf{x}-\mathbf{x}')} \langle \Phi | \hat{a}_{\mathbf{k}}^\dagger \hat{a}_{\mathbf{k}} | \Phi \rangle$$

$$= \int \frac{d^3k}{(2\pi)^3} e^{-i\mathbf{k}\cdot(\mathbf{x}-\mathbf{x}')} n_{\mathbf{k}}, \tag{10.23}$$

where $n_{\mathbf{k}}$ is the density of particles with momentum $\mathbf{k}$ per unit volume. When $\mathbf{x} = \mathbf{x}'$, R is just the number-density. If

$$n_{\mathbf{k}} = n_0 (2\pi)^3 \delta^3(\mathbf{k}) + f(\mathbf{k}), \tag{10.24}$$

where $f(\mathbf{k})$ is a smooth function, then

$$R(\mathbf{x}, \mathbf{x}') = n_0 + \tilde{f}(\mathbf{x} - \mathbf{x}'). \tag{10.25}$$

Here $\tilde{f}$ is the Fourier transform of f. By the Riemann-Lebesgue lemma, $\tilde{f}(\mathbf{x} - \mathbf{x}')$ tends to zero as $\mathbf{x}$ and $\mathbf{x}'$ get far apart. The nonvanishing of $R(\mathbf{x}, \mathbf{x}')$ at large separations of $\mathbf{x}$ and $\mathbf{x}'$ therefore implies that $n_0 \neq 0$, and so is another signal of BEC.

States $|\Phi\rangle$ for which, for any pair of operators $O_1(\mathbf{x})$, $O_2(\mathbf{x})$, we have

$$\langle \Phi | O_1(\mathbf{x}) O_2(\mathbf{x}') | \Phi \rangle \to \langle \Phi | O_1(\mathbf{x}) | \Phi \rangle \langle \Phi | O_2(\mathbf{x}') | \Phi \rangle \tag{10.26}$$

at large separation of $\mathbf{x}$, $\mathbf{x}'$ are said to satisfy the *cluster decomposition property*, or to be *clustering*. Sometimes, because of a slightly confusing analogy with classical statistical mechanics, they are called "pure states." This terminology should not be confused with the conventional usage of the phrase in quantum mechanics. In the present context we have

$$\langle \Phi | \varphi^\dagger(\mathbf{x}) \varphi(\mathbf{x}') | \Phi \rangle \to n_0 \tag{10.27}$$

at large separation, even for states with a fixed number of particles. Since for these states $\langle \Phi | \varphi | \Phi \rangle = 0$, such states do *not* have the clustering property. Coherent states like $|\phi\rangle$ for which $\langle \phi | \varphi | \phi \rangle = (phase)\sqrt{n_0}$ are clustering.

For superfluid ^{4}He at low temperature the quantity n_0 has been estimated to be about 0.08 to 0.11 of the total density so, because of interactions, only a small fraction of the particles actually occupy the lowest single-particle state. In the next section we study a simple model for such interactions.

10.3 Dilute Bose Gas

10.3.1 Bogoliubov Transfomation

The dilute Bose gas was introduced by Bogoliubov as a model for superfluid ^{4}He. It is not a very good model for what is in reality quite a complex liquid,[1] but it does contain enough of the physics to be instructive. It may be a more realistic model for the alkali metal vapor atomic Bose condensates, which are currently under intensive study.

Since we wish to single out the lowest φ_0 mode for special treatment, let us consider a finite system where the momentum takes discrete values. We take as our hamiltonian

$$\hat{H} = \sum_{\mathbf{k}} (\frac{\mathbf{k}^2}{2m} - \mu) \hat{a}_{\mathbf{k}}^{\dagger} \hat{a}_{\mathbf{k}}$$
$$+ \frac{\lambda}{2V} \sum \delta_{\mathbf{k}_1+\mathbf{k}_2, \mathbf{k}_4+\mathbf{k}_3} \hat{a}_{\mathbf{k}_4}^{\dagger} \hat{a}_{\mathbf{k}_3}^{\dagger} \hat{a}_{\mathbf{k}_2} \hat{a}_{\mathbf{k}_1} \qquad (10.28)$$

The interaction term is the discrete momentum-space version of the delta-function repulsion $\frac{\lambda}{2}(\varphi^{\dagger}\varphi)^2$. We will assume that λ is small. The chemical potential μ is a parameter that determines N, the number of particles in the system.

Let us suppose that of the N particles present a macroscopic fraction, N_0, of them are in the $n = 0$ state. We can represent this by constructing a state $|\Phi_0\rangle$ of the form (10.16) with $\sqrt{V}\phi = \sqrt{N_0}$ and using this as the ground state for the non-interacting system. Other states are made by applying $\hat{a}_{\mathbf{k}}^{\dagger}$ to $|\Phi_0\rangle$.

Given the huge preponderance of particles in the $n = 0$ state, nearly all the interactions will be between particles in the $n = 0$ state and at most one other. We can therefore approximate the interaction term in (10.28) as

$$\hat{H}_{int} \approx \frac{\lambda}{2V} \left[\hat{a}_0^{\dagger} \hat{a}_0^{\dagger} \hat{a}_0 \hat{a}_0 + \sum_{\mathbf{k} \neq 0} \left\{ 4\hat{a}_{\mathbf{k}}^{\dagger} \hat{a}_0^{\dagger} \hat{a}_{\mathbf{k}} \hat{a}_0 + \hat{a}_{\mathbf{k}}^{\dagger} \hat{a}_{-\mathbf{k}}^{\dagger} \hat{a}_0 \hat{a}_0 + \hat{a}_0^{\dagger} \hat{a}_0^{\dagger} \hat{a}_{\mathbf{k}} \hat{a}_{-\mathbf{k}} \right\} \right].$$
(10.29)

Between states built by applying $\hat{a}_{\mathbf{k}}^{\dagger}$, $\mathbf{k} \neq 0$, to $|\Phi_0\rangle$ we can replace $\hat{a}_0$ and $\hat{a}_0^{\dagger}$ by their eigenvalue $\sqrt{N_0}$, so

$$\hat{H} \approx \sum_{\mathbf{k} \neq 0} (\frac{\mathbf{k}^2}{2m} - \mu) \hat{a}_{\mathbf{k}}^{\dagger} \hat{a}_{\mathbf{k}} - \mu N_0$$
$$+ \frac{\lambda}{2V} \left\{ N_0^2 + 4N_0 \sum_{\mathbf{k} \neq 0} \hat{a}_{\mathbf{k}}^{\dagger} \hat{a}_{\mathbf{k}} + N_0 \sum_{\mathbf{k} \neq 0} (\hat{a}_{\mathbf{k}}^{\dagger} \hat{a}_{-\mathbf{k}}^{\dagger} + \hat{a}_{\mathbf{k}} \hat{a}_{-\mathbf{k}}) \right\}.$$
(10.30)

---

[1] The main problem is that the assumption of pairwise interactions fails badly for liquid helium.

10.3 Dilute Bose Gas

Note that there is a purely c-number term equal to $\frac{\lambda}{2V} N_0^2 - \mu N_0$. We get a first estimate of N_0 by minimizing this term. The minimum is at $\frac{\lambda}{V} N_0 = \mu$, and we will assume that N_0 has been fixed in terms of μ in this way. Later we will see that this assumption is consistent with our solution of the system.

Our hamiltonian has become

$$\hat{H} \approx \sum_{\mathbf{k} \neq 0} (\epsilon(\mathbf{k}) + \mu) \hat{a}_\mathbf{k}^\dagger \hat{a}_\mathbf{k}$$
$$+ \frac{1}{2} \mu \sum_{\mathbf{k} \neq 0} (\hat{a}_\mathbf{k}^\dagger \hat{a}_{-\mathbf{k}}^\dagger + \hat{a}_\mathbf{k} \hat{a}_{-\mathbf{k}}), \tag{10.31}$$

where $\epsilon(\mathbf{k}) = \mathbf{k}^2 / 2m$.

To diagonalize (10.31) we perform a *canonical*, or *Bogoliubov, transformation* on each of the $\hat{a}_\mathbf{k}$ and $\hat{a}_\mathbf{k}^\dagger$. We write

$$\hat{a}_\mathbf{k} = \hat{b}_\mathbf{k} \cosh \theta + \hat{b}_{-\mathbf{k}}^\dagger \sinh \theta,$$
$$\hat{a}_\mathbf{k}^\dagger = \hat{b}_{-\mathbf{k}} \sinh \theta + \hat{b}_\mathbf{k}^\dagger \cosh \theta, \tag{10.32}$$

where $\theta(\mathbf{k})$ is a real parameter that we will choose later. The form of the Bogoliubov transformation is dictated by the requirement that the $\hat{b}_\mathbf{k}$ and $\hat{b}_\mathbf{k}^\dagger$ have the same commutation relations as the $\hat{a}_\mathbf{k}$ and $\hat{a}_\mathbf{k}^\dagger$. We can in fact find a unitary operator such that $\hat{b}_\mathbf{k} = U \hat{a}_\mathbf{k} U^{-1}$, but we will not display its explicit form here.

With this transformation (10.31) becomes (ignoring c-number from normal-ordering)

$$\sum_{\mathbf{k} \neq 0} \hat{b}_\mathbf{k}^\dagger \hat{b}_\mathbf{k} ((\epsilon(\mathbf{k}) + \mu) \cosh 2\theta + \mu \sinh 2\theta))$$
$$+ \frac{1}{2} (\hat{b}_\mathbf{k}^\dagger \hat{b}_{-\mathbf{k}}^\dagger + \hat{b}_\mathbf{k} \hat{b}_{-\mathbf{k}}) ((\epsilon(\mathbf{k}) + \mu) \sinh 2\theta + \mu \cosh 2\theta). \tag{10.33}$$

We now choose θ so that the last term vanishes. We need

$$\tanh 2\theta(\mathbf{k}) = -\frac{\mu}{\epsilon(\mathbf{k}) + \mu}, \quad \mathbf{k} \neq 0 \tag{10.34}$$

or equivalently

$$\cosh 2\theta(\mathbf{k}) = \frac{(\epsilon(\mathbf{k}) + \mu)}{\sqrt{(\epsilon(\mathbf{k}) + \mu)^2 - \mu^2}}, \quad \sinh 2\theta(\mathbf{k}) = \frac{-\mu}{\sqrt{(\epsilon(\mathbf{k}) + \mu)^2 - \mu^2}}. \tag{10.35}$$

The hamiltonian is now reduced to

$$\hat{H} = \sum_{\mathbf{k} \neq 0} \hat{b}_\mathbf{k}^\dagger \hat{b}_\mathbf{k} \sqrt{(\epsilon(\mathbf{k}) + \mu)^2 - \mu^2}, \tag{10.36}$$

which represents a gas of free bosons having energy

$$E_\mathbf{k} = \sqrt{(\epsilon(\mathbf{k}) + \mu)^2 - \mu^2}. \tag{10.37}$$

Clearly the new ground state is that state which is annihilated by all the $\hat{b}_\mathbf{k}$. Since $\hat{a}_\mathbf{k}|\Phi_0\rangle = 0$, and $\hat{b}_\mathbf{k} = U\hat{a}_\mathbf{k}U^{-1}$, this is

$$|\Phi\rangle = U|\Phi_0\rangle. \tag{10.38}$$

Note what our assumption that $N_0 = V\mu/\lambda$ has bought us. The two μ's in $(\epsilon(\mathbf{k}) + \mu)^2 - \mu^2$ are equal, so $E_\mathbf{k} = 0$ at $\mathbf{k} = 0$. Tracing through the algebra you can see that any other choice of N_0 would have lead to the "μ" added to ϵ being different from the "μ" outside the bracket. The energy of the zero-momentum excitations would not then be zero. This would allow us to lower the energy by either adding or removing particles from the condensate. Clearly we made the correct choice for N_0.

For small $\mathbf{k}$, equation (10.37) reduces to

$$E_\mathbf{k} \approx |\mathbf{k}|\sqrt{\frac{\mu}{m}} = |\mathbf{k}|\sqrt{\frac{\lambda\rho_0}{m}}. \tag{10.39}$$

Here $\rho_0 = N_0/V$ is the density of particles in the condensate. This linear dispersion curve is characteristic of *sound waves* in a fluid. For higher values of $\mathbf{k}$, $E_\mathbf{k}$ crosses over into the usual free-particle expression $E_\mathbf{k} \approx \mathbf{k}^2/2m$.

If the long-wavelength eigenstates are indeed sound waves, there is something slightly odd at work. In the last chapter there was a clear distinction between the quasiparticle states that reduced to single electrons in the noninteracting particle limit, and the collective sound-wave-like excitations, such as plasmons and zero sound, which were composed of electron-hole pairs. Here there appears to be no difference. Both sound waves and single-particle excitations appear on the same dispersion curve, each blending gradually into the other. This merger is a consequence of the spontaneous breakdown of the $U(1)$ number-conservation symmetry. The quasiparticle and collective excitations of the last chapter were distinguished by their quantum numbers: the electron-like quasiparticles carried $U(1)$ charge $+1$, while the collective sound-wave-like modes were $U(1)$ neutral. Once the $U(1)$ symmetry has spontaneously broken, as it is here, the excitations are no longer charge eigenstates because they may gain or lose any such charge to the ground state at will.

We can find a better approximation to the density than $N = N_0$ by calculating

$$\begin{aligned} N &= \langle\Phi|\hat{a}_0^\dagger\hat{a}_0 + \sum_{\mathbf{k}\neq 0}\hat{a}_\mathbf{k}^\dagger\hat{a}_\mathbf{k}|\Phi\rangle \\ &= N_0 + \sinh^2\theta\,\langle\Phi|\hat{b}_\mathbf{k}\hat{b}_\mathbf{k}^\dagger|\Phi\rangle \\ &= N_0 + \frac{1}{2}\sum_{\mathbf{k}\neq 0}\left\{\frac{(\epsilon+\mu)}{\sqrt{(\epsilon+\mu)^2 - \mu^2}} - 1\right\}. \end{aligned} \tag{10.40}$$

We see that the number of particles present in the interacting many-body ground state exceeds the number in the single-particle condensate by a fraction of order λ. This first-order correction be contrasted with the estimate of Penrose and Onsager that in superfluid ^{4}He only about 8% of the helium atoms are in the single-particle ground state.

10.3 Dilute Bose Gas

It is difficult to improve systematically the present approach to the problem by taking into account higher orders in λ. The next section leads to a more suitable method.

10.3.2 Field Equations

We can obtain the same E_k dispersion curve that we found in the last section directly from the classical field equations. Take our lagrangian density as

$$\mathcal{L} = \varphi^\dagger (i\partial_t + \frac{1}{2m}\nabla^2 + \mu)\varphi - \frac{\lambda}{2}(\varphi^\dagger \varphi)^2. \tag{10.41}$$

Here μ is the chemical potential for the bosons.

The potential part of this expression is

$$V(\varphi) = \frac{\lambda}{2}(\varphi^\dagger \varphi)^2 - \mu \varphi^\dagger \varphi. \tag{10.42}$$

This has its minimum at $\varphi^\dagger \varphi = \mu/\lambda$, so the possible stationary solutions have

$$\langle \varphi \rangle = \varphi_c = e^{i\theta}\sqrt{\frac{\mu}{\lambda}}. \tag{10.43}$$

Any choice of θ gives a stationary solution, and is a candidate classical ground state.

Let us look for small oscillations about one of these stationary points. Without loss of generality we may take $\theta = 0$. We set $\varphi = \varphi_c + \eta$ so

$$V(\varphi) = \frac{\lambda}{2}\left(\varphi^\dagger \varphi - \varphi_c^\dagger \varphi_c\right)^2 - \frac{\mu^2}{2\lambda}$$
$$= \frac{\lambda}{2}\left((\varphi_c^\dagger + \eta^\dagger)(\varphi_c + \eta) - \varphi_c^\dagger \varphi_c\right)^2 - \frac{\mu^2}{2\lambda}$$
$$= const. + \mu \eta^\dagger \eta + \frac{\mu}{2}\eta\eta + \frac{\mu}{2}\eta^\dagger \eta^\dagger + O(\eta^3). \tag{10.44}$$

Remembering that $\mu = \lambda \rho_0$, where ρ_0 is the density of particles in the condensate, we see that each of these terms corresponds to the interaction between the η field and the condensate that we saw in the earlier approach.

Fig 1. Diagrams representing the interaction of the particles with the condensate in (10.44).

Keeping only the quadratic terms gives the linearized equations of motion

$$i\partial_t \eta = -\frac{1}{2m}\nabla^2 \eta + \mu\eta + \mu\eta^\dagger$$
$$-i\partial_t \eta^\dagger = -\frac{1}{2m}\nabla^2 \eta^\dagger + \mu\eta^\dagger + \mu\eta. \quad (10.45)$$

If we look for plane-wave solutions of the form

$$\eta = a e^{ikx-i\omega t} + b^\dagger e^{-ikx+i\omega t}, \quad (10.46)$$

we find that

$$\begin{bmatrix} \frac{1}{2m}k^2 - \omega + \mu & \mu \\ \mu & \frac{1}{2m}k^2 + \omega + \mu \end{bmatrix} \begin{bmatrix} a \\ b \end{bmatrix} = 0. \quad (10.47)$$

Setting the determinant of this to zero yields

$$\omega^2 = (\frac{1}{2m}k^2 + \mu)^2 - \mu^2. \quad (10.48)$$

Recalling that $\mu = \lambda|\varphi_c|^2 = \lambda\rho_0$ shows that this is exactly the same dispersion curve as (10.37).

During the wave motion the tip of the φ vector describes an ellipse about the equilibrium postion $\varphi = \varphi_c$. The ratio of the major axes of the ellipse is equal to $(a - b)/(a + b)$, where (a, b) is the solution to (10.47), while the sense in which the ellipse is traversed is determined by the fact that for a sound wave the density, $|\varphi|^2$, is greater during the phase of the wave where the particles are moving in the direction of propagation.

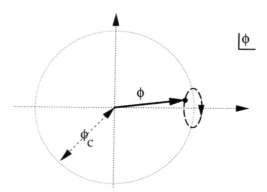

Fig 2. *The orbit of the φ field for solutions of (10.45).*

10.3.3 Quantization

When we quantize the bose field φ, we treat φ_c as a classical constant. Only the η field is an operator with canonical commutation relations

$$[\eta(\mathbf{x}), \eta^\dagger(\mathbf{x}')]_{t=t'} = \delta^3(\mathbf{x} - \mathbf{x}'). \quad (10.49)$$

10.3 Dilute Bose Gas

If we keep only the quadratic terms in the interaction potential (10.44), we can calculate the η field propagators. In addition to $\langle\Phi|T\{\eta^\dagger(x)\eta(x')\}|\Phi\rangle$, we will find that both $\langle\Phi|T\{\eta^\dagger(x)\eta^\dagger(x')\}|\Phi\rangle$ and $\langle\Phi|T\{\eta(x)\eta(x')\}|\Phi\rangle$ are nonzero. This is because the interactions with the condensate mix the η and $\eta^\dagger$ fields. Indeed we will find that in momentum space the possible propagators form a matrix inverse to that in (10.47). Using the abbreviated notation

$$\langle\Phi|T\{\eta^\dagger(x)\eta(x')\}|\Phi\rangle = <\eta^\dagger\eta>, \tag{10.50}$$

etc. we find that

$$\begin{bmatrix} <\eta^\dagger\eta> & <\eta^\dagger\eta^\dagger> \\ <\eta\eta> & <\eta\eta^\dagger> \end{bmatrix} = \frac{1}{(\epsilon+\mu)^2-\mu^2-\omega^2}\begin{bmatrix} \epsilon+\omega+\mu & -\mu \\ -\mu & \epsilon-\omega+\mu \end{bmatrix}. \tag{10.51}$$

We can use these expressions for the propagators to find the lowest-order correction to the density. We compute

$$<\rho> = |\varphi_c|^2 + <\eta^\dagger\eta>$$
$$= |\varphi_c|^2 + \int \frac{d^3k}{(2\pi)^3}\frac{d\omega}{2\pi}\frac{\epsilon+i\omega+\mu}{(\epsilon+\mu)^2-\mu^2+\omega^2}. \tag{10.52}$$

In writing down this expression I have rotated ω into the euclidean region. Performing the ω integral we find

$$<\rho> \stackrel{?}{=} |\varphi_c|^2 + \int \frac{d^3k}{(2\pi)^3}\frac{(\epsilon+\mu)}{\sqrt{(\epsilon+\mu)^2-\mu^2}}. \tag{10.53}$$

This not quite right, because in failing to include a small time-shift to force $\eta^\dagger(x)$ to lie to the left of $\eta(x)$ in the time ordered product, we have inadvertently computed $\frac{1}{2}<\eta^\dagger\eta+\eta\eta^\dagger>$ instead of $<\eta^\dagger\eta>$. Fixing this error by subtracting the commutator $\frac{1}{2}\delta^3(0)$ gives

$$<\rho> = |\varphi_c|^2 + \int \frac{d^3k}{(2\pi)^3}\left\{\frac{(\epsilon+\mu)}{\sqrt{(\epsilon+\mu)^2-\mu^2}}-1\right\}, \tag{10.54}$$

which coincides with (10.40).

If we keep higher-order terms in 10.44, then vertices such as $\eta^\dagger\eta\varphi_c^\dagger\eta$ allow a single particle to evolve into an particle-antiparticle pair. As alluded to above, this is the reason that there is no distinction between quasiparticles and sound-wave-like collective modes in this system.

Identifying the sound waves in our Bose gas with those in liquid helium allows us to try to fix some of the parameters. The mass m should be taken to be that of a helium atom (6.647×10^{-27} kg). The number density ρ_0 is then determined by the density of liquid helium ($m\rho_0 = 145$ kg m^{-3}), and the parameter λ is selected to fit the measured speed of sound ($c = \sqrt{\lambda\rho_0/m} = 230$ ms^{-1}) in the fluid. Perhaps surprisingly this choice of parameters yields properties for quantities like the healing length that are not far from measured values. Unfortunately, many other properties are not well described by this model.

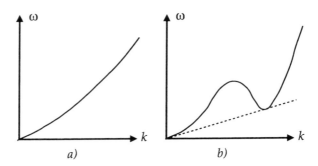

Fig 3. The dispersion curves for a) the dilute Bose gas, b) real superfluid ^{4}He. The dotted line in b) is the curve $\omega = v_0 k$, which is tangent to the roton minimum.

In particular, the dispersion curve for sound waves in real helium has a pronounced "roton minimum," no sign of which appears in the simple model.

10.3.4 Landau Criterion for Superfluidity

There is an interesting argument, due to Landau, relating the phonon dispersion curve to the existence of superfluidity. Consider a body of large mass M moving at velocity $\mathbf{V}$ through the fluid. If phonons are the only excitations possible in the fluid, then the only way for the body to experience any retarding force is for it to emit some of these phonons. In doing so it will loose energy, δE, and momentum, $\delta \mathbf{k}$, where

$$\delta E = \delta \mathbf{k} \cdot \mathbf{V} + O(\frac{1}{M}). \tag{10.55}$$

Now the energy and momentum lost by the body must equal the energy and momentum gained by the phonons

$$\delta E = -\sum \omega(\mathbf{k}_i), \qquad \delta \mathbf{k} = -\sum \mathbf{k}_i. \tag{10.56}$$

Next suppose that the phonons have a nonzero minimum phase velocity

$$v_0 = \inf \left(\frac{\omega(\mathbf{k})}{|\mathbf{k}|} \right). \tag{10.57}$$

(See Fig. 3b.) Exploiting this minumum phase velocity in (10.55) we establish a chain of inequalities

$$|\delta E| = \sum \omega(\mathbf{k}_i) \geq v_0 \sum |\mathbf{k}_i| \geq v_0 |\sum \mathbf{k}_i| = v_0 |\delta \mathbf{k}|. \tag{10.58}$$

Thus $|\delta \mathbf{k} \cdot \mathbf{V}| \geq v_0 |\delta \mathbf{k}|$ for any permitted $\delta \mathbf{k}$. Such a $\delta \mathbf{k}$ can only exist if $|\mathbf{V}| \geq v_0$. The body must exceed this minimum velocity before it can experience any drag.

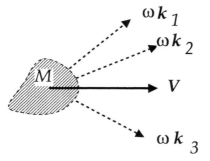

Fig 4. A body emitting phonons.

This *Landau criterion* for the critical velocity below which no dissipation occurs sounds convincing, but the argument must be flawed. The experimental dispersion curve for real helium gives $v_0 \approx 60$ m s^{-1}, while the observed critical velocities depend sensitively on the experimental geometry, and are much less than this value. Before we attempt to explain away this embarrassment, we should note that the condition $|\mathbf{V}| \geq v_0$ has a classical interpretation. When a ship moves through water, one of the principal sources of drag is the formation of a wave train or *wake* of surface waves emitted by the ship. We must all have noted that the pattern of wave crests and troughs maintains a constant form when seen from the frame of the ship. In other words, the component of *phase velocity* of the waves in the direction of the ship's motion is equal to the ship's forward speed. This is always true of radiation emitted by a body in uniform motion through a medium. The Landau criterion is simply the condition that sound waves exist whose phase velocity can match that of the body and form such a wake.

The problem with Landau's argument lies in his assumption that phonons are the only low-energy excitations in the superfluid. As we will see in the next section, there also exist vortex-line configurations. These can be shed, or preexisting ones stretched and entangled, in the course of the fluid flow, and all these processes drain energy from the bulk motion and give rise to friction.

10.3.5 *Normal and Superfluid Densities*

When we calculated the dispersion relation for the phonon excitations in the dilute Bose gas, we did so in a frame in which the condensate was at rest. To find the spectrum when the fluid is in motion, we need to understand how energy and momentum transform under a Galilean transformation. The easiest way to do this is to consider some external object (a neutron perhaps) of mass m impacting the fluid and creating a single phonon. If, in the rest frame of the fluid, the impacting particle starts with energy $\frac{1}{2}mv^2$ and momentum $m\mathbf{v}$, and ends with energy $\frac{1}{2}m(v')^2$ and momentum $m\mathbf{v}'$, then it has created an phonon with energy $E = \frac{1}{2}mv^2 - \frac{1}{2}m(v')^2$ and momentum $\mathbf{p} = m(\mathbf{v} - \mathbf{v}')$.

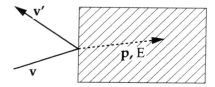

Fig 5. A particle scattering and creating a phonon.

Viewed from a frame in which the fluid is moving at speed **V**, the incoming particle has energy $\frac{1}{2}m|\mathbf{v} + \mathbf{V}|^2$ and its outgoing energy is $\frac{1}{2}m|\mathbf{v}' + \mathbf{V}|^2$. The delivered energy is therefore

$$\tilde{E} = \frac{1}{2}m|\mathbf{v} + \mathbf{V}|^2 - \frac{1}{2}m|\mathbf{v}' + \mathbf{V}|^2 = E + m(\mathbf{v} - \mathbf{v}') \cdot \mathbf{V}, \qquad (10.59)$$

while the momentum transfer is

$$m\left((\mathbf{v} + \mathbf{V}) - (\mathbf{v}' + \mathbf{V})\right) = m(\mathbf{v} - \mathbf{v}') = \mathbf{p}, \qquad (10.60)$$

which is the same as in the rest frame. Since the identical excitation is being created, its energy in the moving frame must be

$$\tilde{E}(\mathbf{p}) = E(\mathbf{p}) + \mathbf{p} \cdot \mathbf{V}. \qquad (10.61)$$

An alternative way to interpret this transformation law is to recognize E as $\hbar\omega$ and **p** as $\hbar\mathbf{k}$, where ω is the frequency of the sound wave and **k** its wavenumber. The phonon dispersion relation thus transforms as

$$\omega(\mathbf{k}) \to \tilde{\omega}(\mathbf{k}) = \omega(\mathbf{k}) + \mathbf{k} \cdot \mathbf{V} \qquad (10.62)$$

This is just the Doppler shift of the frequency as the sound-wave is carried past the observer by the medium moving at velocity **V**.

This transformation law has an important consequence at nonzero temperatures. Consider our Bose fluid flowing through a stationary capillary tube. Thermally excited phonons will scatter off the wall of the tube and will come into equilibrium with it. Because phonons traveling with the flow have higher energies than those opposing the flow, there will be more phonons traveling downstream, and this will tend to reduce the total momentum in the flow. At temperature T the momentum-density due to the phonons will be

$$\mathbf{P}_{phonon} = \int \frac{d^3k}{(2\pi)^3} (\hbar\mathbf{k}) n_\beta(\omega(\mathbf{k}) + \mathbf{k} \cdot \mathbf{V}), \qquad (10.63)$$

where $\beta = 1/kT$ and

$$n_\beta(\omega) = \frac{1}{e^{\beta(\hbar\omega)} - 1} \qquad (10.64)$$

is the Bose distribution function. For small **V** we can write this as

$$\begin{aligned}\mathbf{P}_{phonon} &= \int \frac{d^3k}{(2\pi)^3} (\hbar\mathbf{k})(\mathbf{k} \cdot \mathbf{V}) \frac{\partial n_\beta}{\partial \omega} \\ &= -\rho_n \mathbf{V},\end{aligned} \qquad (10.65)$$

where

$$\rho_n(T) = -\frac{\hbar}{3}\int \frac{d^3k}{(2\pi)^3} k^2 \frac{\partial n_\beta}{\partial \omega}. \tag{10.66}$$

This is a positive quantity because the derivative of n_β is everywhere negative. The total momentum density associated with the flow is therefore

$$\mathbf{P}_{tot} = \rho\mathbf{V} - \rho_n\mathbf{V} = \rho_s\mathbf{V}. \tag{10.67}$$

(Here, contrary to our usage elsewhere, ρ denotes the mass density. This is the usual convention in the superfluid literature.) The last equality serves to define the *superfluid mass density*, $\rho_s = \rho - \rho_n$. Because the momentum density is also the mass current (see the appendix for a discussion of this), we see that a flow of the background condensate at speed $\mathbf{V}$, which we will now denote by $\mathbf{v}_s$, the *superfluid velocity*, only transports mass at a rate of

$$\mathbf{j} = \rho_s \mathbf{v}_s. \tag{10.68}$$

A fraction, ρ_n, apparently remains stationary because of friction with the walls of the capillary.

If the walls are themselves moving at a velocity $\mathbf{v}_n$ then the last line of (10.65) is replaced by

$$\mathbf{P}_{phonon} = \rho_n(\mathbf{v}_s - \mathbf{v}_n), \tag{10.69}$$

and, consequently, the mass current becomes

$$\mathbf{j} = \rho\mathbf{v}_s - \rho_n(\mathbf{v}_s - \mathbf{v}_n) = \rho_s\mathbf{v}_s + \rho_n\mathbf{v}_n. \tag{10.70}$$

This equation is the basis of the *two fluid model* of Tisza, London, and Landau. One envisions that a fraction ρ_s of the fluid takes part in the superflow and the remaining part, the *normal component* with density ρ_n, is a normal viscous fluid. Although the normal component is due to phonons, do not fall into the trap of thinking that $\mathbf{v}_n$ is in any sense their mean velocity. On the contrary, as we have seen above, even when $\mathbf{v}_n = 0$, the phonons tend to stream in the opposite direction to $\mathbf{v}_s$. The normal-component velocity $\mathbf{v}_n$ is simply that of the frame in which the fluid is locally in thermal equilibrium.

10.4 Charged Bosons

10.4.1 Gross-Pitaevskii Equation

The nonlinear Schrödinger equation we get by varying the action (10.41) is called the *Gross-Pitaevskii equation*. It is often used as a model for the motion of the condensate in a superfluid. As we show here it can also model the electrodynamics of a superconductor. In the latter case, of course, we are imagining that the bosons are Cooper pairs of electrons. This means that in the following we should set $m = 2m_e^*$ and take the charge of the particles to be $e \to q = -2e$.

If we restore the $\hbar$'s and include an interaction with an external electromagnetic field, the Gross-Pitaevskii equation takes its general form as

$$i\hbar(\partial_t - ieA_0/\hbar)\varphi = -\frac{\hbar^2}{2m}\sum_{a=1}^{3}(\partial_a - ieA_a/\hbar)^2\varphi + \lambda(|\varphi|^2 - \rho_0)\varphi. \quad (10.71)$$

By using the *Madelung transformation*, (10.71) can be recast as the equation of motion of a charged, compressible fluid having equilibrium particle-number density ρ_0. Madelung sets $\varphi = \sqrt{\rho}e^{i\theta}$ and defines a velocity field **v** in such a way that the number current,

$$\mathbf{j} = \frac{\hbar}{2mi}\left(\varphi^*(\nabla - ie\mathbf{A}/\hbar)\varphi - ((\nabla + ie\mathbf{A}/\hbar)\varphi^*)\varphi\right) \quad (10.72)$$

may be written as $\mathbf{j} = \rho\mathbf{v}$. This requires

$$\mathbf{v} = \frac{\hbar}{m}(\nabla\theta - e\mathbf{A}/\hbar). \quad (10.73)$$

In the absence of vortex singularities in φ, the vorticity, $\omega = \nabla \wedge \mathbf{v}$, is completely determined by the gauge field to be $\omega = -\frac{e}{m}\nabla \wedge \mathbf{A} = -\frac{e\mathbf{B}}{m}$, i.e.,

$$m\omega + e\mathbf{B} = 0. \quad (10.74)$$

For neutral superfluids (10.74) implies *irrotational motion* and hence, at low velocities, where the effects of compressibility can be ignored, leads to D'Alembert's paradox (the absence of drag forces). For charged superfluids equation (10.74) gives rise to the *Meissner effect*: A penetrating uniform **B** field would require uniform vorticity, i.e., rigid rotation. A rigidly rotating body possesses a kinetic energy that grows faster than the volume of the system, and so is impossible in the thermodynamic limit. Alternatively, taking the curl of the Maxwell equation $\nabla \wedge \mathbf{B} = e\mathbf{j}$ and using $\mathbf{j} = \rho\mathbf{v}_s$ together with (10.74) implies that

$$\nabla^2\mathbf{B} - \frac{e^2\rho}{m}\mathbf{B} = 0, \quad (10.75)$$

which leads to flux screening. Note that $\hbar$ does not appear in (10.74) or in the *London penetration depth* $(e^2\rho/m)^{-1/2}$.

Because the fluid is charged and compressible, it is also able to screen static electric charges. The combined effect of the Meissner effect and the Debye screening is to give an effective mass to the photon. This is a nonrelativistic version of the *Higgs-Kibble* mechanism.

10.4.2 Vortices

The Gross-Pitaevskii equation, for both the charged and neutral Bose fluids, has singular vortex-line solutions. Here the phase of the φ field winds through 2π as one encircles the vortex line. The magnitude of φ must be zero in the core of the vortex, otherwise the field configuration would be ill-defined there.

In the neutral case, the 2π phase winding means that the *circulation* about the vortex

$$\kappa = \oint \mathbf{v} \cdot d\mathbf{r} \tag{10.76}$$

takes the value

$$\kappa = \oint \frac{\hbar}{m} \nabla\theta \cdot d\mathbf{r} = 2\pi \frac{\hbar}{m}. \tag{10.77}$$

Quantization of circulation was first suggested by Onsager[2] in a remark at a conference in 1949 — but he apparently had in mind that a rotating container of superfluid would contain a series of concentric differentially rotating cylinders. It was Feynman in 1955 who realized that the lowest-energy configuration of a such a rotating superfluid would be an array of discrete vortices.

Since the expression (10.77) for the circulation quantum, κ, has been obtained from a classical approximation to a crude model, it is an interesting question as to whether the observed value is consistent with the m appearing in (10.77) being the bare mass of the helium atom, or whether there are many-body corrections. For a *uniform* flow in a homogeneous fluid, galilean invariance requires the m appearing in (10.73) to be the mass of the helium atom, but this does not rule out corrections when the velocity and density depend on position. The observed circulation gives $m = M_{He}$ to within about 1%, which is the precision of the best experiments.

For a straight vortex the velocity field decreases as $1/r$ as we move away from it, just as with a classical vortex line. This means that an isolated vortex in an infinite system would have a logarithmically divergent kinetic energy. There is no exact solution for the solution of (10.71) for the vortex, but a reasonable approximation is

$$|\varphi(r)| \approx \left(\frac{r^2}{r^2 + 2\xi^2} \right)^{\frac{1}{2}}, \tag{10.78}$$

where

$$\xi = \hbar/\sqrt{2m\lambda\rho_0} \tag{10.79}$$

is the *healing length*.

In the charged-fluid case, the circulating current of charged particles creates a magnetic field, and this, via the relation (10.74), sets up an extended anti-vortex whose total vorticity is equal and opposite to that of the singular core. Because of the screening by this antivortex, the velocity field falls off exponentially fast, and there is no circulation at infinity. The magnetic field decreases in accordance with (10.75). The total magnetic flux can be found by integrating round a countour at infinity, where we can set $\mathbf{v} = 0$ in (10.73). Assuming the vortex is parallel to the

---

[2] Lars Onsager. Born Novemeber 27, 1903, Kristiania, Norway. Died October 5, 1976, Coral Gables, FL, Nobel Prize for Chemistry 1968.

z axis, this gives

$$\int B_z \, dx dy = \oint \mathbf{A} \cdot d\mathbf{r} = \frac{\hbar}{e} \oint \nabla \theta \cdot d\mathbf{r} = 2\pi \frac{\hbar}{e} \tag{10.80}$$

(remember that $e \to e^* = 2e$ when we apply this to a fluid of Cooper pairs). These tubes of quantized magnetic flux are called *Abrikosov vortices*. In order for these flux-tube vortices to be stable, we need the penetration length λ to be larger than $\sqrt{2}$ times the healing length ξ. If this inequality is not satisfied, the vortices attract one another and the magnetic field and the superfluid simply phase-separate. When $\lambda > \sqrt{2}\xi$, the superconductor is said to be *type II*. Otherwise it is *type I*.

It is worth noting that the flux-quantization argument depends only on the velocity being a function of a gauge covariant combination of the vector potential and the phase of the order parameter. It therefore depends only on the gauge transformation properties of the order parameter, and is robust against many-body corrections.

10.4.3 Connection with Fluid Mechanics

With the definition (10.72), the imaginary and real parts of (10.71) become, respectively, the continuity equation

$$\partial_t \rho + \nabla \cdot \rho \mathbf{v} = 0, \tag{10.81}$$

and the Euler equation governing the flow of a charged barotropic fluid

$$m(\partial_t \mathbf{v} + \mathbf{v} \cdot \nabla \mathbf{v}) = e(\mathbf{E} + \mathbf{v} \wedge \mathbf{B}) - \nabla \zeta. \tag{10.82}$$

The word *barotropic* refers to the simplifying property that the pressure term $\frac{1}{\rho} \nabla P$, which occurs on the right-hand side of the conventional Euler equation, is here combined into the gradient of a potential

$$\zeta \equiv \lambda(\rho - \rho_0) - \frac{\hbar^2}{2m} \frac{\nabla^2 \sqrt{\rho}}{\sqrt{\rho}}. \tag{10.83}$$

The potential ζ contains the expected compressibility pressure, depending on the deviation from the equilibrium density, together with a correction involving gradients of ρ. This correction, the *quantum pressure*, is the only place that $\hbar$ appears in the flow equations. It sets the length scale $\xi = \hbar/\sqrt{2m\lambda\rho_0}$ over which the superfluid density heals after being forced to zero by a boundary or a vortex singularity. (Of course $\hbar$ also manifests itself in the quantum of circulation, $\kappa = 2\pi\hbar/m$.)

With the parameters chosen as suggested above, we find $\xi = 0.487$ Å.

The Euler equation (10.82) may be derived by first taking the gradient of (10.71) and interpreting the result as the Bernoulli equation,

$$m(\partial_t \mathbf{v} - \mathbf{v} \wedge \boldsymbol{\omega}) = e(\mathbf{E} + \mathbf{v} \wedge \mathbf{B}) - \nabla \left(\frac{1}{2} m v^2 + \zeta\right), \tag{10.84}$$

which is equivalent to (10.82).

In (10.84) a cancellation of the $m\mathbf{v} \wedge \boldsymbol{\omega}$ term against the $e\mathbf{v} \wedge \mathbf{B}$ term is evident upon use of (10.74). It is after this cancellation, and so without reference to either **B** or

ω, that the hydrodynamic picture of superconductivity is conventionally displayed. To me it seems preferable to keep both ω and B in (10.84) and rewrite it as (10.82). By doing this one can see that the only difference between superfluid dynamics and classical fluid dynamics lies in the constraint (10.74).

11
Finite Temperature

11.1 Partition Functions

The equilibrium thermodynamic properties of a many-body system at temperature T are determined by the grand canonical partition function

$$\mathcal{Z} = \text{Tr}\left\{e^{\beta(\mu \hat{N} - \hat{H})}\right\}. \tag{11.1}$$

Here $\beta = 1/kT$ and μ is the chemical potential. For a system of noninteracting bosons whose single-particle energy eigenvalues are E_n, we can expand this out in terms of the number of particles occupying each E_n eigenstate as

$$\begin{aligned}\mathcal{Z} &= \prod_n \left(1 + e^{\beta(\mu - E_n)} + e^{2\beta(\mu - E_n)} + \ldots\right) \\ &= \prod_n \left(1 - e^{\beta(\mu - E_n)}\right)^{-1}.\end{aligned} \tag{11.2}$$

We must have $\mu < E_{min}$ for the sum to make sense. In the particular case of free, nonrelativistic, bosons of mass m, $\mathcal{Z}$ becomes

$$\mathcal{Z} = \exp\left\{-(Vol)\int \frac{d^3k}{(2\pi)^3} \ln(1 - e^{\beta(\mu - k^2/2m)})\right\}. \tag{11.3}$$

Defining the *grand potential* Ω by

$$\mathcal{Z} = e^{-\beta\Omega}, \tag{11.4}$$

we have

$$\Omega = (Vol)\frac{1}{\beta} \int \frac{d^3k}{(2\pi)^3} \ln(1 - e^{\beta(\mu - k^2/2m)}). \tag{11.5}$$

For fermions, the occupation numbers of each state can only be zero or one, so only the first two terms in the occupation number sum can occur. Therefore,

$$\mathcal{Z} = \prod_n \left(1 + e^{\beta(\mu - E_n)}\right). \tag{11.6}$$

For free, nonrelativistic fermions this reduces to

$$\mathcal{Z} = \exp\left\{+Vol \int \frac{d^3k}{(2\pi)^3} \ln(1 + e^{\beta(\mu - k^2/2m)})\right\}, \tag{11.7}$$

so

$$\Omega = -(Vol)\frac{1}{\beta} \int \frac{d^3k}{(2\pi)^3} \ln(1 + e^{\beta(\mu - k^2/2m)}). \tag{11.8}$$

For any many-body system, thermodynamics tells us that the grand potential Ω is equal to $-P(Vol)$, where P is the pressure. Bearing in mind that $\ln(x)$ is negative when $x < 1$, we see that the signs in the right-hand sides of (11.8) and (11.5) ensure that the pressure is positive.

The combinatorics required to count the many-particle energies in terms of the single-particle energy levels are somewhat different in the Bose and Fermi cases — yet the obvious similarities between (11.3) and (11.7) suggests that a unified derivation is possible. We will provide this in the next section.

11.2 Worldlines

In the previous chapters the statistics of second-quantized bosons and fermions have been accounted for via the commutation relations of the fields that create them. It is sometimes easier, and perhaps more intuitive, to choose a different approach and to work directly with a picture of the worldlines of the particles. The arguments of this section are due to Feynman and will lead us part of the way toward his "sum over paths" reformulation of quantum mechanics.

Suppose $G(x, y, \tau)$ is the amplitude for one of a set of identical, nonrelativistic, particles to go from y to x in Euclidean "time" τ. When we study functional integrals, we will see that it can be obtained by summing over the possible paths from y to x, but, however we obtain it, it obeys an imaginary-time version of the Schrödinger equation

$$\frac{1}{2m}\partial_x^2 G(x, y, \tau) = \partial_\tau G(x, y, \tau), \tag{11.9}$$

with initial condition

$$G(x, y, \tau) \to \delta(x - y) \quad \text{as} \quad \tau \to 0_+. \tag{11.10}$$

11. Finite Temperature

To find the grand canonical partition function,

$$\mathcal{Z} = \text{Tr}\left[e^{\beta\mu\hat{N}} e^{-\beta\hat{H}}\right], \tag{11.11}$$

of a system of fermions, for example, at temperature $T = 1/\beta$, we should sum over histories that are periodic in euclidean time with period β. Now "periodic" means that the *state* of the system is the same at time 0 and β but, because of the identity of the particles, the final particle arrangement may well be a *permutation* of the starting one. For fermions Feynman tells us to perform a sum over the elements of S_n, the permutation group of n objects, and include a $\pm$ sign for each permutation. Thus

$$\mathcal{Z} = \sum_n \frac{1}{n!} e^{\beta\mu n} \sum_{P \in S_n} \text{sgn}(P) \int d^3x_1 \ldots d^3x_n G(x_{P(1)}, x_1) \ldots G(x_{P(n)}, x_n). \tag{11.12}$$

Any permutation in S_n can be factored into *cycles* with r_1 1-cycles, r_2 2-cycles etc., and $\sum m r_m = n$. For example,

$$P = (14632)(78) \in S_8 \tag{11.13}$$

has one 1-cycle (5 is left alone by the permutation), one 2-cycle, and one 5-cycle.

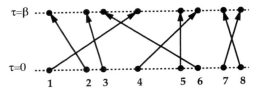

Fig 1. *The permutation* $(14632)(78) \in S_8$.

The sign of the permutation sgn (P) is defined by its effect on the polynomial

$$f(x_1, x_2, \ldots, x_n) = \prod_{i<j}(x_i - x_j). \tag{11.14}$$

The permutation is said to be *odd* [and sgn (P) is set to -1], if the mapping $P : \{x_n\} \to \{x_{P(n)}\}$ changes the sign of f. It is said to be *even* [and sgn $(P) = +1$], if it leaves f alone. The set of even permutations forms a normal subgroup of S_n. One easily sees that

$$\text{sgn}(P) = (+1)^{r_1}(-1)^{r_2}(+1)^{r_3} \ldots. \tag{11.15}$$

Two permutations P_1, P_2 with the same pattern of cycles are in the same *congugacy class*, i.e. there is some P such that $P_1 = P^{-1} P_2 P$. The number of cycles in the conjugacy class labeled by the r_i is

$$N(r_1, r_2 \ldots r_m) = \frac{n!}{1^{r_1} 2^{r_2} \ldots m^{r_m} r_1! r_2! \ldots r_m!}. \tag{11.16}$$

We can use this information about permutations to reorganize the sum in (11.12) as

$$\mathcal{Z} = \sum_n e^{\mu\beta n} \frac{1}{n!} \left[(\text{Tr}\, G)^{r_1} (-\text{Tr}\, G^2)^{r_2} \cdots ((-1)^{m+1} \text{Tr}\, G^m)^{r_m} \right] N(r_1, \ldots r_m), \tag{11.17}$$

where

$$\text{Tr}\, G^2 = \int d^3 x_1 d^3 x_2 G(x_1, x_2) G(x_2, x_1), \quad \text{etc.} \tag{11.18}$$

The sum on n does away with the $\sum m r_m = n$ constraint. This is precisely why chemical potentials and grand canonical partition functions are so useful. Now

$$\mathcal{Z} = \sum_{r_i} \frac{1}{r_1!} \left[e^{\mu\beta} \text{Tr}\, G(+1) \right]^{r_1} \cdots \frac{1}{r_m!} \left[\frac{1}{m} e^{m\mu\beta} \text{Tr}\, G^m (-1)^{m+1} \right]^{r_m} \cdots \tag{11.19}$$

$$= \exp \sum_m \frac{1}{m} e^{\mu\beta m} (-1)^{m+1} \text{Tr}\, G^m \tag{11.20}$$

$$= \exp \left[\text{Tr}\, \ln(1 + e^{\mu\beta} G) \right]. \tag{11.21}$$

We can evaluate the trace over x in any basis for the Hilbert space of functions on $\mathbf{R}^3$. In the plane-wave basis G is diagonal with elements $\exp(-\beta k^2/2m)$, so we have

$$\mathcal{Z} = \exp(Vol) \int \frac{d^3 k}{(2\pi)^3} \ln(1 + e^{\mu\beta} e^{-\beta k^2/2m}). \tag{11.22}$$

Eq (11.22) should be recognizable as the correct answer for free fermions.

The quantity $G(x, y, \tau)$ is of course the Green function for the heat equation. The higher powers of G that appear in the expressions above can be thought of as coming from the "image sources" we need when solving the heat equation on a cylinder of circumference β. Each G^{n+1} comes from "heat" that has wound n times around the cylinder on its journey from y to x. The minus signs that go with these powers of G can be thought of coming from taking *antiperiodic* boundary conditions for the solutions of the heat equation on this cylinder, so each time some heat "winds" it changes sign. Later we will see that the partition function could also be obtained from a path integral over Grassmann fields with antiperiodic boundary conditions. This directly gives $\det(1 + e^\mu G)$ and the same combinatoric excercise as above then serves to provide a demonstration, via an expansion of the determinant in powers of e^μ, of the well-known identity $\ln \det(1 + e^\mu G) = \text{Tr}\, \ln(1 + e^\mu G)$.

For bosons we would omit the sgn(P) and find

$$\mathcal{Z} = \exp\left[-\text{Tr}\, \ln(1 - e^{\mu\beta} G)\right]. \tag{11.23}$$

To include interactions we would write $G(x, y, \tau)$ as a path integral and introduce interactions between the paths.

At low temperatures the particles have the most "time" to wander and loose track of their identity, so it is at low temperatures that the relative signs associated

140   11. Finite Temperature

with different permuations are significant. For free bosons μ must be negative or zero for the trace to make sense, and the formation of a Bose condensate at $\mu = 0$ corresponds to the occurrence of long loops of worldline wrapping arbitrarily many times around the time torus. When this occurs any Bose particle could be the same as any other and a wordline created at x by a field $\varphi^\dagger(x)$ may easily terminate in the field $\varphi(x')$ at a distant x'. This leads to the *long range-order* in the $\langle \varphi^\dagger \varphi \rangle$ correlation function. Fermions behave differently. Here μ can be positive. Although with $\mu > 0$ the series no longer converges, the expression for the sum still makes sense and describes a degenerate Fermi gas.

11.3 Matsubara Sums

Recall from Chapter 5 how the vacuum-energy density for a bosonic field could be written as a euclidean-space momentum integral

$$\mathcal{E}(m^2) = \frac{1}{2} \int \frac{d^4k}{(2\pi)^4} \ln(k^2 + m^2). \tag{11.24}$$

To relate this to the usual $\frac{1}{2}\hbar\omega$ expresion we first differentiated with respect to m^2

$$\frac{\partial \mathcal{E}(m^2)}{\partial m^2} = \frac{1}{2} \int \frac{d^4k}{(2\pi)^4} \frac{1}{k^2 + m^2}, \tag{11.25}$$

then we performed the k_0 integral

$$\frac{\partial \mathcal{E}(m^2)}{\partial m^2} = \frac{1}{4} \int \frac{d^3k}{(2\pi)^3} \frac{1}{\sqrt{\mathbf{k}^2 + m^2}}, \tag{11.26}$$

and finally integrated with repect to m^2, to find

$$\mathcal{E}(m^2) = \frac{1}{2} \int \frac{d^3k}{(2\pi)^3} \sqrt{\mathbf{k}^2 + m^2}, \tag{11.27}$$

up to an m-independent constant.

At nonzero temperature the results of the previous section suggest that we regard the euclidean time τ as being periodic $\tau \sim \tau + \beta$. The finite period requires that the Fourier integral over k_0 be replaced by a Fourier sum over a set of discrete *Matsubara frequencies* $k_0 \to \omega_n = 2\pi n/\beta$. The integral becomes the sum

$$\int \frac{dk_0}{2\pi} f(k_0) \to \frac{1}{\beta} \sum_n f(\omega_n). \tag{11.28}$$

Using the identity

$$\sum_{-\infty}^{\infty} \frac{1}{n^2 + z^2} = \frac{\pi}{z} \coth \pi z \tag{11.29}$$

we find that

$$\sum_n \frac{1}{\left(\frac{2\pi n}{\beta}\right)^2 + \mathbf{k}^2 + m^2} = \frac{\beta}{2} \frac{1}{\sqrt{\mathbf{k}^2 + m^2}} \coth \frac{\beta}{2} \sqrt{\mathbf{k}^2 + m^2}. \qquad (11.30)$$

This is the derivative with repect to m^2 of

$$\ln \sinh \frac{\beta}{2} \sqrt{\mathbf{k}^2 + m^2} =$$
$$= \frac{\beta}{2} \sqrt{\mathbf{k}^2 + m^2} + \ln(1 - e^{-\beta\sqrt{\mathbf{k}^2 + m^2}}). \qquad (11.31)$$

The finite-temperature version of the "vacuum energy" integral has become the sum of the zero-point energy of the vacuum and the grand potential of a gas of relativistic bosons.

In the case of Dirac fermions, the zero-temperature vacuum energy is (after Wick rotation and with euclidean gamma matrices)

$$\mathcal{E}(m) = -\int \frac{d^4k}{(2\pi)^4} \operatorname{tr} \ln(i\slashed{k} + m). \qquad (11.32)$$

Here the trace is over the spinor indices in the gamma matrices, and the overall minus sign is from the factor of (minus 1) from the closed fermion loop. There is no factor of 1/2 because the fermion propagators in the ouroboros diagram have an arrow on them. To evaluate this we note that

$$\operatorname{tr}\{\ln(i\slashed{k} + m)\} = \operatorname{tr}\{\gamma_5^2 \ln(i\slashed{k} + m)\}$$
$$= \operatorname{tr}\{\gamma_5 \ln(i\slashed{k} + m)\gamma_5\} = \operatorname{tr}\{\ln(-i\slashed{k} + m)\} \qquad (11.33)$$

and

$$\ln(i\slashed{k} + m) + \ln(-i\slashed{k} + m) = \ln(i\slashed{k} + m)(-i\slashed{k} + m) = \ln(k^2 + m^2). \qquad (11.34)$$

The trace over the identity matrix that is tacit in this equation gives a factor of 4, so

$$\mathcal{E}(m) = -2 \int \frac{d^4k}{(2\pi)^4} \ln(k^2 + m^2). \qquad (11.35)$$

Treating this in the same manner as the zero-temperature Bose expression gives

$$\mathcal{E}(m) = -2 \int \frac{d^3k}{(2\pi)^3} \sqrt{\mathbf{k}^2 + m^2}. \qquad (11.36)$$

This is interpreted as being the sum of the (negative) energies of the states in the Dirac sea.[1] The factor of 2 counts the two possible spin orientations.

---

[1] There is complete symmetry between particles and antiparticles. This means that we cannot tell whether positrons are holes in an infinite sea of electrons, or electrons holes in an infinite sea of positrons. Whichever point of view we take, the vacuum-energy density comes out the same.

11. Finite Temperature

At finite temperature we replace the k_0 integral by a Matsubara sum, but now we take the discrete set of frequencies as those appropriate for *antiperiodic* boundary conditions. This is in accord with the observation in the previous section that the minus signs from the Fermi statistics are equivalent to inverting the sign of the image charges as we go around the τ cylinder. Thus $k_0 \to \omega_n = 2\pi(n+\frac{1}{2})/\beta$. To evaluate the sum, we need the identity

$$\sum_{n=-\infty}^{\infty} \frac{1}{(n+\frac{1}{2})^2 + z^2} = \frac{\pi}{z} \tanh \pi z. \tag{11.37}$$

We find

$$\frac{1}{Vol}\Omega = -\frac{4}{\beta} \int \frac{d^3k}{(2\pi)^3} \ln \cosh \frac{\beta}{2}\sqrt{\mathbf{k}^2 + m^2} \tag{11.38}$$

or

$$\frac{1}{Vol}\Omega = -4\frac{1}{\beta}\int \frac{d^3k}{(2\pi)^3}\left\{\frac{\beta}{2}\sqrt{\mathbf{k}^2+m^2} + \ln(1 + e^{-\beta\sqrt{\mathbf{k}^2+m^2}})\right\}. \tag{11.39}$$

This again combines the zero-temperature vacuum-energy contribution with the grand potential of the two spin species for each of two charges of relativistic particles.

12
Path Integrals

In this chapter we will introduce the Feynman path integral. We will derive the simplest path integral from conventional quantum mechanics. In doing so we may perhaps give the impression that path integration is merely a useful trick for solving a certain class of quantum mechanics problems. As familiarity with the technique grows, however, the gentle reader may come to follow Feynman in his belief that the path integral captures the fundamental physics, and that hamiltonians and Hilbert space are merely useful mathematical methods for evaluating path integrals.

12.1 Quantum Mechanics of a Particle

12.1.1 Real Time

In the quantum mechanics of a single particle, the propagator function $G(x_2, x_1, t)$ is the amplitude for the particle to start from the point x_1, and to be found a time t later at the point x_2. When the hamiltonian takes the simple form

$$\hat{H} = -\frac{1}{2m}\partial_x^2 + V(x), \tag{12.1}$$

G obeys the time-dependent Schrödinger equation

$$i\partial_t G(x_2, x_1, t) = \left(-\frac{1}{2m}\partial_{x_2}^2 + V(x_2)\right) G(x_2, x_1, t) \tag{12.2}$$

with initial condition

$$G(x_2, x_1, 0) = \delta(x_1 - x_2). \tag{12.3}$$

144   12. Path Integrals

The solution to this initial-value problem may be written in terms of the normalized energy eigenfunctions $\varphi_n(x)$ as

$$G(x_2, x_1, t) = \sum_n \varphi_n(x_2)\varphi_n^*(x_1)e^{-iE_n t}. \tag{12.4}$$

Rewriting this using Dirac notation, where $\varphi_n(x) \equiv \langle x|n\rangle$, we have

$$G(x_2, x_1, t) = \sum_n \langle x_2|n\rangle e^{-iE_n t}\langle n|x_1\rangle$$

$$= \langle x_2|e^{-i\hat{H}t}|x_1\rangle$$

$$= \langle x_2, t|x_1, 0\rangle. \tag{12.5}$$

There is a subtlety here. Despite its t-dependence, the symbol $|x, t\rangle$ stands for a *Heisenberg-picture* state. To be precise, we define the ket $|x, t\rangle$ to be the eigenstate of the Heisenberg operator

$$\hat{x}(t) = e^{i\hat{H}t}\hat{x}(0)e^{-i\hat{H}t} \tag{12.6}$$

having eigenvalue x. To maintain this property the $|x, t\rangle$ for different t must be related by

$$|x, t\rangle = e^{i\hat{H}t}|x, 0\rangle. \tag{12.7}$$

Note that the exponent has the opposite sign to that we might naïvely have expected. We should think of this t-dependence not as the time evolution of a particular state, but as the relation between members of a family of t-labeled, time-independent, states that have simple descriptions as eigenstates of $\hat{x}(t')$ only at the moment $t = t'$. In this regard they are similar to the $|\mathbf{k}; in\rangle$ and $|\mathbf{p}; out\rangle$ states of Chapter 6. The "backwards time-evolution" in (12.7) should be compared with the superficially paradoxical $|\mathbf{k}; in\rangle = S|\mathbf{p}; out\rangle$ we found there.

We have

$$|x, t\rangle = \sum_n e^{iE_n t}|n\rangle\langle n|x\rangle \tag{12.8}$$

and

$$\langle x, t| = \sum_n e^{-iE_n t}\langle x|n\rangle\langle n|, \tag{12.9}$$

so, for example,[1]

$$\langle x, t|n\rangle = \varphi_n(x)e^{-iE_n t}. \tag{12.11}$$

---

[1] Compare the present notation with the Schrödinger-picture version of the same equation. This would read

$$\langle x|n, t\rangle_S = \varphi_n(x)e^{-iE_n t}. \tag{12.10}$$

12.1 Quantum Mechanics of a Particle

The Feynman path integral provides an expression for $\langle x_2, t | x_1, 0\rangle$ in terms of a sum over all possible ways of getting from x_1 to x_2 that take time t to complete. To derive the basic form of the path-integral formula we again assume that the hamiltonian is of the simple form

$$\hat{H} = \frac{1}{2m}\hat{p}^2 + V(\hat{x}). \tag{12.12}$$

We break the time interval $(t, 0)$ into N small intervals of duration δt, and write

$$\langle x_N | e^{-i\hat{H}t} | x_0 \rangle = \langle x_N | e^{-i\delta t \hat{H}} e^{-i\delta t \hat{H}} \ldots e^{-i\delta t \hat{H}} | x_0 \rangle$$

$$= \int dx_{N-1} dx_{N-2} \ldots dx_1 \times$$

$$\langle x_N | e^{-i\hat{H}\delta t} | x_{N-1}\rangle \langle x_{N-1} | e^{-i\hat{H}\delta t} | x_{N-2}\rangle \ldots \langle x_1 | e^{-i\hat{H}\delta t} | x_0 \rangle. \tag{12.13}$$

Let us focus on one of the factors — say $\langle x_1 | e^{-i\hat{H}\delta t} | x_0\rangle$. We break it apart as

$$e^{-i(\frac{1}{2m}\hat{p}^2 + V(\hat{x}))\delta t} = e^{-i\delta t(\frac{1}{2m}\hat{p}^2)} e^{-i\delta t V(\hat{x})} e^{-iO(\delta t^2)}. \tag{12.14}$$

For the moment ignore the $O(\delta t^2)$ piece and evaluate

$$\langle x_1 | e^{-i\delta t(\frac{1}{2m}\hat{p}^2)} e^{-i\delta t V(\hat{x})} | x_0\rangle$$

$$= \int dx \int \frac{dp}{2\pi} \int \frac{dp'}{2\pi} \langle x_1 | p\rangle \langle p | e^{-i\delta t(\frac{1}{2m}\hat{p}^2)} | p'\rangle \langle p' | x\rangle \langle x | e^{-i\delta t V(\hat{x})} | x_0\rangle$$

$$= \int \frac{dp}{2\pi} \langle x_1 | p\rangle \langle p | x_0\rangle e^{-i\delta t(\frac{1}{2m}p^2)} e^{-i\delta t V(x_0)}$$

$$= \int \frac{dp}{2\pi} e^{ip(x_1 - x_0)} e^{-i\delta t(\frac{1}{2m}p^2)} e^{-i\delta t V(x_0)}. \tag{12.15}$$

When we perform the Fresnel integral, we get

$$\langle x_1 | e^{-i\hat{H}\delta t} | x_0\rangle = e^{-i\frac{1}{4}\pi} \frac{1}{2\pi} \sqrt{\frac{2m\pi}{\delta t}} \exp i\delta t \left\{ \frac{m(x_1 - x_0)^2}{2(\delta t)^2} - V(x_0) \right\}. \tag{12.16}$$

Thus the propagator is approximately

$$\langle x_N, t | x_0, 0\rangle \approx \mathcal{N} \int dx_{N-1} dx_{N-2} \ldots dx_1 e^{i\delta t \sum_i \left\{ \frac{m(x_{i+1} - x_i)^2}{2(\delta t)^2} - V(x_i)\right\}}. \tag{12.17}$$

Here $\mathcal{N}$ is a normalization constant containing a product of N factors of $\sqrt{m/2\pi\delta t}$.

The approximation becomes better as δt becomes smaller, and provided the $O(\delta t^2)$ errors do not accumulate pathologically as the number of factors becomes large, we can take the limit that $N \to \infty$, $\delta t \to 0$ while keeping the value of the endpoint x_N fixed at $x_N = x_t$. If the trajectory mapped out by the sequence x_i were smooth, then we would have

$$i\delta t \sum_i \left\{ \frac{m(x_{i+1} - x_i)^2}{2(\delta t)^2} - V(x_i) \right\} \to i \int_0^t dt \left\{ \frac{m\dot{x}^2}{2} - V(x) \right\}. \quad (12.18)$$

This motivates the notation for the infinite-dimensional integral

$$\langle x_t, t | x_0, 0 \rangle = \mathcal{N} \int_{x_0}^{x_t} d[x(t)] e^{i \int_0^t dt \left\{ \frac{m\dot{x}^2}{2} - V(x) \right\}}. \quad (12.19)$$

Here $d[x(t)]$ is shorthand for

$$\lim_{N \to \infty} dx_{N-1} dx_{N-2} \ldots dx_1, \quad (12.20)$$

and may be regarded as the volume element or "measure" on the space of paths $x(t)$. After the limit is taken the normalization constant $\mathcal{N}$ contains infinitely many factors, and is divergent. We can either absorb these factors into the definition of the symbol $d[x(t)]$, in which case we will omit it, or we can observe that $\mathcal{N}$ will cancel out of any physical quantity we may care to compute.

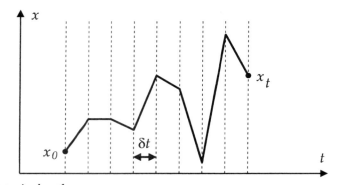

Fig 1. A typical path.

Restoring $\hbar$, we can write the path-integral expression for (12.4) as

$$\langle x_2, t | x_1, 0 \rangle = \mathcal{N} \int_{x_1(0)}^{x_2(t)} d[x(t)] e^{\frac{i}{\hbar} S[x(t)]}, \quad (12.21)$$

where $S[x(t)]$ is the *classical* action for each path $x(t)$ that occurs in the sum.

The occurrence of the classical action in the exponent strongly suggests that the classical principle of least action comes about because as $\hbar$ becomes small, the path integral will come to be dominated by a single stationary-phase trajectory.

12.1.2 Euclidean Time

The i in the Fresnel integrals causes problems with the formal measure theory on the space of paths. It also makes numerical work with the path integral next-to-impossible. For many applications it is convenient to replace the Minkowski-space propagator by its euclidean version. We simply replace t by $-i\tau$ where we intend the euclidean time τ to be real. Then

$$\langle x_2|e^{-\hat{H}\tau}|x_1\rangle = \sum_n \varphi_n(x_2)\varphi_n^*(x_1)e^{-E_n\tau}$$

$$= \mathcal{N}\int_{x_1(0)}^{x_2(\tau)} d[x(\tau)]e^{-\int_0^\tau d\tau\left\{\frac{m\dot{x}^2}{2}+V(x)\right\}}. \tag{12.22}$$

Now $\dot{x}$ stands for $dx/d\tau$. Note how the relative sign of the kinetic and potential terms in the action have changed.

Equation (12.22) is much better behaved mathematically than (12.19). In particular the new exponent tends to supress very jagged paths. Even so, the "typical" path $x(\tau)$ is merely continous and not differentiable. This means that that the $\dot{x}^2$ appearing in the exponent should not be taken too literally. The formal expression (12.22) is to be understood only as a mnemonic shorthand for the multiple integral.

Try to resist the temptation (which can be rather strong) toward metaphysical speculation about the meaning of imaginary time. Think of (12.22) as merely a formula from which we can extract various quantities such as wavefunctions and energies which have meanings in the conventional, real time, world.

One immediate application is the *Feynman-Kac formula* for the ground-state energy

$$E_0 = \lim_{T\to\infty}\left\{-\frac{1}{T}\ln\int_{x_1(0)}^{x_2(T)} d[x(\tau)]e^{-\int_0^T d\tau\left\{\frac{m\dot{x}^2}{2}+V(x)\right\}}\right\}. \tag{12.23}$$

This formula works because, when T is large, the lowest energy state dominates the sum $\sum_n \varphi_n(x_2)\varphi_n^*(x_1)\exp\{-E_n\tau\}$. The particular choices of the end-point values of x_1, x_2 become irrelevant once the logarithm is taken. Remember that the E_0 that we have found here is the ground-state energy of the conventional *real time* Schrödinger equation. Imaginary time was merely a convenient trick for obtaining it.

Note that in the euclidean world the time evolution of the Heisenberg operator $\hat{x}(\tau)$ is defined by

$$\hat{x}(\tau) = e^{\hat{H}\tau}\hat{x}(0)e^{-\hat{H}\tau}. \tag{12.24}$$

The corresponding τ dependence of the family of τ-labeled Heisenberg states is therefore

$$\langle x,\tau| = \langle x,0|e^{-\hat{H}\tau}, \qquad |x,\tau\rangle = e^{\hat{H}\tau}|x,0\rangle. \tag{12.25}$$

This τ evolution is not unitary and the states $|x,\tau\rangle$ are not normalized. They still provide a resolution of the identity, however, since

$$\int dx|x,\tau\rangle\langle x,\tau| = \int dx e^{\hat{H}\tau}|x,0\rangle\langle x,0|e^{-\hat{H}\tau} = e^{\hat{H}\tau}\cdot I\cdot e^{-\hat{H}\tau} = I. \tag{12.26}$$

If we write

$$d\mu[x(\tau)] = d[x(\tau)]e^{-\int d\tau\frac{m\dot{x}^2}{2}}, \tag{12.27}$$

148   12. Path Integrals

then $d\mu(x(\tau))$ is called the *Wiener measure*[2] on the space of paths. The theory of Wiener measure was originally developed to deal with Langevin equations and their connection with random walks and diffusion. We will discuss this later.

12.2   Gauge Invariance and Operator Ordering

The path integral makes things seem too easy. It seems to imply that for any classical action there is a unique quantum mechanics. This is certainly not true. One way to see that there are easily missed subtleties is to consider generalizing our path integral to include an external magnetic field. The classical lagrangian for a charge e particle in a magnetic field is

$$\frac{1}{2}\dot{x}^2 - V(x) + e\mathbf{A}\cdot\dot{\mathbf{x}}. \tag{12.28}$$

Let us try to derive the (euclidean) Schrödinger equation for the particle by using the path integral. We will look at the equation obeyed by the quantity

$$\psi(x,\tau) = \mathcal{N}\int_{x(0)}^{x(\tau)} d[x(\tau)] e^{-\int d\tau\left\{\frac{\dot{x}^2}{2}+V(x)-ie\sum_\mu A_\mu\dot{x}_\mu\right\}}. \tag{12.29}$$

Note that in this expression the coupling to the gauge field remains imaginary even though we are in euclidean time. This is because

$$i\int dt\,\mathbf{A}\cdot\frac{d\mathbf{x}}{dt} = i\int d\mathbf{x}\cdot\mathbf{A} = i\int d\tau\,\mathbf{A}\cdot\frac{d\mathbf{x}}{d\tau} \tag{12.30}$$

Now we discretize the path integral as

$$\lim_{\delta\tau\to 0}\int\prod_i\left[\frac{d^3x_i}{(2\pi\delta\tau)^{3/2}}\right] e^{-\sum_i\left\{\frac{(x_{i+1}-x_i)^2}{2\delta\tau}+V(x_i)\delta\tau-ie(x_{i+1}-x_i)^\mu A_\mu(\frac{1}{2}(x_{i-1}+x_i))\right\}}. \tag{12.31}$$

Note how I have evaluated A_μ at the *midpoint* of the interval (x_{i+1}, x_i). This will turn out to be crucial. The point selected for the evaluation of V is not so important. Now if we integrate over all intermediate x_i except the last one, we find

$$\psi(x,\tau+\delta\tau) = \int\frac{d^3\eta}{(2\pi\delta\tau)^{3/2}} e^{-\frac{\eta^2}{2\delta\tau}-V(x)\delta\tau-ie\eta^\mu A_\mu(x+\frac{1}{2}\eta)}\psi(x+\eta,\tau) \tag{12.32}$$

Therefore

$$\psi(x,\tau+\delta\tau) = \int\frac{d^3\eta}{(2\pi\delta\tau)^{3/2}} e^{-\frac{\eta^2}{2\delta\tau}}\times$$

$$\times\left(\psi(x,\tau)+\eta_\mu\partial_\mu\psi+\frac{1}{2}\eta_\mu\eta_\nu\partial^2_{\mu\nu}\psi\ldots\right)$$

---

[2]Norbert Wiener. Born November 26, 1894, Columbia, MO. Died March 18, 1964, Stockholm, Sweden.

12.2 Gauge Invariance and Operator Ordering

$$\times (1 - V(x)\delta\tau + \ldots)$$

$$\times \left(1 - ie\eta^\mu A_\mu + \frac{(ie)^2}{2}\eta^\mu A_\mu \eta^\nu A_\nu - ie\eta^\mu \frac{\eta^\nu}{2}\partial_\nu A_\mu + \ldots\right) \quad (12.33)$$

The $\frac{1}{2}$ in the very last displayed term comes from the *midpoint rule* we adopted above for the discretization of the A_μ-field term. In writing down the above expression I have kept all terms of $O(\delta\tau)$ — bearing in mind that $O(\delta\tau) \sim O(\eta^2)$ because of the way the gaussian integral restrains η from getting too large. Now

$$\int \frac{d^3\eta}{(2\pi\delta\tau)^{3/2}} e^{-\frac{\eta^2}{2\delta\tau}} = 1, \quad (12.34)$$

$$\int \frac{d^3\eta}{(2\pi\delta\tau)^{3/2}} e^{-\frac{\eta^2}{2\delta\tau}} \eta^\mu \eta^\nu = \delta\tau\, \delta^{\mu\nu}, \quad (12.35)$$

so

$$\psi(x, \tau+\delta\tau) = \psi(x,\tau) +$$
$$+ \left\{\frac{1}{2}\nabla^2\psi - V\psi + \frac{(ie)^2}{2}A_\mu A_\mu \psi - ieA_\mu \partial_\mu \psi - \frac{ie}{2}(\partial_\mu A_\mu)\psi\right\}\delta\tau.$$

$$(12.36)$$

In other words,

$$\frac{\partial\psi}{\partial\tau} = \frac{1}{2}\left(\partial_\mu - ieA_\mu\right)^2\psi - V\psi. \quad (12.37)$$

This is the gauge-invariant euclidean Schrödinger equation. Note how crucial it was that we chose to evaluate $A_\mu(x)$ at the *midpoint* of the interval. Without this rule we do not get the organization of the terms into gauge-covariant derivatives.

The reason we need the midpoint rule is that we require a gauge transformation $A_\mu \to A_\mu + \partial_\mu \Lambda$ to give

$$\int d[x(\tau)] e^{-\int d\tau\left\{\frac{\dot{x}^2}{2} - ie(A_\mu + \partial_\mu \Lambda)\dot{x}_\mu\right\}}$$

$$= \int d[x(\tau)] e^{-\int\left\{\frac{\dot{x}^2}{2} - ieA_\mu \dot{x}_\mu\right\}d\tau + ie\int \partial_\mu \Lambda dx^\mu}$$

$$\int d[x(\tau)] e^{-\int\left\{\frac{\dot{x}^2}{2} - ieA_\mu \dot{x}_\mu\right\}d\tau} e^{ie\{\Lambda(x,\tau)-\Lambda(0,0)\}}. \quad (12.38)$$

In general, however, the naive calculus manipulation

$$\int_{\tau_1}^{\tau_2} \frac{d\Lambda}{dx}\dot{x}\,d\tau = \int_{x(\tau_1)}^{x(\tau_2)} \frac{d\Lambda}{dx}dx = \Lambda(x(\tau_2)) - \Lambda(x(\tau_1)) \quad (12.39)$$

fails because $\delta x = O(\sqrt{\delta\tau})$ in a diffusion process. If, however, we use the midpoint, we have

$$\Lambda(x+a) = \Lambda(x) + a \left.\frac{d\Lambda}{dx}\right|_{x+\frac{1}{2}a} + O(a^3) \qquad (12.40)$$

and the absence of an $O(a^2)$ error makes this an accurate enough estimate for things to work.

In general, different discretization recipes for the path integral may correspond to different orderings of the noncommuting operators in the hamiltonian. Given a novel action to quantize, the guiding principle should be to try to select a discretization that preserves as many of the desired symmetries as possible.

12.3 Correlation Functions

In field-theory the "Green functions" are the vacuum expectation values of (euclidean or Minkowski) time-ordered products of Heisenberg operators:

$$G^{(n)}(x_1, \ldots, x_n) = \langle 0|T\{\hat{\varphi}(x_1)\ldots\hat{\varphi}(x_n)\}|0\rangle. \qquad (12.41)$$

Here the fields live in a space with $d-1$ space and 1 time dimension. Quantum mechanics can be regarded as a field theory in 0 space and 1 time dimension. With this identification the variable $\hat{x}(\tau)$ plays the role of the field $\hat{\varphi}(x)$. The quantum-mechanics analogs of the Green functions are therefore the quantities

$$G^{(n)}(\tau_1, \ldots, \tau_n) = \langle 0|T\{\hat{x}(\tau_1)\ldots\hat{x}(\tau_n)\}|0\rangle, \qquad (12.42)$$

where $|0\rangle$ is the ground state.

Our path integral gives a recipe for calculating quantities like

$$\langle x_T, T|T\{\hat{x}(\tau_1)\ldots\hat{x}(\tau_n)\}|x_{-T}, -T\rangle, \qquad (12.43)$$

where the ground state has been replaced by eigenstates of $\hat{x}$ with eigenvalue x_T, x_{-T}. To find the recipe we work with euclidean time and assume that $T > \tau_n > \tau_{n-1} > \ldots > \tau_1 > -T$.

$\langle x_T, T|T\{\hat{x}(\tau_1)\ldots\hat{x}(\tau_n)\}|x_{-T}, -T\rangle$

$= \int dx(\tau_n)\ldots dx(\tau_1) \times$

$\times \langle x_T, T|\hat{x}(\tau_n)|x(\tau_n), \tau_n\rangle\langle x(\tau_n), \tau_n|\hat{x}(\tau_{n-1})|x(\tau_{n-1}), \tau_{n-1}\rangle\langle x(\tau_{n-1}), \tau_{n-1}|\ldots|x_0\rangle.$
$\qquad (12.44)$

Note how we have put in intermediate states that are eigenstates of the $\hat{x}(\tau)$ operators that immediately precede them. We therefore replace the operators by their

12.3 Correlation Functions

eigenvalues and get

$$\int dx(\tau_n) \ldots dx(\tau_1) \times$$

$$\times \langle x_T, T | x(\tau_n), \tau_n \rangle x(\tau_n) \langle x(\tau_n), \tau_n | x(\tau_{n-1}), \tau_{n-1} \rangle x(\tau_{n-1}) \langle x(\tau_{n-1}), \tau_{n-1} | \ldots | x_0 \rangle$$

$$= \mathcal{N} \int_{x_{-T}}^{x_T} d[x(\tau)] \, \{x(\tau_n) x(\tau_{n-1}) \ldots x(\tau_1)\} \, e^{-\int d\tau \left\{ \frac{m\dot{x}^2}{2} + V(x) \right\}}. \tag{12.45}$$

In the last line we have replaced the various $\langle x(\tau_i), \tau_i | x(\tau_{i-1}), \tau_{i-1} \rangle$ by their path integral expressions, and observed that the integrations over the intermediate $x(\tau_i)$ serve to sew the $n+1$ separate segments together to make one overall integral over paths from x_{-T} to x_T.

Note that, since the $x(\tau_m)$ appearing in the path integral are now *numbers*, the order in which they are written is immaterial.

Now

$$|x_{-T}, -T\rangle = e^{-\hat{H}T} |x_{-T}\rangle = \sum_n |n\rangle \langle n | x_{-T} \rangle e^{-E_n T}, \tag{12.46}$$

$$\langle x_T, T| = e^{-\hat{H}T} \langle x_T| = \sum_n e^{-E_n T} \langle x_T | n \rangle \langle n |, \tag{12.47}$$

so when we take $T \to \infty$, the ground state will dominate the sum over the initial and final states. Thus

$$\langle x_T, T | T\{\hat{x}(\tau_1) \ldots \hat{x}(\tau_n)\} | x_{-T}, -T \rangle$$

$$\approx \langle x_T | 0 \rangle \langle 0 | T\{\hat{x}(\tau_1) \ldots \hat{x}(\tau_n)\} | 0 \rangle \langle 0 | x_{-T} \rangle e^{-2E_0 T}. \tag{12.48}$$

Division by $\langle x_T, T | x_{-T}, -T \rangle$ serves a double role of removing the $\langle x_T | 0 \rangle$ and $\langle 0 | x_{-T} \rangle$ wavefuctions and cancelling the unwanted vacuum energy factor. Doing this gives

$$\langle 0 | T\{\hat{x}(\tau_1) \ldots \hat{x}(\tau_n)\} | 0 \rangle = \lim_{T \to \infty} \frac{\langle x_T, T | T\{\hat{x}(\tau_1) \ldots \hat{x}(\tau_n)\} | x_{-T}, -T \rangle}{\langle x_T, T | x_{-T}, -T \rangle}. \tag{12.49}$$

We have therefore proved:

Theorem:

The euclidean Green functions are given by

$$\langle 0 | T\{\hat{x}(\tau_1) \ldots \hat{x}(\tau_n)\} | 0 \rangle$$

$$= \frac{1}{Z} \int_{x(-\infty)}^{x(\infty)} d[x(\tau)] \, \{x(\tau_n) x(\tau_{n-1}) \ldots x(\tau_0)\} \, e^{-S[x]}, \tag{12.50}$$

where

$$S[x] = \int_{-\infty}^{\infty} d\tau \left\{ \frac{m\dot{x}^2}{2} + V(x) \right\} \tag{12.51}$$

and

$$Z = \int_{x_\infty}^{x_0} d[x(\tau)] e^{-S[x]}. \tag{12.52}$$

COMMENT: Compare this with the expression for the thermal average of variables in statistical mechanics

$$\langle X_1 X_2 \ldots X_N \rangle = \frac{1}{Z} \sum \{X_1 X_2 \ldots X_N\} e^{-\beta H}. \tag{12.53}$$

We see that all the complicated operators have disappeared and been replaced by the classical statistical mechanics of an elastic string.

12.4 Fields

The extension of the quantum-mechanics results of the previous section to d-dimensional field theories is formally trivial — although full of mathematical and physical subtleties that we will occupy us in later sections. We use a "Schrödinger representation" for the fields where a state $|\Psi\rangle$ is represented by a wave-functional $\Psi(\varphi)$. Here

$$\Psi(\varphi) = \langle \varphi | \Psi \rangle, \qquad \hat{\varphi}(\mathbf{x}) | \varphi \rangle = \varphi(\mathbf{x}) | \varphi \rangle. \tag{12.54}$$

In the basis where $\hat{\varphi}(\mathbf{x})$ is diagonal, the operator $\hat{\varphi}(\mathbf{x})$ and its canonical conjugate $\hat{\pi}(\mathbf{x})$ become

$$\hat{\varphi}(\mathbf{x}) \to \varphi(\mathbf{x}), \qquad \hat{\pi}(\mathbf{x}) \to -i \frac{\delta}{\delta \varphi(\mathbf{x})}. \tag{12.55}$$

In other words $\hat{\varphi}(\mathbf{x})$ corresponds to multiplication by the function $\varphi(\mathbf{x})$. Clearly

$$[\hat{\varphi}(\mathbf{x}), \hat{\pi}(\mathbf{x}')] = i \frac{\delta \varphi(\mathbf{x})}{\delta \varphi(\mathbf{x}')} = i \delta^{d-1}(\mathbf{x} - \mathbf{x}'). \tag{12.56}$$

With this formalism in place we simply transcribe the results of the previous section and obtain

$$\langle 0 | T\{\hat{\varphi}(x_1) \ldots \hat{\varphi}(x_n)\} | 0 \rangle$$

$$= \frac{1}{Z} \int_{\varphi(-\infty)}^{\varphi(\infty)} d[\varphi(x)] \{\varphi(x_n) \varphi(x_{n-1}) \ldots \varphi(x_0)\} e^{-S[x]}, \tag{12.57}$$

where

$$S[\varphi] = \int_{-\infty}^{\infty} d^d x \left\{ \frac{1}{2} (\partial_\mu \varphi)^2 + V(\varphi) \right\} \tag{12.58}$$

and

$$Z = \int_{\varphi(-\infty)}^{\varphi(\infty)} d[\varphi(x)] e^{-S[x]}. \tag{12.59}$$

This appears as the d-dimensional classical statistical mechanics of an elastic *membrane* with a tension energy term $\frac{1}{2}(\partial\varphi)^2$ and a displacement potential $V(\varphi(x))$.

Note how these formulae make no reference to which of the d dimensions we chose to be "time." Indeed the final path-integral expressions are completely rotationally invariant. In Minkowski space rotation invariance becomes Lorentz invariance. The manifest Lorentz or ratoation invariance is one of the great advantages of the path integral formalism. It comes about because the path-integral formula contains the Lagrangian density, which is a Lorentz scalar. The hamiltonian density, on the other hand, is the 00 component of the energy-momentum tensor.

A word of warning: the "typical" configurations in the sum over fields are even less smooth than the continuous, but nondifferentiable, paths that appear in the quantum-mechanics path integral. It is this lack of smoothness that is responsible for the appearance of divergent integrals in the perturbation series. Taming this anfractuosity sufficiently for us to take the n-dimensional analog of $\delta\tau \to 0$ (the *continuum limit*) will require some deep ideas.

12.5 Gaussian Integrals and Free Fields

12.5.1 Real Fields

Free fields correspond to actions that are quadratic in the field variables φ. All integrals involving such quadratic actions can be reduced to the fundamental Gaussian integral

$$\int_{-\infty}^{\infty} e^{-\frac{1}{2}ax^2} dx = \sqrt{\frac{2\pi}{a}}, \tag{12.60}$$

or to its simple variant

$$\int_{-\infty}^{\infty} e^{-\frac{1}{2}ax^2+ibx} dx = \sqrt{\frac{2\pi}{a}} e^{-\frac{1}{2}\frac{b^2}{a}}. \tag{12.61}$$

The latter is obtained from the former by completing the square

$$-\frac{1}{2}ax^2 + ibx = -\frac{1}{2}a(x - ib/a)^2 - \frac{1}{2}\frac{b^2}{a}, \tag{12.62}$$

and shifting the variable of integration from x to $x - ib/a$.

We generalize these integrals to many x_i's by using an orthogonal transformation $x_i \to y_i$, which diagonalizes the symmetric matrix A_{ij}

$$\int \frac{d^N x}{(\sqrt{2\pi})^N} e^{-\frac{1}{2}x^i A_{ij} x^j} = \int \frac{d^N y}{(\sqrt{2\pi})^N} e^{-\frac{1}{2}\lambda_i y_i^2} = (\prod_i \lambda_i)^{-\frac{1}{2}} = \det{}^{-\frac{1}{2}} A. \tag{12.63}$$

Here the λ_i are the eigenvalues of the matrix A_{ij}. Similarly

$$\int \frac{d^N x}{(\sqrt{2\pi})^N} e^{-\frac{1}{2} x^i A_{ij} x^j + i b_i x^i} = (\det A)^{-\frac{1}{2}} e^{-\frac{1}{2} b_i A_{ij}^{-1} b_j}. \qquad (12.64)$$

Algebraically the generalization is straightforward. In extending further to integrals with a continuous set of x^i we should take note of a technical, but important, subtlety. The exponent in the integral involves a *quadratic form*, $A(x,x) = A_{ij} x^i x^j$. In other words it depends on a symmetric, doubly covariant tensor A_{ij}. The right-hand side involves the determinant of the matrix A — but in order to have an invariantly defined determinant, the matrix has to be interpreted as an *endomorphism* of a vector space V. That is, as a mapping $A: V \to V$. In any particular basis an endomorphism is represented by a mixed tensor $A^i{}_j$ acting as $(x')^i = A^i{}_j x^j$. To relate the two we require an inner product $\langle x, y \rangle = g_{ij} x^i y^j$. We write

$$A(x,x) = x_i A_{ij} x_j = \langle x, Ax \rangle = x^i g_{ij} A^j{}_k x^k. \qquad (12.65)$$

The integral tacitly depends on this inner product because, in chosing the measure to be

$$\frac{d^N x}{(\sqrt{2\pi})^N} \qquad (12.66)$$

in (12.63) and (12.64), we are treating the x^i as orthogonal coordinates, and the notion of orthogonality requires a choice of inner product. In many cases the choice of inner product will be obvious, but in other contexts (such as quantum gravity or string theory) some effort is required to find the right one. In all cases the symmetry of the quadratic form is reflected in the operator A being self-adjoint with respect to the selected inner product.

For free fields the quadratic form in our path integral is

$$A(\varphi, \varphi) = 2S[\varphi] = \int d^d x\, \varphi(x) \left(-\partial^2 + m^2\right) \varphi(x), \qquad (12.67)$$

so $A \to -\partial^2 + m^2$. When we interpret this in the usual way as an operator

$$-\partial^2 + m^2 : L^2(\mathbf{R}^d) \to L^2(\mathbf{R}^d), \qquad (12.68)$$

we are implicitly selecting the $L^2(\mathbf{R}^d)$ inner product

$$\langle \varphi_1, \varphi_2 \rangle = \int d^d x\, \varphi_1(x) \varphi_2(x). \qquad (12.69)$$

The A^{-1} appearing in (12.63) and (12.64) is the Green function for $-\partial^2 + m^2$

$$(-\partial^2 + m^2)^{-1}_{xy} = G(x,y) = \int \frac{d^d x}{(2\pi)^d} \frac{e^{ik(x-y)}}{k^2 + m^2}. \qquad (12.70)$$

The quantity $\det A$ must now be interpreted as the *Fredholm determinant* of the differential operator. This is a more complicated object, because unless the operator can be written in the form $A = 1 + B$ where B is *trace-class*, the determinant will be divergent and will need regularization and renormalization. For the moment we define it heuristically via the identity (valid for finite matrices)

$$\ln \det A = \ln \prod_i \lambda_i = \sum_i \ln \lambda_i = \operatorname{tr} \ln A. \tag{12.71}$$

We met the identity

$$\ln \det A = \operatorname{tr} \ln A, \tag{12.72}$$

in Chapter 11. In the present case

$$\ln \det (-\partial^2 + m^2) = (Vol) \int \frac{d^d k}{(2\pi)^d} \ln(k^2 + m^2). \tag{12.73}$$

where (Vol) referes to the volume of space-time. Coupled with the Feynman-Kac formula, this gives the same expression for the ground-state energy density that we derived in Chapter 6.

We also transcribe (12.64) to read

$$\frac{1}{Z} \int d[\varphi(x)] e^{-\int d^d x \{\frac{1}{2}\varphi(x)(-\partial^2 + m^2))\varphi(x) + i J(x)\varphi(x)\}}$$
$$= \langle 0 | T \{ e^{i \int d^d x\, J(x)\varphi(x)} \} | 0 \rangle$$
$$= \exp\left\{ -\frac{1}{2} \int d^d x\, d^d y\, J(x) G(x,y) J(y) \right\}. \tag{12.74}$$

We have thus recovered Wick's theorem as an almost trivial consequence of the formalism.

12.5.2 Complex Fields

So far we have used real fields and real numbers. We can extend our results to complex fields by introducing the complex measure

$$dz^* dz = (dx - i dy)(dx + i dy) = 2i\, dx\, dy \tag{12.75}$$

(I am using the notation of differential forms so dx and dy anticommute) so

$$\frac{dz^* dz}{2i} = dx\, dy. \tag{12.76}$$

The basic gaussian integral now becomes

$$\int \frac{dz^* dz}{2\pi i} e^{-|z|^2} = \int \frac{dx\, dy}{\pi} e^{-(x^2+y^2)} = 1. \tag{12.77}$$

This implies, for real a,

$$\int \frac{dz^* dz}{2\pi i} e^{-a|z|^2 + b^* z + z^* b}$$
$$= \int \frac{dz^* dz}{2\pi i} e^{-a(z - b/a)(z^* - b^*/a) + |b|^2/a}$$

$$= \frac{1}{a} \exp\left\{\frac{1}{a}|b|^2\right\}. \tag{12.78}$$

For multiple integrals

$$\int \prod_i \left[\frac{dz_i^* dz_i}{2\pi i}\right] e^{-z_i^* A_{ij} z_j + b_i^* z_i + z_i^* b_i} = (\det A)^{-1} \exp\left\{b_i^* A_{ij}^{-1} b_j\right\}. \tag{12.79}$$

We apply this to fields and find

$$\mathcal{N} \int d[\varphi^*][d\varphi] e^{-\int d^d x \varphi^*(-\partial^2 + m^2)\varphi + J^*(x)\varphi(x) + J(x)\varphi^*(x)}$$
$$= \left(\det(-\partial^2 + m^2)\right)^{-1} \exp\left\{\int d^d x d^d y J^*(x) G(x,y) J(y)\right\}. \tag{12.80}$$

From this we find

$$\langle 0|T\{\hat{\varphi}(x)\hat{\varphi}^\dagger(y)\}|0\rangle = G(x,y), \tag{12.81}$$

and all the other results we derived earlier using operators.

12.6 Perturbation Theory

To obtain the perturbation expansion we simply expand-out the exponential of the interaction term in the action, and integrate term by term.

To state this formally we write

$$\mathcal{Z}(J) = \int d[\varphi] e^{-\int d^d x \left\{\frac{1}{2}(\partial\varphi)^2 + \frac{m^2}{2}\varphi^2 + V(\varphi) - J(x)\varphi(x)\right\}}$$
$$= \sum_{n=0}^{\infty} \int d[\varphi] \frac{1}{n!} \left[-\int d^d x \, V(\varphi(x))\right]^n e^{-\int d^d x \left\{\frac{1}{2}(\partial\varphi)^2 + \frac{m^2}{2}\varphi^2 + J(x)\varphi(x)\right\}}$$
$$= \sum_{n=0}^{\infty} \frac{1}{n!} \left[-\int d^d x \, V\left(\frac{\delta}{\delta J(x)}\right)\right]^n e^{\frac{1}{2}\int d^d x d^d y J(x) G(x,y) J(y)}. \tag{12.82}$$

This is essentially the same perturbation expansion that we obtained from the Dyson expansion and Wick's theorem. The only difference is that the terms in the expansion are simpler because we are in euclidean space — there are no factors of i to keep track of. All we need is to note that, for $V = \frac{1}{4!}\lambda\varphi^4$, for example, that there is a factor of $-\lambda$ for each vertex.

Using the euclidean path integral allows us to address the issue of the convergence or otherwise of the perturbation series. As a toy model for this consider the zero dimensional "field theory" defined by the ordinary integral

$$\mathcal{Z}(\lambda) = \int_{-\infty}^{\infty} dx \, e^{-x^2 - \lambda x^4}. \tag{12.83}$$

12.6 Perturbation Theory   157

Let us expand out the nonlinear term in the exponent and integrate term by term. We find

$$\mathcal{Z}(\lambda) = \int_{-\infty}^{\infty} dx \, e^{-x^2 - \lambda x^4}$$

$$= \int_{-\infty}^{\infty} dx \sum_{n=0}^{\infty} \frac{(-\lambda)^n}{n!} x^{4n} e^{-x^2}$$

$$\stackrel{?}{=} \sum_{n=0}^{\infty} (-1)^n \lambda^n \frac{1}{n!} \Gamma(2n + \frac{1}{2}). \tag{12.84}$$

Something is not right here. The warning signal for the error is seen in the growth of the expansion coefficients at large n. Since $\Gamma(2n + \frac{1}{2})/n! \approx 4^n n!$, the radius of convergence of the perturbation expansion is zero. Reviewing our manipulations, we see that the power series in λ fails to converge uniformly on the domain of integration, so the interchange of sum and integral in the last line cannot be justified.

It is not hard to find a more informative reason for the divergence of the perturbation series. The radius of convergence of any power series is given by the distance from the origin of the expansion to the nearest singularity of the function that is being expanded. In the case of our $\mathcal{Z}(\lambda)$, that nearest singularity is located at $\lambda = 0$. Although $\mathcal{Z}(\lambda)$ is an analytic function with no singularities in the half-plane $\text{Re } \lambda > 0$, it is clear that the integral diverges once λ becomes negative. We can define $\mathcal{Z}(\lambda)$ for $\text{Re } \lambda < 0$ by analytic continuation, but the resulting function has a branch-cut running along the negative real axis. The origin is a branch point therefore, and no expansion about that point can be convergent.

This is a generic problem in field theory. For example, $\lambda \varphi^4$ theory has an unstable vacuum when $\lambda < 0$. This can be seen either from realizing that the φ field can tunnel from the neigborhood $\varphi \approx 0$ to regions where φ is large and has arbitrarily negative energy, or equivalently by realizing that the attractive force between the φ particles implied by the negative coupling constant means that a bound state of n particles has binding energy that goes as n^2. Thus, for large enough n, the binding energy of an n-particle composite exceeds the total rest mass of the particles that compose it. The vacuum will spontaneously decay into these bound states. Decay implies an imaginary part for the ground state energy, and this implies a branch cut in $\mathcal{Z}$ at $\lambda = 0$. Similarly for QED with $e^2 < 0$. Once $e^2 < 0$, like particles attract and unlike repel. This makes the vacuum unstable to decay into electron-positron pairs. Again there will be branch cut at $e^2 = 0$.

All is not lost however. Watson's lemma guarantees that the series (12.84) is at least an *asymptotic expansion* for $\mathcal{Z}(\lambda)$. In other words, if we keep a fixed number of terms in the series and then take $\lambda \to 0$, the error in truncating the series is of the order of the first term omitted, and this may be small. It is hoped, but usually not known for certain, that field-theory perturbation series are also asymptotic expansions. It must, however, be borne in mind that many different functions may have the same asymptotic expansion. The perturbation series does not therefore serve to define the theory.

13

Functional Methods

In this chapter we will introduce some useful functionals that are analogous to thermodynamic potentials. We will use them to investigate spontaneous symmetry breaking and Goldstone's theorem.

13.1 Generating Functionals

Starting from the action

$$S[\varphi] = \int d^d x \left\{ \frac{1}{2}(\partial\varphi)^2 + \frac{m^2}{2}\varphi^2 + V(\varphi) + J(x)\varphi(x) \right\}, \tag{13.1}$$

define the quantities $\mathcal{Z}(J)$ and $W(J)$ by

$$\mathcal{Z}(J) = e^{-W(J)} = \int d[\varphi] e^{-S[\varphi] - \int d^d x\, J(x)\varphi(x)}. \tag{13.2}$$

In these expressions we should think of $J(x)$ as playing the role of probe that we use to extract the Green functions. Writing $\langle \hat{\varphi}(x)\hat{\varphi}(y) \rangle$ as shorthand for $\langle 0|T\{\hat{\varphi}(x)\hat{\varphi}(y)\}|0\rangle$ etc., we have

$$\langle \hat{\varphi}(x) \rangle = -\frac{1}{\mathcal{Z}}\frac{\delta \mathcal{Z}}{\delta J(x)}\bigg|_{J=0} = \frac{\delta W}{\delta J(x)}\bigg|_{J=0}. \tag{13.3}$$

The general n-point Green function is given by

$$\langle \hat{\varphi}(x_1) \ldots \hat{\varphi}(x_n) \rangle = (-1)^n \frac{\delta^n \mathcal{Z}}{\delta J(x_1) \ldots \delta J(x_n)}\bigg|_{J=0}. \tag{13.4}$$

13.1 Generating Functionals

We can regard $\mathcal{Z}(J)$ as a *generating functional* for the Green functions and write

$$\mathcal{Z}(J) = \sum_{n=0}^{\infty} \frac{(-1)^n}{n!} \int dx_1 \ldots dx_n \, \langle \hat{\varphi}(x_1) \ldots \hat{\varphi}(x_n) \rangle J(x_1) \ldots J(x_n). \tag{13.5}$$

Similarly $W(J)$ provides a generating functional for the *connected Green functions*. This follows from the fact, established in Chapter 4, that the perturbation expansion for W contains only connected diagrams.

We can confirm this by considering

$$-\frac{\delta^2 W}{\delta J(x)\delta J(x)} = \langle \hat{\varphi}(x)\hat{\varphi}(y) \rangle - \langle \hat{\varphi}(x) \rangle \langle \hat{\varphi}(y) \rangle \stackrel{def}{=} \langle \hat{\varphi}(x)\hat{\varphi}(y) \rangle_{con}. \tag{13.6}$$

Similarly,

$$\langle \hat{\varphi}(x)\hat{\varphi}(y)\hat{\varphi}(z) \rangle_{con} = \frac{\delta^3 W}{\delta J(x)\delta J(x)\delta J(z)}\bigg|_{J=0}$$
$$= \langle \hat{\varphi}(x)\hat{\varphi}(y)\hat{\varphi}(z) \rangle - \langle \hat{\varphi}(x)\hat{\varphi}(y) \rangle_{con}\langle \hat{\varphi}(z) \rangle - \langle \hat{\varphi}(y)\hat{\varphi}(z) \rangle_{con}\langle \hat{\varphi}(x) \rangle$$
$$- \langle \hat{\varphi}(z)\hat{\varphi}(x) \rangle_{con}\langle \hat{\varphi}(y) \rangle - \langle \hat{\varphi}(x) \rangle\langle \hat{\varphi}(y) \rangle\langle \hat{\varphi}(z) \rangle. \tag{13.7}$$

In each case the terms with minus signs remove the disconnected pieces from the correlation function.

The connected n-point function is given by

$$(-1)^{n+1}\frac{\delta^n W(J)}{\delta J(x_1)\ldots\delta J(x_n)} = \langle \hat{\varphi}(x_1)\ldots\hat{\varphi}(x_n) \rangle_{con} \tag{13.8}$$

and

$$W(J) = \sum_{n=0}^{\infty} \frac{(-1)^{n+1}}{n!} \int dx_1 \ldots dx_n \, \langle \hat{\varphi}(x_1) \ldots \hat{\varphi}(x_n) \rangle_{con} J(x_1) \ldots J(x_n). \tag{13.9}$$

The connected n-point functions are sometimes called *cumulants* after similar quantities in statistics.

To get some insight into the diagrammatic expansion of $W(J)$, consider the purely classical problem of determining φ from J. We would have to solve the equation

$$(-\partial^2 + m^2)\varphi(x) + \frac{\lambda}{3!}\varphi^3(x) + J(x) = 0. \tag{13.10}$$

At lowest order we would simply ignore the nonlinear term and write

$$\varphi(x) = -\int G(x, y)J(y)d^d y + O(\lambda), \tag{13.11}$$

where $G(x, y) = (-\partial^2 + m^2)^{-1}_{xy}$ is, as usual, the propagator. Diagramatically

$$x \bullet \!\!\!\!\!-\!\!\!-\!\!\!-\!\!\!-\!\!\!-\!\!\!-\!\!\!-\!\!\!-\!\!\!-\!\!\!\times J(y)$$

13. Functional Methods

This would come from functionally differentiating

$$W(J)^{(0)}_{classical} = -\frac{1}{2} \int J(x) G(x,y) J(y) d^d x\, d^d y, \tag{13.12}$$

with respect to $J(x)$ We represent this expression by the diagram

$$J(x) \times\!\!\text{------}\!\!\times\, J(y)$$

where the crosses represent the location of the source J. At next order we would take into account the first-order value of φ in the nonlinear term to obtain

$$\varphi(x) = -\int G(x,y) J(y) d^d y$$
$$+ \frac{\lambda}{3!} \int d^d y\, d^d x_1 d^d x_2 d^d x_3 G(x,y) G(y,x_1) G(y,x_2) G(y,x_3) + O(\lambda^2). \tag{13.13}$$

Diagrammatically, the second term is

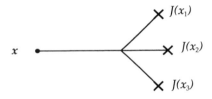

This comes from functionally differentiating

$$W(J)^{(1)}_{clasical} = \frac{\lambda}{4!} \int d^d y\, d^d x_1 d^d x_2 d^d x_3 d^d x_4 J(x_1) J(x_2) J(x_3) J(x_4)$$
$$\times G(y,x_1) G(y,x_2) G(y,x_3) G(y,x_4), \tag{13.14}$$

which we represent as

13.1 Generating Functionals

If we continue, we find that $W(J)_{classical}$ is given by the sum over all *tree diagrams* with $J(x)$ factors at the end of each branch. These same tree diagrams appear as a subset of the terms appearing in the full quantum-mechanical $W(J)$, but the quantum perturbation expansion also contains diagrams with closed loops. Closed loops are therefore characteristic of quantum corrections.

13.1.1 Effective Action

As in thermodynamics, it is often convenient to make a Legendre transformation which interchanges the role of $\langle \hat{\varphi}(x) \rangle$ and $J(x)$. We define a new functional

$$\Gamma(\varphi_c) = W(J) - \int d^d x\, J(x) \varphi_c(x), \tag{13.15}$$

where $\varphi_c = \langle \hat{\varphi} \rangle$. $\Gamma(\varphi_c)$ is regarded as a function of φ_c in the same way that the hamiltonian $H = p\dot{q} - L$ is regarded as function of p, and not as a function of $\dot{q}$.
The source $J(x)$ can be recovered from $\Gamma(\varphi_c)$ by noting that

$$\frac{\delta \Gamma}{\delta \varphi_c(y)} = \frac{\delta W}{\delta \varphi_c(y)} - \int d^d x \frac{\delta J(x)}{\delta \varphi_c(y)} \varphi_c(x) - \int d^d x \frac{\delta \varphi_c(x)}{\delta \varphi_c(y)} J(x)$$

$$= \int d^d x \frac{\delta W}{\delta J(x)} \frac{\delta J(x)}{\delta \varphi_c(y)} - \int d^d x \frac{\delta J(x)}{\delta \varphi_c(y)} \varphi_c(x) - J(y)$$

$$= -J(y). \tag{13.16}$$

In the penultimate line we used

$$\frac{\delta W}{\delta J(x)} = \langle \hat{\varphi}(x) \rangle = \varphi_c. \tag{13.17}$$

The quantity $\Gamma(\varphi_c)$ is called the *effective action*. This is because it plays exactly the same role in determining the exact $\varphi_c(x)$ via the equation

$$\frac{\delta \Gamma(\varphi_c)}{\delta \varphi_c(x)} + J(x) = 0 \tag{13.18}$$

as the classical action does in determining the classical value of $\varphi(x)$ through the action principle

$$\frac{\delta S[\varphi]}{\delta \varphi(x)} + J(x) = 0. \tag{13.19}$$

Effective Potential

We already know that, when J is a constant, $W(J)$ has the physical interpretation as $(Vol)(Time)\mathcal{E}_0(J)$, where $\mathcal{E}_0(J)$ is the ground-state energy density in the theory with the source J. When φ_c is constant, $\Gamma(\varphi_c)$ has a similar energy-density interpretation. To see this let

$$\mathcal{V}(\varphi_c) = \frac{1}{(Vol)} \inf \langle \Psi | \hat{H} | \Psi \rangle, \tag{13.20}$$

where the infimum is taken over all normalized states $|\Psi\rangle$, subject to the condition

$$\frac{1}{(Vol)} \int d^3x \langle \Psi | \hat{\varphi} | \Psi \rangle = \varphi_c. \tag{13.21}$$

By introducing Lagrange multipliers E and J to impose the constraints, we can find an equation for the state $|\Psi\rangle$ that minimizes $\langle \Psi | \hat{H} | \Psi \rangle$ subject to these conditions. We find that stationarity under variations of $|\Psi\rangle$ requires

$$\left(\hat{H} + J \int d^3x \, \hat{\varphi} - E \right) |\Psi\rangle = 0. \tag{13.22}$$

The minimum is therefore achieved if $|\Psi\rangle$ is chosen to be the eigenstate of $\hat{H} + J \int d^3x \, \hat{\varphi}$ corresponding to the lowest-energy $E = E_0(J)$. The value of J must be chosen so that (13.21) is satisfied. Since the Lagrange multiplier J is independent of position, we expect the corresponding φ_c to be a constant. If it were not, then there would be a spontaneous breakdown of translational invariance; barring this, we can write

$$\mathcal{V}(\varphi_c) = \mathcal{E}_0(J) - J\varphi_c. \tag{13.23}$$

Given the connection between W and $\mathcal{E}_0$, this implies that

$$(Vol)(Time)\mathcal{V}(\varphi_c) = \Gamma(\varphi_c). \tag{13.24}$$

The function $\mathcal{V}(\varphi_c)$ is called the *effective potential* of the theory. Since

$$\frac{\partial \mathcal{V}}{\partial \varphi_c} + J = 0, \tag{13.25}$$

the stationary points of $\mathcal{V}(\varphi_c)$ will tell us possible values of $\langle \hat{\varphi} \rangle$ in the theory with the probe J set to zero.

Spontaneous Symmetry Breaking

As described in Chapter 10, spontaneous symmetry breaking is said to occur when the ground state (vacuum) is not invariant under the action of some symmetry group. This is usually signalled by the appearance of a non-zero vacuum expectation value (VEV) for a field that transforms nontrivially under the group. For example, if one starts with a classical potential that is invariant under the discrete transformation $\varphi \to -\varphi$, or a potential that is a function only of $|\varphi|$, then a nonzero value for $\langle \hat{\varphi} \rangle$ indicates that the symmetry is broken.

Spontaneous symmetry breaking will occur at the classical level if one starts with a potential whose minima are away from the origin. Examples are a $\varphi \to -\varphi$ symmetric double-well, or a rotationally invariant Mexican-hat potential. Quantum corrections will of course alter the value of $\langle \hat{\varphi} \rangle$ in the true ground state, and may even restore the broken symmetry.[1]

---

[1] This latter effect will always occur in simple quantum-mechanical systems since the ground state is always nondegenerate. To illustrate this point consider a particle of mass

One should be a little careful when using the *effective potential* to explore these effects. We have

$$\frac{\partial^2 \mathcal{E}_0}{\partial J^2} = -\langle \hat{\varphi}^2 - \varphi_c^2 \rangle$$
$$= -\langle (\hat{\varphi} - \varphi_c)^2 \rangle. \tag{13.26}$$

When we evaluate $\langle (\hat{\varphi} - \varphi_c)^2 \rangle$ via the path integral, the integrand is strictly positive. Indeed e^{-S} is positive, as is $(\varphi - \varphi_c)^2$. The last line in (13.26) is therefore guaranteed to be negative. Thus

$$\frac{\partial^2 \mathcal{E}_0}{\partial J^2} < 0. \tag{13.27}$$

This means that the curve obtained by plotting $\mathcal{E}_0$ against J is everywhere *concave*. Since

$$\frac{\partial^2 \mathcal{E}_0}{\partial J^2} = \frac{\partial \varphi_c}{\partial J} < 0, \tag{13.28}$$

we have

$$\frac{\partial^2 \mathcal{V}}{\partial \varphi_c^2} \equiv -\frac{\partial J}{\partial \varphi_c} > 0. \tag{13.29}$$

The corresponding curve of $\mathcal{V}(\varphi_c)$ against φ_c is therefore everywhere *convex*. This convexity property is often violated in a perturbative expansion for Γ, but must hold true for the full expression for $\mathcal{V}(\varphi_c)$.

The full effective potential can never have a double well therefore. The reason for this is not hard to find. If we have a vacuum state $|\varphi_c\rangle$ with $\langle \hat{\varphi} \rangle = \varphi_c$, then there must exist a state $|-\varphi_c\rangle$ with the same energy and $\langle \hat{\varphi} \rangle = -\varphi_c$. The state

$$|\alpha\rangle = \sqrt{\alpha}|\varphi_c\rangle + \sqrt{1-\alpha}|-\varphi_c\rangle, \tag{13.30}$$

with $0 \leq \alpha \leq 1$, is another possible vacuum with $\langle \alpha | \varphi | \alpha \rangle = (2\alpha - 1)\varphi_c$. Any number between the two extremes, $\pm\varphi_c$, is therefore the VEV for some vacuum state. Even at the classical level, in order to construct the "true" effective potential from a given double-well potential, we must construct the *convex hull* of the potential by joining the two minima by a horizontal line, and discarding the barrier between the two wells. In statistical mechanics, where $\mathcal{E}_0(J)$ corresponds to the Helmholtz free energy density, and the effective potential to the Gibbs free energy density, this yields the "Maxwell construction" for mixed phases. In this language, the thermodynamic states beween the extrema correspond to "mixed phases" (a mixture of ice and water, say) while the extrema are "pure phases" (only ice or only water). Only the pure phases obey the cluster decomposition property.

M free to move on a circle. The eigenfunctions are $\langle \theta | m \rangle = e^{im\theta}$ corresponding to $E_m = m^2/2M$. A state corresponding to the particle sitting at a definite angle $|\theta\rangle = \sum e^{-im\theta}|m\rangle$ can only be an eigenstate if $M = \infty$. Similarly, tunneling will restore the symmetry in the discrete double-well case.

Diagram Expansion of $\Gamma(\varphi_c)$

When we expand Γ out in powers of φ_c

$$\Gamma(\varphi_c) = \sum_{n=0}^{\infty} \frac{1}{n!} \int dx_1 \ldots dx_n \, \Gamma^{(n)}(x_1, \ldots, x_n) \varphi_c(x_1) \ldots \varphi_c(x_n), \tag{13.31}$$

the coefficients $\Gamma^{(n)}$ turn out to be given by *one-particle irreducible* (1PI) graphs. These are diagrams that cannot be separated into two pieces by cutting a single line. Also the propagators on the external legs have been amputated.

The perturbation expansion of $\Gamma(\varphi_c)$ or $V(\varphi_c)$ is best derived by setting $\varphi(x) = \varphi_c(x) + \eta(x)$ and adjusting the source $J(x)$ to make $\langle \hat{\eta}(x) \rangle = 0$. For simplicity let us discuss the expansion of $V(\varphi_c)$. Here φ_c is a constant so the shift only affects the polynomial part of the action:

$$\frac{1}{2}\varphi^2 + \frac{\lambda}{4!}\varphi^4 + J\varphi$$
$$\to \frac{1}{2}\varphi_c^2 + \frac{\lambda}{4!}\varphi_c^4 + J\varphi_c + \eta(J + m^2 + \frac{\lambda}{3!}\varphi_c^3)$$
$$+ \frac{1}{2}\eta^2(m^2 + \frac{\lambda}{2}\varphi_c^2) + \frac{\lambda}{3!}\eta^3\varphi_c + \frac{\lambda}{4!}\eta^4. \tag{13.32}$$

The shifted theory therefore has a φ_c dependent mass term and contains η^3 vertices as well as the original quartic interactions.

The diagrams contributing to $W(J)$ for the shifted theory can be arranged organized in families such as

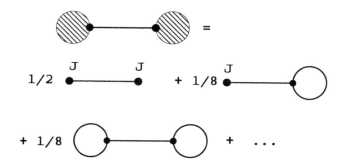

All propagators that would disconnect the graph were they cut are explicitly shown, and the others are lumped together into the "blobs."

The sources $J(x)$ occur only in blobs terminating lines, and then only this form $J(x) + \Gamma_1(x)$ where $\Gamma_1(x)$ is the sum of all one particle irreducible graphs with a marked point. If we select J so that $0 = J(x) + \Gamma_1(x)$, all graphs that are one-particle reducible vanish. A little thought shows that this choice of J is exactly what is needed to make $\langle \eta \rangle = 0$, and so

$$\Gamma(\varphi_c) = S[\varphi_c] + \frac{1}{2}\ln\det\left(\frac{\delta^2 S}{\delta\varphi(x)\delta\varphi(y)}\right) + \text{sum of 1PI graphs.} \tag{13.33}$$

The first term in this expression is simply the classical action that appears in the path integral while the second is the Ouroboros graph of the shifted action. When

$$S[\varphi] = \int_{-\infty}^{\infty} d^d x \left\{\frac{1}{2}(\partial_\mu \varphi)^2 + \frac{1}{2}m^2\varphi^2 + \frac{\lambda}{4!}\varphi^4\right\}, \tag{13.34}$$

and φ_c is chosen to be independent of space and time, the Ouroboros graph contributes

$$+\frac{1}{2}\int \frac{d^d k}{(2\pi)^d}\ln(k^2 + m^2 + \frac{\lambda}{2}\varphi_c^2) \tag{13.35}$$

to $\mathcal{V}(\varphi_c)$. As we know, this is equal to the sum of the $\frac{1}{2}\hbar\omega$'s from the zero-point motion of field modes in the theory with mass^2 $m^2 + \frac{\lambda}{2}\varphi_c^2$.

Since the propagator in these graphs is the inverse of

$$-\partial^2 + m^2 + \frac{1}{2}\lambda\varphi_c^2, \tag{13.36}$$

λ appears in the denominator of every propagator. The diagram-by-diagram expansion is therefore no longer an expansion in powers of λ. It is best to organize the expansion by counting the number of loops in the graphs. This is equivalent to an expansion in powers of $\hbar$. Restoring $\hbar$ makes $L \to \hbar^{-1}L$, so each propagator is proportional to $\hbar$ while each vertex has a factor of $\hbar^{-1}$. A diagram with V vertices and P propagators is therefore proportional to $\hbar^{P-V}$. The number of loops in the diagram, however, is $P - V + 1$. Now with the $\hbar$'s included $W(J) = (Vol)(Time)\mathcal{E}_0/\hbar$. Taking this into account we see that the $S[\varphi_c]$ contribution to $\mathcal{V}(\varphi_c)$ is of order $O(\hbar^0)$, while the Ouroboros graph is $O(\hbar)$. We should not think of $\hbar$ as a small quantity — it is not dimensionless and is indeed equal to one equal to 1 in our units, but the loop expansion is unaffected by shifts of fields and by any re-assignment of terms in the lagrangian to the free and interacting parts.

To 1-loop accuracy,

$$\mathcal{V}(\varphi_c) = V(\varphi_c) + \frac{1}{2}\int \frac{d^d k}{(2\pi)^d}\ln(k^2 + m^2 + \frac{\lambda}{2}\varphi_c^2). \tag{13.37}$$

This expression is not convex. The only hint of the convexity problem is that the 1-loop term has an imaginary part when φ_c takes values where the effective mass-squared, $m^2 + \frac{\lambda}{2}\varphi_c^2$, is negative. A more careful calculation, taking into account non-perturbative vacuum decay effects would reveal an exponentially small imaginary part for all values of φ_c where the convexity inequality is violated.

13.2 Ward Identities

Recall how Noether's theorem works for complex fields with

$$S = \int d^d x \left\{ \partial_\mu \varphi^* \partial_\mu \varphi + m^2 |\varphi|^2 + V(|\varphi|^2) \right\}. \tag{13.38}$$

We observe that S is unchanged when we make the replacement

$$\varphi \to e^{i\alpha} \varphi, \qquad \varphi^* \to e^{-i\alpha} \varphi^*, \tag{13.39}$$

with α is independent of x. When we permit α to be a function of x we find that

$$\delta S = \int d^d x \, \partial_\mu \alpha(x) \left\{ \frac{1}{i} [\varphi^* \partial_\mu \varphi - \varphi \partial_\mu \varphi^*] \right\}. \tag{13.40}$$

This means that

$$\frac{\delta S}{\delta \alpha(x)} = -\partial_\mu \left\{ \frac{1}{i} [\varphi^* \partial_\mu \varphi - \varphi \partial_\mu \varphi^*] \right\} = -\partial_\mu j_\mu = 0, \tag{13.41}$$

where the last equality is a consequence of δS being zero for *any* variation — including, in particular, the variation (13.39) — provided the fields satisfy the classical equations of motion. Now the fields being integrated over in the path integral do *not* in general satisfy the equations of motion, but a similar strategy of x dependent rotations will produce useful consequences.

Consider the path integral

$$I = \int d[\varphi^*] d[\varphi] \left\{ \varphi(x_1) \ldots \varphi(x_n) \varphi^*(y_1) \ldots \varphi^*(y_m) \right\} e^{-S[\varphi, \varphi^*]}, \tag{13.42}$$

and regard the transformation (13.39) as a change of variables. Bear in mind that the value of an integral is not altered by a change of variables. Now note that the measure $d[\varphi^*] d[\varphi]$ is also unchanged by (13.39) — even when α is x dependent. Putting together these two observations, we see that under the variation

$$\varphi(x) \to \varphi(x)(1 + i\delta\alpha(x)), \qquad \varphi^*(x) \to \varphi^*(x)(1 - i\delta\alpha(x)), \tag{13.43}$$

we must have

$$0 = \delta I = \int d[\varphi^*] d[\varphi] \left(\delta \left\{ \varphi(x_1) \ldots \varphi^*(y_m) \right\} e^{-S} + \left\{ \varphi(x_1) \ldots \varphi^*(y_m) \right\} \delta e^{-S} \right)$$

$$= \int d[\varphi^*] d[\varphi] \left(\sum_{i=1}^{n} \int i\delta\alpha(x) \delta(x - x_i) \left\{ \varphi(x_1) \ldots \varphi^*(y_m) \right\} dx \right.$$

$$\left. - \sum_{i=1}^{m} \int i\delta\alpha(x) \delta(x - y_i) \left\{ \varphi(x_1) \ldots \varphi^*(y_m) \right\} dx \right) e^{-S}$$

$$+ \int d[\varphi^*] d[\varphi] \left\{ \varphi(x_1) \ldots \varphi^*(y_m) \int dx \, \delta\alpha(x) \partial_\mu j_\mu(x) \right\} e^{-S} \tag{13.44}$$

for any $\delta\alpha(x)$. The coefficient of $\delta\alpha(x)$ must therefore be zero.

Identifying this coefficient, we find that we have proved that

$$i \sum_{i=1}^{n} \delta(x - x_i) \langle 0| T \left\{ \hat{\varphi}(x_1) \ldots \hat{\varphi}^\dagger(y_m) \right\} |0\rangle$$

$$- i \sum_{i=1}^{m} \delta(x - y_i) \langle 0| T \left\{ \hat{\varphi}(x_1) \ldots \hat{\varphi}^\dagger(y_m) \right\} |0\rangle$$

$$= -\partial_\mu \langle 0| T \left\{ \hat{j}_\mu(x) \hat{\varphi}(x_1) \ldots \hat{\varphi}^\dagger(y_m) \right\} |0\rangle. \quad (13.45)$$

This is similar to the Ward identity we derived in Chapter 8 by the use of commutators. We may establish analogous identities for any continuous symmetry.

Note that the ∂_μ lies *outside* the time ordering. Our theorem relating the path integral to the vacuum expectation value of time-ordered products applies to polynomials in $\hat{\varphi}$ only. It does not apply to operators containing $\pi(x)$, the field canonically conjugate to φ. In particular, we must write

$$\frac{1}{\mathcal{Z}} \int d[\varphi] \left\{ \left(\partial_{x_0} \varphi(x) \right) \varphi(y) \right\} e^{-S} = \frac{1}{\mathcal{Z}} \partial_{x_0} \int d[\varphi] \left\{ \varphi(x) \varphi(y) \right\} e^{-S}$$

$$= \partial_{x_0} \langle 0| T \{ \hat{\varphi}(x) \hat{\varphi}(y) \} |0\rangle. \quad (13.46)$$

A shorthand notation for the Ward identity is

$$\partial_\mu \langle 0| T \left\{ \hat{j}_\mu(x) \hat{\varphi}(x_1) \ldots \hat{\varphi}^\dagger(y_m) \right\} |0\rangle = -\delta \langle 0| T \left\{ \hat{\varphi}(x_1) \ldots \hat{\varphi}^\dagger(y_m) \right\} |0\rangle. \quad (13.47)$$

13.2.1 Goldstone's Theorem

If we integrate (13.45) over a region Ω that contains all the x_i, y_i, we find

$$\int_{\partial\Omega} dS^\mu \langle 0| T \left\{ \hat{j}_\mu(x) \hat{\varphi}(x_1) \ldots \hat{\varphi}^\dagger(y_m) \right\} |0\rangle$$

$$= (n - m) \langle 0| T \left\{ \hat{\varphi}(x_1) \ldots \hat{\varphi}^\dagger(y_m) \right\} |0\rangle, \quad (13.48)$$

where dS^μ is the surface element with outward normal on the boundary, $\partial\Omega$, of the region Ω and n, m, are the number of $\hat{\varphi}(x)$'s and $\hat{\varphi}^\dagger(y)$'s, respectively. Under conditions where we can ignore the boundary term (for example, if space-time is a sphere, or if the boundary is at infinity, all particles are massive), then we must have $n = m$, or else $\langle 0| T \left\{ \hat{\varphi}(x_1) \ldots \hat{\varphi}^\dagger(y_m) \right\} |0\rangle = 0$.

An important circumstance where we cannot ignore the boundary terms occurs when the continous symmetry that gives rise to the Ward identity is spontaneously broken and $\langle 0|\hat{\varphi}|0\rangle \neq 0$. Then we will have a nonvanishing right-hand side to (13.48) even when $n \neq m$. Therefore we must have a nonvanishing contribution from the boundary term on the LHS, however far away it is. This contribution is possible because the spontaneous breakdown of a continuous symmetry produces massless particles — *Goldstone bosons* — which make the system very sensitive to boundary conditions.

If space-time were large, but finite, with no boundary (a sphere or torus, for example) the Ward identity tells us that $\langle \hat{\varphi} \rangle$ ought to be zero. This is indeed what the path integral gives us, since the sum over all space-time field configurations includes summing over all possible directions of symmetry breaking, and so φ averages to zero. In the hamiltonian language, the sum over periodic boundary conditions in the time direction means that we are taking a trace over all states in the Hilbert space, and again $\langle\hat\varphi\rangle$ averages to zero. The symmetry may still be broken however. We are simply computing Green functions that do not satisfy cluster decompostion. We can detect the spontaneous symetry breaking by introducing a perturbation that *explicitly* breaks the symmetry. Then spontaneous symmetry breaking will be signalled by the failure to commute of the large volume limit and the removal of the explicit symmetry breaking term.

Let us consider an explicit example. Let us break the field φ into its real and imaginary components $\varphi = \varphi_1 + i\varphi_2$, and apply the Ward identity to

$$\langle\hat\varphi_a(y)\rangle = \frac{1}{Z}\int d[\varphi_1]d[\varphi_2]\varphi_a(y)e^{-S[\varphi_1,\varphi_2]}, \tag{13.49}$$

where

$$S[\varphi_1,\varphi_2] = \int d^dx \left\{\frac{1}{2}(\partial\varphi_1)^2 + \frac{1}{2}(\partial\varphi_2)^2 + \frac{\lambda}{2}(\varphi_1^2 + \varphi_2^2 - \kappa)^2 - \epsilon\varphi_1\right\}. \tag{13.50}$$

The form of the potential we have chosen is an example of a "Mexican hat" potential, which, for $\epsilon = 0$, has a degenerate classical minimum at any point on the circle $\varphi_1^2 + \varphi_2^2 = \kappa$. The term $-\epsilon\varphi_1$ serves to explicitly break the rotational symmetry and select a point near $\varphi_1 = \kappa$, $\varphi_2 = 0$ as the absolute classical minimum. Quantum fluctuations will modify the exact value of $\langle\hat\varphi_1\rangle$ but should not affect $\langle\hat\varphi_2\rangle = 0$. Classically, in the limit $\epsilon \to 0$, there will be a gapless mode corresponding to vibrations of the field along the floor of the potential valley. If this mode remains gapless after quantization, the associated particle will be the Goldstone boson of the spontaneously broken symmetry. The purely radial mode will remain massive.[2] We will use the Ward identity to show that broken symmetry does indeed always give rise to a Goldstone boson.

Under the rotation

$$\delta\varphi_1(x) = -\alpha(x)\varphi_2(y), \qquad \delta\varphi_2(x) = \alpha(x)\varphi_1(y), \tag{13.51}$$

we have

$$\delta S = \int d^dx \left\{\partial_\mu\alpha(\varphi_1\partial_\mu\varphi_2 - \varphi_2\partial_\mu\varphi_1) + \alpha\epsilon\varphi_2\right\}. \tag{13.52}$$

The equation of motion therefore implies

$$\partial_\mu(\varphi_1\partial_\mu\varphi_2 - \varphi_2\partial_\mu\varphi_1) = \epsilon\varphi_2. \tag{13.53}$$

When $\epsilon = 0$, the current

$$j_\mu(x) = (\varphi_1(x)\partial_\mu\varphi_2(x) - \varphi_2(x)\partial_\mu\varphi_1(x)) \tag{13.54}$$

---

[2] Note the difference between the relativistically invariant sytem we are considering here and the nonrelativistic Bose system we considered in chapter 12. The latter does not have distinct radial and tangential modes.

is conserved.

Inserting the variation into the path integral for $\langle\varphi_2\rangle$ gives the Ward identity

$$\int d^d x \alpha(x) \left\{ \delta^d(x-y)\langle\hat\varphi_1\rangle + \partial_\mu \langle \hat\jmath_\mu(x)\hat\varphi_2(y)\rangle - \epsilon \langle \hat\varphi_2(x)\hat\varphi_2(y)\rangle \right\} = 0. \quad (13.55)$$

There are some immediate consequences of this.

If we keep $\epsilon \neq 0$, then all particles should be massive and we can ignore the boundary terms when we set $\alpha = 1$. We then find for the zero-momentum Fourier component of $\langle\hat\varphi_2\hat\varphi_2\rangle$

$$\langle\hat\varphi_2\hat\varphi_2\rangle_{p=0} = \frac{1}{\epsilon}\langle\hat\varphi_1\rangle. \quad (13.56)$$

This must be true whether or not the symmetry is broken. If the symmetry is broken, the LHS is singular as $\epsilon \to 0$.

To understand the nature of this singularity, we make the approximation of replacing $\hat\varphi_1$ by its vacuum expectation value $\langle\hat\varphi_1\rangle$. Then we find that

$$\delta^d(x-y)\langle\hat\varphi_1\rangle + \langle\hat\varphi_1\rangle\partial_x^2 \langle\hat\varphi_2(x)\hat\varphi_2(y)\rangle - \epsilon \langle\hat\varphi_2(x)\hat\varphi_2(y)\rangle = 0. \quad (13.57)$$

In other words $\langle\hat\varphi_2(x)\hat\varphi_2(y)\rangle$ satisfies the free Klein-Gordon equation for $m^2 = \frac{1}{\epsilon}\langle\hat\varphi_1\rangle$ and

$$\langle\hat\varphi_2\hat\varphi_2\rangle_p = \frac{1}{p^2 + \frac{\epsilon}{\langle\hat\varphi_1\rangle}}. \quad (13.58)$$

We see that the particle created by the $\hat\varphi_2$ field becomes massless as $\epsilon \to 0$. This is the Goldstone boson.

Jeffrey Goldstone gave a formal argument for the existence of his boson that does not depend on the uncontrolled approximation of replacing $\hat\varphi_1$ by its vacuum expectation value. He considered the two-point function $\langle \hat\jmath_\mu(x)\hat\varphi_2(y)\rangle$ in the case when $\epsilon = 0$, and the current is conserved. Suppose that $\hat\jmath_\mu(x)$ couples the vacuum to some state with momentum p. Lorentz invariance requires

$$\langle p|\hat\jmath_\mu(x)|0\rangle = N p_\mu e^{ipx}. \quad (13.59)$$

When the state is massive ($p^2 \neq 0$), the operator equation $\partial_\mu \hat\jmath_\mu(x) = 0$ requires that $N = 0$. When the state is massless ($p^2 = 0$), however, N may be nonzero. Let us show that $\langle\hat\varphi_1\rangle \neq 0$ *requires* such a coupling to exist. We write a Lehmann expansion for $\langle \hat\jmath_\mu(x)\hat\varphi_2(y)\rangle$:

$$\langle \hat\jmath_\mu(x)\hat\varphi_2(y)\rangle_p = \int dm^2\, \rho_2(m^2) \frac{p_\mu}{p^2+m^2}. \quad (13.60)$$

Inserting this into (13.55) gives

$$\langle\hat\varphi_1\rangle = \int dm^2\, \rho_2(m^2) \frac{p^2}{p^2+m^2}. \quad (13.61)$$

The left-hand side of this equation is p independent. The only way the right-hand side can agree with this is if $\rho_2(m^2) = \langle\hat\varphi_1\rangle \delta(m^2)$. The spectral function therefore contains only the contribution of a zero-mass state — the Goldstone boson.

It is possible for this argument to fail. Long-range gauge fields can make the limit $p \to 0$ singular. We are familiar with at least one example of this — the plasmon mode in an electron gas. As we saw in Chapter 9, this mode is a sound wave (and thus the Goldstone mode of spontaneously broken translation invariance) that has been pushed up to the plasma frequency by the long-range Coulomb interactions. The dispersion curve for such modes look like $\omega^2(\mathbf{p}) = c_{sound}^2 \mathbf{p}^2 + \omega_p^2$, for all $\mathbf{p} \neq 0$. For *exactly* zero momentum the mode reduces to a global translation of the system. Such a translation produces no restoring force, and so has zero frequency. Similarly, for relativistic systems, the Higgs-Kibble mechanism combines the Goldstone boson with the transverse massless gauge fields to produce a massive vector gauge field. The physics of this is identical with the Meissner effect.

14
Path Integrals for Fermions

In order to have path-integral expressions for Fermi fields it is necessary to integrate over anticommuting c-number fields. The "integrals" we define for such fields are entirely formal objects, but the notation, which is due to Berezin, is remarkable fruitful.

14.1   Berezin Integrals

We first introduce a set of anticommuting symbols called *Grassmann variables* — although they are not really variables. Suppose there are N of them, which we denote by ξ_i. The only properties we require is that they be linearly independent and that

$$\xi_i \xi_j + \xi_j \xi_i = 0. \tag{14.1}$$

We combine the ξ_i with a coefficient field (either **R** or **C**) and form the algebra $\mathcal{A}$ consisting of all sums of products of the ξ_i. A typical element of the algebra would be

$$A = a + a_i \xi_i + \frac{1}{2} a_{ij} \xi_i \xi_j + \frac{1}{3!} a_{ijk} \xi_i \xi_j \xi_k + \ldots, \tag{14.2}$$

where the $a_{ij\ldots}$ are elements of the coefficient field. We assume that they are antisymmetric under interchange of pairs of indices. The sophisticated reader will recognize that we are constructing the exterior algebra of the vector space spanned by the symbols ξ_i. The series (14.2) terminates at the N-th term because any monomial with a repeated ξ_i is equal to minus itself, and therefore zero. Expressions containing only terms with an even number of ξ_i factors commute with all elements of the algebra and are called *even* or *bosonic* elements. Those with odd numbers of factors anticommute with one another, and are said to be *odd* or *fermionic*. In physics contexts we will find ourselves only adding even elements to even elements, and odd elements to odd elements, but this is not a mathematical requirement in the algebra.

We define the "integral" as a linear functional taking elements of the algebra to elements of the coefficient field. In particular, we set

$$\int d\xi_N \ldots d\xi_2 d\xi_1 \, \xi_1 \xi_2 \ldots \xi_N = 1. \tag{14.3}$$

If any of the ξ_i appearing in the measure are absent from the integrand, the integral is defined to be zero. The order of the variables in the integrand matters, of course, and

$$\int d\xi_N \ldots d\xi_2 d\xi_1 (a_{i_1 i_2 \ldots i_N} \xi_{i_1} \xi_{i_2} \ldots \xi_{i_N})$$
$$= \epsilon_{i_1 i_2 \ldots i_N} a_{i_1 i_2 \ldots i_N}$$
$$= N! a_{12 \ldots N}. \tag{14.4}$$

The integral therefore takes in an element of the algebra and returns the coefficient of the "top form." This definition is purely algbraic. There is no real notion of the analytic process of integration — if anything, what we are doing is suggestive of differentiation — but the notation is justified by its mnemonic usefulness.

To deal with Dirac (charged) fermions we double the number of generators in the algebra and define an involution that takes an element ξ_i to an asociated element ξ_i^*, inverts the orders of products, and takes the complex conjugate of the coefficients. The term "involution" means that if we perform the mapping twice, we get back the original element, i.e. $(\xi_i^*)^* = \xi_i$. Despite the similarity of this procedure to the operation of hermitian conjugation in matrix algebra, the variable ξ_i^* should be regarded as being an object quite independent from ξ_i.

Following the rule from above, we have

$$\int d\xi d\xi^* \, e^{\xi^* a \xi} = \int d\xi d\xi^* (1 + \xi^* a \xi) = a. \tag{14.5}$$

The exponential series terminated after the second term because $\xi^2 = (\xi^*)^2 = 0$. More generally (using the symbol $d[\xi]d[\xi^*]$ as shorthand for $d\xi_N d\xi_N^* \ldots d\xi_1 d\xi_1^*$) we have

$$\int d[\xi]d[\xi^*] e^{\xi_i^* A_{ij} \xi_j}$$
$$= \int d[\xi]d[\xi^*] A_{1i_1} A_{2i_2} \ldots A_{N i_N} \xi_1^* \xi_{i_1} \xi_2^* \xi_{i_2} \ldots \xi_N^* \xi_{i_N}$$
$$= \det A. \tag{14.6}$$

Here only the N-th term in the exponential series contributed. Its factor of $1/N!$ was eaten by the $N!$ identical terms we get when we arrange the ξ_i^* factors in their natural order.

14.1 Berezin Integrals

We can also define an exponential integral with "real" ξ_i. If A_{ij} is a $2N \times 2N$ antisymmetric matrix, then

$$\int d[\xi] e^{\frac{1}{2}\xi_i A_{ij} \xi_j}$$

$$\frac{1}{2^N N!} \int d[\xi] A_{i_1 i_2} A_{i_3 i_3} \ldots A_{i_{N2-1} i_{2N}} \xi_{i_1} \xi_{i_2} \ldots \xi_{i_{2N}}$$

$$= \frac{1}{2^N N!} \epsilon_{i_1 i_2 \ldots i_{2N}} A_{i_1 i_2} A_{i_3 i_4} \ldots A_{i_{N2-1} i_{2N}}$$

$$= \text{Pf } A. \tag{14.7}$$

The last two lines serve as the definition of the *Pfaffian* of A. It is easy to prove that for antisymmetric A we have

$$(\text{Pf } A)^2 = \det A. \tag{14.8}$$

One way to establish this is to note that $\text{Pf } A = \text{Pf}(O^T A O)$ where O is an orthogonal matix. Then we note that for nearly all antisymmetric A there exist orthogonal matrices O such that $O^T A O$ is reduced to block diagonal form where the blocks are 2×2 matrices of the form

$$\begin{pmatrix} 0 & -\lambda \\ \lambda & 0 \end{pmatrix}. \tag{14.9}$$

For such matrices, the theorem is immediate.

To complete the analogy with gaussian integration, we should define and evaluate integrals of the form

$$\int d[\xi] e^{\frac{1}{2}\xi_i A_{ij} \xi_j + \eta_i \xi_i}. \tag{14.10}$$

To do this we must embed the algebra in a larger one where the η_i form a set of elements that anticommute with each other and with the ξ_i. They serve as "constants" that will not be integrated over. To evaluate the integral we must complete the square in the exponent and shift the variable of integration. To make the shift we must show that the integral is not affected by it, just as a convergent integral of a real function over $[-\infty, +\infty]$ is unaffected by a shift in the integration variable. No "domain of integration" appears in (14.3), but we see that, if we set $\xi' = \xi + \eta$, then

$$\int d\xi' \, \xi' = 1 = \int d\xi \, (\xi + \eta), \tag{14.11}$$

so the one-dimensional integral is invariant under shifts. The same holds for the general shift $\xi_i \to \xi_i + \eta_i$, since η_i will appear only in monomials lacking ξ_i, and these will integrate to zero.

Since the inverse of an antisymmetric matrix is itself antisymmetric, we can write

$$\int d[\xi] e^{\frac{1}{2}\xi_i A_{ij} \xi_j + \xi_i \eta_i}$$

174   14. Path Integrals for Fermions

$$= \int d[\xi] e^{\frac{1}{2}(\xi_i + A_{ik}^{-1}\eta_k)A_{ij}(\xi_j + A_{jl}^{-1}\eta_l) + \frac{1}{2}\eta_i A_{ij}^{-1}\eta_j}$$

$$= (\text{Pf } A) e^{+\frac{1}{2}\eta_i A_{ij}^{-1}\eta_j} \propto \sqrt{\det A} e^{+\frac{1}{2}\eta_i A_{ij}^{-1}\eta_j}. \tag{14.12}$$

This is exactly like the real x gaussian integral — except that the determinant is in the numerator instead of the denominator.

A related observation is that if we make a linear change of variables $\xi_i' = M_{ij}\xi_j$, the measure changes by

$$d[\xi'] = (\det M)^{-1} d[\xi]. \tag{14.13}$$

The jacobian determinant therefore appears to the *inverse* power, again the opposite of the real-variable measure.

For "complex" Grassmann variables we have

$$\int d[\xi] d[\xi^*] e^{\xi_i^* A_{ij} \xi_j + \xi_i^* \eta_i + \eta_i^* \xi_i}$$

$$= \int d[\xi] d[\xi^*] e^{(\xi_i^* + \eta_k^* A_{ki}^{-1}) A_{ij} (\xi_j + A_{jl}^{-1}\eta_l) - \eta_i^* A_{ij}^{-1}\eta_j}$$

$$= (\det A) e^{-\eta_i^* A_{ij}^{-1}\eta_j}. \tag{14.14}$$

14.1.1 A Simple Supersymmetry

Consider the gaussian integral

$$I = \int_{-\infty}^{\infty} d[\mathcal{F}] e^{-\frac{1}{2} g_{ij} \mathcal{F}^i \mathcal{F}^j}, \tag{14.15}$$

where $d[\mathcal{F}]$ represents $\prod_{i=1}^{N} d\mathcal{F}^i / (2\pi)^{N/2}$. The integral I is of course equal to $\det^{-\frac{1}{2}}|g_{ij}|$. We can, however, disguise this simple formula by setting $\mathcal{F}^i = \mathcal{F}^i(\varphi^\mu)$, where the φ^μ are another set of N variables. Let us assume that $\mathcal{F}(\varphi)$ provides an invertible map $\mathcal{F}: \mathbf{R}^N \to \mathbf{R}^N$. Then the usual change of variables formula gives

$$I = \int_{-\infty}^{\infty} d[\varphi] J(\varphi) e^{-\frac{1}{2} g_{ij} \mathcal{F}^i(\varphi) \mathcal{F}^j(\varphi)}, \tag{14.16}$$

where

$$J = \det\left(\frac{\partial \mathcal{F}^i}{\partial \varphi^\mu}\right) \tag{14.17}$$

is the jacobian determinant of the transformation.

We now introduce two sets of N Grassmann variables, $\bar{\psi}_i$ and ψ^μ, and use them to express the jacobian as a Berezin integral. We find

$$I = \int_{-\infty}^{\infty} d[\varphi] d[\psi] d[\bar{\psi}] e^{\bar{\psi}_i M_\mu^i \psi^\mu - \frac{1}{2} g_{ij} \mathcal{F}^i(\varphi) \mathcal{F}^j(\varphi)}. \tag{14.18}$$

Here

$$M_\mu^i = \frac{\partial \mathcal{F}^i}{\partial \varphi^\mu}.$$

I've made a distinction between Greek and Roman indices because the variables they label are playing rather different roles, and because I want stress that $\bar\psi$ should not be thought of as the complex conjugate of ψ.

Lastly we introduce another set of N commuting variables ω_i, and write

$$I = \det{}^{-\frac{1}{2}}|g_{ij}| \int d[\varphi]d[\psi]d[\bar\psi]d[\omega] e^{\bar\psi_i M^i_\mu \psi^\mu + \frac{1}{2} g^{ij}\omega_i \omega_j - \omega_i \mathcal{F}^i(\varphi)}.$$

Here g^{ij} is the matrix inverse to g_{ij}, and for the integrals to converge, we should integrate the ω_i along the imaginary axis. Cancelling the $\det{}^{-\frac{1}{2}}|g_{ij}|$ from both sides we have

$$1 = \int d[\varphi]d[\psi]d[\bar\psi]d[\omega] e^{\bar\psi_i M^i_\mu \psi^\mu + \frac{1}{2} g^{ij}\omega_i \omega_j - \omega_i \mathcal{F}^i(\varphi)}. \tag{14.19}$$

This seems like a particularly perverse way to write the number 1 — but it is rather remarkable that this, now thoroughly disguised, integral is independent of the form of the functions $\mathcal{F}^i$, and of the matrix g_{ij}.

The integrand possesses another remarkable property. It is invariant under a *supersymmetry* transformation. Define the infinitesimal supersymmetry transformation by

$$\delta\varphi^\mu = \epsilon\psi^\mu, \quad \delta\psi^\mu = 0,$$
$$\delta\bar\psi_i = \epsilon\omega_i, \quad \delta\omega_i = 0. \tag{14.20}$$

The transformation mixes the commuting and anticommuting variables so we have included an anticommuting "infinitesimal" constant ϵ to ensure that we add commuting variations to commuting variables, and anticommuting variations to anticommuting variables.

Now look at the variation of the exponent in the integral

$$\delta\left\{\bar\psi_i M^i_\mu \psi^\mu + \frac{1}{2} g^{ij}\omega_i \omega_j - \omega_i \mathcal{F}^i(\varphi)\right\}$$
$$= \epsilon\left\{\omega_i M^i_\mu \psi^\mu - \bar\psi_i \frac{\partial M^i_\mu}{\partial \varphi^\nu}\psi^\nu \psi^\mu - \omega_i \frac{\partial \mathcal{F}^i}{\partial \varphi^\mu}\psi^\mu\right\}$$
$$= 0. \tag{14.21}$$

(The second term vanishes because of the antisymmetry of the product $\psi^\nu \psi^\mu$ coupled with symmetry of the mixed partial derivative.) The "action" is therefore invariant under the supersymmetry.

This particular supersymmetry is usually called a BRS transformation after Becchi, Rouet, and Stora who introduced it in the context of gauge theories.

Note that with the ϵ included we have $\delta(AB) = (\delta A)B + A(\delta B)$ for any A, B is the algebra. We can avoid the need for the ϵ by defining an operation s by

$$s\varphi^\mu = \psi^\mu, \quad s\psi^\mu = 0,$$
$$s\bar\psi_i = \omega_i, \quad s\omega_i = 0. \tag{14.22}$$

176   14. Path Integrals for Fermions

Now, however, we have to keep track of the commuting and anticommuting properties of A and B by hand. To do this we will assign a *grading* to an expression. We will set $g(A) = 0$ if A is an even expression (containing an even number of Grassmann variables), and $g(A) = 1$ if A is an odd expression (containing an odd number of Grassmann variables). With this notation we must take

$$s(AB) = s(A)B + (-1)^{g(A)} A s(B). \tag{14.23}$$

An operation that obeys this form of Leibnitz rule is called an *antiderivation*.[1] The antiderivation s obeys $s^2 = 0$. This is strongly reminiscent of the properties of the exterior derivative d or the operation of taking a boundary (∂) in homological algebra. Indeed, supersymmetry is intimately related to homology and cohomology.

The identity $s^2 = 0$ gives us some insight as to why the action is supersymmetric. It turns out that the action is the s of something. Indeed

$$\left\{ \bar{\psi}_i M^i_\mu \psi^\mu + \frac{1}{2} g^{ij} \omega_i \omega_j - \omega_i \mathcal{F}^i(\varphi) \right\}$$
$$= s \left\{ \bar{\psi}_i \left(\frac{1}{2} g^{ij} \omega_j - \mathcal{F}^i(\varphi) \right) \right\}, \tag{14.24}$$

and since the s of the s of anything vanishes, s applied to the action gives zero.

To gain further insight into the origin of the supersymmetry, let us assume that there is a function, called a *superpotential*, $W(\varphi)$ such that

$$\mathcal{F}^i = \frac{\partial W}{\partial \varphi_i}. \tag{14.25}$$

In this case it makes sense to drop the distinction between the Greek and Roman indices and to combine all the variables, φ_i, ψ_i, $\bar{\psi}_i$, and ω_i into a single set of *superfields* $\Phi_i(\theta, \bar{\theta})$, where

$$\Phi_i(\theta, \bar{\theta}) = \varphi_i + \bar{\theta} \psi_i + \bar{\psi}_i \theta + \bar{\theta} \theta \omega_i. \tag{14.26}$$

Here $\bar{\theta}$ and θ are a pair of Grassmann variables. We think of them as anticommuting coordinates on *superspace*, and of (14.26) as the Taylor expansion of the superfield about the origin in this superspace. Of course, because $\theta^2 = \bar{\theta}^2 = 0$, the Taylor expansion soon terminates.

We note that

$$S = \int d\theta d\bar{\theta} \left\{ \frac{1}{2} g^{ij} \Phi_i \frac{\partial^2}{\partial \theta \partial \bar{\theta}} \Phi_j - W(\Phi) \right\}$$
$$= \frac{1}{2} g^{ij} \omega_i \omega_j - \frac{\partial W}{\partial \varphi_i} \omega_i + \bar{\psi} \frac{\partial^2 W}{\partial \varphi_i \partial \varphi_j} \psi_j$$

---

[1] Any operation that obeys the conventional Leibnitz rule is called a *derivation*. Such operations need not involve derivatives. An example is the operation Ad_M on matrices. This is defined by $\text{Ad}_M A = [M, A]$. Leibnitz' rule follows from the Jacobi identity.

$$= \frac{1}{2}g^{ij}\omega_i\omega_j - \mathcal{F}^i\omega_i + \bar{\psi}_i \frac{\partial^2 W}{\partial \varphi_i \partial \varphi_j}\psi_j. \tag{14.27}$$

(Partial derivatives such as $\partial_\theta \Phi$ are computed by commuting θ to the extreme left of any monomial containing it, and then suppressing it. This makes ∂_θ an antiderivation.) This is our previous action, in the special circumstance that $\mathcal{F}$ is derived from a superpotential.

Now if we let $\bar{\theta} \to \bar{\theta} + \epsilon$, we have

$$\Phi_i \to (\varphi_i + \epsilon\psi_i) + \bar{\theta}\psi_i + (\bar{\psi}_i + \epsilon\omega_i)\theta + \bar{\theta}\theta\omega_i. \tag{14.28}$$

Clearly the effect of this supertranslation coincides with our earlier supersymmetry transformation on the component fields. The action is therefore invariant under it. However, it should no longer be surprising that the action in (14.27) is invariant under such a translation. The action has after all been obtained by integrating something over all superspace!

14.2 Fermionic Coherent States

Now we must make the connection with the Hilbert space description of the quantum mechanics. We start with a pair of fermion annihilation and creation operators $\hat{a}, \hat{a}^\dagger$ obeying the familar relation $\{\hat{a}, \hat{a}^\dagger\} = 1$. We let ξ and ξ^* be Grassmann variables anticommuting with both $\hat{a}$ and $\hat{a}^\dagger$. Note that this implies that ξ commutes with $|0\rangle$ and anticommutes with $|1\rangle = \hat{a}^\dagger|0\rangle$.

We define a fermionic *coherent state*

$$|\xi\rangle = e^{\hat{a}^\dagger \xi}|0\rangle = (1 + \hat{a}^\dagger \xi)|0\rangle = |0\rangle - \xi|1\rangle. \tag{14.29}$$

This is an eigenstate of $\hat{a}$ with eigenvalue ξ, because

$$\hat{a}|\xi\rangle = \hat{a}|0\rangle - \hat{a}\xi|1\rangle = \xi(|0\rangle - \xi|1\rangle) = \xi|\xi\rangle. \tag{14.30}$$

We also define the operation of hermitian conjugation as acting in the usual way on $\hat{a}, \hat{a}^\dagger$, and as the $\xi \to \xi^*$ involution on the algebra of Grassmann variables. We find, therefore, that

$$(e^{\hat{a}^\dagger \xi})^\dagger = e^{\xi^* \hat{a}}. \tag{14.31}$$

From this we find

$$\langle \xi_1 | \xi_2 \rangle = \langle 0 | e^{\xi_1^* \hat{a}} e^{\hat{a}^\dagger \xi_2} | 0 \rangle = \langle 0 | (1 + \xi_1^* \hat{a})(1 + \hat{a}^\dagger \xi_2 | 0 \rangle$$
$$= \langle 0 | 0 \rangle + \langle 0 | \xi_1^* \hat{a} \hat{a}^\dagger \xi_2 | 0 \rangle$$
$$1 + \xi_1^* \xi_2 = e^{\xi_1^* \xi_2}. \tag{14.32}$$

Finally we find a completeness relation

$$\int d\xi^* d\xi \, e^{-\xi^* \xi} |\xi\rangle\langle\xi|$$

178   14. Path Integrals for Fermions

$$= \int d\xi^* d\xi (1 - \xi^*\xi)(|0\rangle - \xi|1\rangle)(\langle 0| + \xi^*\langle 1|)$$

$$= \int d\xi^* d\xi (1 - \xi^*\xi)(|0\rangle\langle 0| + \xi\xi^*|1\rangle\langle 1|)$$

$$= |0\rangle\langle 0| + |1\rangle\langle 1| = I. \quad (14.33)$$

Now let us use this to obtain a path integral expression for the partition function $\text{Tr}\{e^{-\beta\hat{H}}\}$. We start with the simplest possible hamiltonian $\hat{H} = \hat{a}^\dagger h \hat{a}$, where h is a number. In this case the eigenvalues of $\hat{H}$ are 0 or h so

$$\text{Tr}\{e^{-\beta\hat{H}}\} = 1 + e^{-\beta h}. \quad (14.34)$$

Now we try to obtain this by inserting sets of intermediate states

$$\text{Tr}\{e^{-\beta\hat{H}}\} = \sum_{n_i=0,1} \langle n_N | e^{-\frac{\beta}{N}\hat{H}} | n_{N-1}\rangle \langle n_{N-1} | e^{-\frac{\beta}{N}\hat{H}} | n_{N-2}\rangle \ldots \langle n_1 | e^{-\frac{\beta}{N}\hat{H}} | n_N\rangle. \quad (14.35)$$

This is approximately

$$\sum_{n_i=0,1} \langle n_N | \left(1 - \frac{\beta}{N}\hat{H}\right) | n_{N-1}\rangle \quad \langle n_{N-1} | \left(1 - \frac{\beta}{N}\hat{H}\right) | n_{N-2}\rangle \ldots$$

$$\ldots \langle n_1 | \left(1 - \frac{\beta}{N}\mathcal{H}\right) | n_N\rangle. \quad (14.36)$$

We use the completeness relation to replace

$$\sum_{n=0,1} |n\rangle\langle n| \to \int d\xi^* d\xi e^{-\xi^*\xi} |\xi\rangle\langle\xi| \quad (14.37)$$

and

$$\langle \xi_{i+1} | \left(1 - \frac{\beta}{N}\hat{H}\right) | \xi_i\rangle \to \langle \xi_{i+1} | \xi_i\rangle \left(1 - \frac{\beta}{N}\xi^*_{i+1} h \xi_i\right) = e^{\xi^*_{i+1}\xi_i - \frac{\beta}{N}\xi^*_{i+1} h \xi_i}. \quad (14.38)$$

Thus we are tempted to assert (there is an error we will catch in a moment)

$$\text{Tr}\{e^{-\beta\hat{H}}\} \stackrel{?}{=} \int d[\xi^*]d[\xi] e^{\left\{-\sum_i (\xi^*_{i+1}(\xi_{i+1}-\xi_i) + \frac{\beta}{N}\xi^*_{i+1} h \xi_i)\right\}}. \quad (14.39)$$

The integral on the right is of the form we have considered earlier and is equal to

$$\begin{vmatrix} 1 & \left(\frac{\beta h}{N} - 1\right) & 0 & \cdots & 0 \\ 0 & 1 & \left(\frac{\beta h}{N} - 1\right) & \cdots & 0 \\ 0 & 0 & 1 & \cdots & 0 \\ \vdots & \vdots & \vdots & \ddots & \vdots \\ \left(\frac{\beta h}{N} - 1\right) & 0 & 0 & \cdots & 1 \end{vmatrix}, \quad (14.40)$$

and this is in turn equal to

$$1 + (-1)^{N+1}\left(\frac{\beta h}{N} - 1\right)^N = 1 - \left(1 - \frac{\beta h}{N}\right)^N \approx 1 - e^{-\beta h}. \tag{14.41}$$

This has a sign wrong! The problem is due to the outermost integration. Assuming that $\hat{A}$ is even (as it is in the current application), we have

$$\int d\xi^* d\xi \, \langle \xi | \hat{A} | \xi \rangle e^{-\xi^* \xi}$$
$$= \int d\xi^* d\xi \, \left(\langle 0 | + \xi^* \langle 1 | \right) \hat{A} \left(|0\rangle - \xi |1\rangle \right)(1 - \xi^* \xi)$$
$$= \int d\xi^* d\xi \, \left(\langle 0 | \hat{A} | 0 \rangle + \xi^* \xi \langle 1 | \hat{A} | 1 \rangle \right)(1 - \xi^* \xi)$$
$$= \langle 0 | \hat{A} | 0 \rangle - \langle 1 | \hat{A} | 1 \rangle = \operatorname{tr}\left\{ (-1)^{\hat{F}} \hat{A} \right\}. \tag{14.42}$$

Here $\hat{F}$ is the operator that counts the number of fermions present. If we want the trace of $\hat{A}$, we must use

$$\operatorname{tr}\left\{\hat{A}\right\} = \int d\xi^* d\xi \, \langle \xi | \hat{A} | (-\xi) \rangle e^{-\xi^* \xi}. \tag{14.43}$$

This means that we should impose *antiperiodic* boundary conditions on our Grassmann variables. We have met this antiperiodicity requirement before in Chapter 11.

Extending our result to a matrix H is trivial. If $\hat{H} = \hat{a}_\alpha^\dagger H_{\alpha\beta} \hat{a}_\beta$, we have

$$\operatorname{tr}\left\{ e^{-\beta \hat{H}} \right\} = \int d[\xi] d[\xi^*] e^{-\int_0^\beta (\xi_\alpha^* \partial_\tau \xi_\alpha + \xi_\alpha^* H_{\alpha\beta} \xi_\beta)} \tag{14.44}$$

with the understanding that we have antiperiodic boundary conditions on the ξ.

14.3 Superconductors

As a physics application of the Berezin integral we will explore the BCS theory of superconductivity.

The partition function of a gas of nonrelativistic spin-$\frac{1}{2}$ fermions may be written as a Berezin path integral

$$\mathcal{Z} = \operatorname{Tr}(e^{-\beta H})$$
$$= \int d[\psi] d[\psi^\dagger] e^{\left(-\int_0^\beta d^3x d\tau \left\{ \sum_{\alpha=1}^2 \psi_\alpha^\dagger (\partial_\tau - \frac{1}{2m}\nabla^2 - \mu)\psi_\alpha - g\psi_1^\dagger \psi_2^\dagger \psi_2 \psi_1 \right\} \right)}. \tag{14.45}$$

The indices $\alpha = 1, 2$ refer to the two components $\uparrow, \downarrow$, of spin. The Grassmann-valued Fermi fields are to be taken antiperiodic under the shift $\tau \to \tau + \beta$. There is no problem with the nonquadratic term $\psi_1^\dagger \psi_2^\dagger \psi_2 \psi_1$ because between coherent states all operators such as $\hat{\psi}_1^\dagger$ may be replaced by their Grassmann-valued eigenvalues.

I have included a short-range interaction, $g\psi_1^\dagger\psi_2^\dagger\psi_2\psi_1$, in (14.45). A positive value for g corresponds to an attractive potential. Given an attractive interaction, and a low enough temperature, the system should be unstable with respect to the onset of superconductivity. To detect this instability we introduce an ancillary complex scalar field Δ, which will become the superconducting order parameter. We use it to decouple the interaction by writing

$$\mathcal{Z} = \int d[\psi]d[\psi^\dagger]d[\Delta]d[\Delta^*] \exp\left(-\int_0^\beta d^3x d\tau \left\{ \sum_{\alpha=1}^2 \psi_\alpha^\dagger(\partial_\tau - \frac{1}{2m}\nabla^2 - \mu)\psi_\alpha \right.\right.$$
$$\left.\left. -\Delta^*\psi_2\psi_1 - \Delta\psi_1^\dagger\psi_2^\dagger + \frac{1}{g}|\Delta|^2 \right\}\right). \tag{14.46}$$

The equation of motion for Δ shows us that $\Delta \equiv g\psi_2\psi_1$.

We may now integrate out the fermions to find an effective action for the Δ field. From this point on we will set the temperature, β^{-1}, to zero.

Taking note of the anticommutativity of the Grassmann fields, the quadratic form in the exponent can be arranged as a matrix:

$$S = \int d^3x d\tau (\psi_1^\dagger, \psi_2) \begin{pmatrix} \partial_\tau - \frac{\nabla^2}{2m} - \mu & \Delta \\ \Delta^* & \partial_\tau + \frac{\nabla^2}{2m} + \mu \end{pmatrix} \begin{pmatrix} \psi_1 \\ \psi_2^\dagger \end{pmatrix}. \tag{14.47}$$

The fermion contribution to the effective action is the logarithm of the Fredholm determinant of this matrix of differential operators:

$$S_F = -\ln \text{Det} \begin{pmatrix} \partial_\tau - \frac{\nabla^2}{2m} - \mu & \Delta \\ \Delta^* & \partial_\tau + \frac{\nabla^2}{2m} + \mu \end{pmatrix}. \tag{14.48}$$

We begin by first assuming that Δ is a constant. Under these circumstances S_F is given by

$$S_F = \int d^3x d\tau \left\{ -\int \frac{d^3k}{(2\pi)^3} \frac{d\omega}{2\pi} \text{tr} \ln \begin{pmatrix} i\omega + \frac{k^2}{2m} - \mu & \Delta \\ \Delta^* & i\omega - \frac{k^2}{2m} + \mu \end{pmatrix} \right\}. \tag{14.49}$$

It is convenient to introduce the notation $\epsilon = k^2/2m - \mu$, whence the momentum integral becomes

$$I = \int \frac{d^3k}{(2\pi)^3} \frac{d\omega}{2\pi} \ln(\omega^2 + |\Delta|^2 + \epsilon^2). \tag{14.50}$$

Since everything is spherically symmetric, we can replace the integration over k by an integration over ϵ

$$I = \int \rho(\epsilon) d\epsilon \frac{d\omega}{2\pi} \ln(\omega^2 + |\Delta|^2 + \epsilon^2), \tag{14.51}$$

at the expense of introducing the density of states $\rho(\epsilon)$. We evaluate (14.51) by first differentiating with respect to Δ:

$$\frac{dI}{d\Delta} = \int d\epsilon \frac{d\omega}{2\pi} \rho(0) \frac{\Delta^*}{\omega^2 + |\Delta|^2 + \epsilon^2}. \tag{14.52}$$

In (14.52) we have approximated $\rho(\epsilon)$ by $\rho(0)$. This approximation is reasonable because the integrand is peaked near the Fermi surface $\epsilon = 0$. To obtain a finite value for the integral we must introduce a cut-off. In phonon-mediated BCS superconductivity the interaction naturally dies out at the energy ϵ_D corresponding to the Debye frequency. Taking this value for the cut-off we find

$$\frac{dI}{d\Delta} = \frac{\rho(0)}{2} \int_{-\epsilon_D}^{\epsilon_D} \frac{\Delta^* d\epsilon}{\sqrt{\epsilon^2 + |\Delta|^2}} = \rho(0)\Delta^* \sinh^{-1}\frac{\epsilon_D}{|\Delta|} \approx \rho(0)\Delta^* \ln\frac{\epsilon_D}{|\Delta|}. \tag{14.53}$$

Putting this together with the $|\Delta|^2$ part of the exponent, we find that the effective potential (the action per unit volume space-time) for Δ is minimized when

$$\frac{dV_{eff}}{d\Delta} = \left(\frac{\Delta^*}{g} + \rho(0)\Delta^* \ln\frac{|\Delta|}{\epsilon_D}\right) = 0, \tag{14.54}$$

or when

$$|\Delta| = |\Delta_0| = \epsilon_D \exp\left\{-\frac{1}{g\rho(0)}\right\}. \tag{14.55}$$

The effective potential itself is

$$V_{eff} = \frac{1}{g}|\Delta|^2 + \frac{\rho(0)}{2}\left\{|\Delta|^2 \ln\frac{|\Delta|^2}{\epsilon_D^2} - |\Delta|^2\right\}. \tag{14.56}$$

14.3.1 Effective Action

Having found the equilibrium value of Δ to 1-loop order, we should set $\Delta = \langle\Delta\rangle + \eta$ and perform the η integral. In practice, however, the fluctuations are small and most of the physics of traditional superconductors can be understood by replacing Δ by its mean-field value and foregoing the fluctuation integral. Nontheless, it is still useful to understand how the superconductor responds to small deviations from equilibrium, and this can be approached by computing the Fredholm determinant, or 1-loop effective action $S_F(\Delta)$ for Δ fields that have some dependence on space and time. Because the *phase* of Δ is the Goldstone boson for the spontaneously broken symmetry, it will be this field that will be most important in determining the *low-energy effective action*. We will be able to find most of the terms in this effective action by appeals to the symmetries of the system.

We begin by considering the effect of uniform twists. Suppose that the phase of the order parameter varies linearly with position, i.e. $\Delta(x) = e^{2ik_s x}\Delta_0$. This should correspond to a uniform superflow with velocity $v_s = k_s/m$.

We need to find the Fredholm determinant of

$$K = \begin{pmatrix} \partial_\tau - \frac{\nabla^2}{2m} - \mu & \Delta_0 e^{2ik_s x} \\ \Delta_0^* e^{-2ik_s x} & \partial_\tau + \frac{\nabla^2}{2m} + \mu \end{pmatrix}. \tag{14.57}$$

182   14. Path Integrals for Fermions

Now if

$$U = \begin{pmatrix} e^{ik_s x} & 0 \\ 0 & e^{-ik_s x} \end{pmatrix}, \qquad (14.58)$$

then

$$U^{-1}KU = \begin{pmatrix} \partial_\tau - \frac{(\nabla + ik_s)^2}{2m} - \mu & \Delta_0 \\ \Delta_0^* & \partial_\tau + \frac{(\nabla - ik_s)^2}{2m} + \mu \end{pmatrix}. \qquad (14.59)$$

The determinant is not affected by such a unitary transformation[2] and we can now evaluate $\ln \text{Det } K$ by Fourier transforming

$$\frac{1}{VT} \ln \text{Det } K = \int \frac{d^3k \, d\omega}{(2\pi)^3 \, 2\pi} \text{tr} \ln \begin{pmatrix} i\omega + \frac{(k+k_s)^2}{2m} - \mu & \Delta_0 \\ \Delta_0^* & i\omega - \frac{(k-k_s)^2}{2m} + \mu \end{pmatrix} \qquad (14.60)$$

$$= \int \frac{d^3k \, d\omega}{(2\pi)^3 \, 2\pi} \text{tr} \ln \begin{pmatrix} (i\omega + v_s k) + \frac{k^2}{2m} - (\mu - \frac{k_s^2}{2m}) & \Delta_0 \\ \Delta_0^* & (i\omega + v_s k) - \frac{k^2}{2m} + (\mu - \frac{k_s^2}{2m}) \end{pmatrix} \qquad (14.61)$$

The net effect has been to shift $i\omega \to i\omega + v_s k$ and $\mu \to \mu - k_s^2/2m$.

If we momentarily forget about the shift in the ω variable, we find that, for small k,

$$S_F(\Delta_0 e^{2ik_s x}) = S_F(\Delta_0) - (VT) \frac{k_s^2}{2m} \frac{\partial S_F}{\partial \mu}. \qquad (14.62)$$

Now $-\partial S_F/\partial \mu$ is the number density of the system, so

$$S_F(\Delta_0 e^{2ik_s x}) = S_F(\Delta_0) + (VT)\rho \frac{k_s^2}{2m}. \qquad (14.63)$$

This is what we would expect for a steady uniform flow of the entire fluid with $v_s = k_s/m$.

Let us now ask for the consequences of the sideways translation of the ω contour. In evaluating the density in (14.63) it is the ω contour integral that determines the occupation numbers of the various quasiparticle modes. These modes manifest themselves as poles in the $\langle \psi^\dagger \psi \rangle$ Green function at $\omega = \pm\sqrt{\epsilon^2 + |\Delta|^2}$, and the contour integral is to be closed in such a manner that the negative energy states are encircled, and thus counted as occupied. These negative energy states are particle-like for $k \ll k_f$ and hole-like for $k \gg k_f$. If v_s is sufficiently large, poles that were within the contour before the $iv_s \cdot k$ shift may no longer be enclosed, and conversely, previously unoccupied states may be occupied. The physical reason is that, when seen from the rest frame, the quasiparticle energies in the moving fluid are Doppler-shifted from $\omega(k)$ to $\omega(k) + v_s k$. The use of a chemical potential μ to determine the average number of particles implies equilibrium with the stationary walls of the container, and it is the negative energy states as seen from the container

---

[2] This is not true for relativistic systems where the determinant *is* altered by such chiral transformations because of anomalies. There are no anomalies for nonrelativistic systems.

frame that are occupied. There is a range of v_s for which the occupation numbers are unchanged, and, consequently, a critical value of v_s below which no "normal" fluid will be created. Above v_{crit} we need to replace the ρ in (14.63) by an effective ρ_s. Since the alteration of occupation numbers tends to reduce the momentum in the system, we have $\rho_s < \rho$. Only the superfluid fraction is flowing. The normal component stays at rest.

We now consider the time variation $\Delta = e^{2i\Omega t}\Delta_0$. This shifts $\mu \to \mu - i\Omega$ so, for small Ω,

$$S_F(e^{2i\Omega t}\Delta_0) = S_F(\Delta_0) + (VT)i\rho\Omega. \tag{14.64}$$

Thus we know that the euclidean action S_F will contain the terms

$$S_F(\Delta) \approx S_F(\Delta_0) + \int d^3x dt \left\{ i\rho\partial_\tau\phi/2 + \frac{\rho}{2m}(\partial_x\phi/2)^2 \right\}, \tag{14.65}$$

where $\Delta = \Delta_0 e^{i\phi(x,t)}$.

Note that if we work in real time (as opposed to the Matsubara imaginary time that we have been using) then, at least for $v_s < v_{crit}$, S_F is invariant under the combined phase transformation $\Delta \to (e^{imv_s x - i\frac{1}{2}mv_s^2 t})^2 \Delta$. This corresponds to a simultaneous galilean transformation acting on all the particles in the fluid. When the fluid and its container are moved together, the occupation numbers should remain unchanged even if $v_s > v_{crit}$. In this circumstance the contour should be arranged so that the $v_s k$ translation has no effect on the occupation numbers.

If we work harder at evaluating the low-energy effective action, including the effects of changing the density and an external gauge field, but *not* including fluctuations in the magnitude[3] of Δ, we will arrive at the action

$$S = \int d^3x dt \left\{ -\rho \left(\frac{\partial \phi/2}{\partial t} - eA_0 \right) - \frac{\rho}{2m}(\nabla\phi/2 - e\mathbf{A})^2 - \lambda(\rho - \rho_0)^2 \right\}, \tag{14.66}$$

Here λ parameterizes the response of the system to a uniform change of density.

Now *define* a field $\Psi = \sqrt{\frac{\rho}{2}}e^{i\phi}$. For mnemonic purposes Ψ can be thought of as a "wavefunction" of the Cooper pairs.[4] We also define the symbol $e^* = 2e$, which can be though of as the Cooper-pair charge, and the symbol $m^* = 2m$, which can be thought of as the Cooper-pair mass. Finally we set $\rho_0^* = \rho_0/2$, and this can be thought of as the Cooper-pair density.

---

[3]These do not correspond to excitation of Goldstone modes, and therefore will not appear in the low-energy effective action.

[4]Some texts give the impression that Δ can be regarded as the condensate wavefunction. I believe that this is misleading. The magnitude of the wavefunction should be related to the density, and there is, in general, no simple relation between $|\Delta|$ and $\sqrt{\rho}$.

184   14. Path Integrals for Fermions

With these definitions we can write (14.66) in a appealingly simple form. Up to higher-order gradients of ρ^*, it is equivalent to

$$S = \int d^x dt \left\{ i\Psi^\dagger (\partial_t - ie^* A_0)\Psi - \frac{1}{2m^*}|(\nabla - ie^* \mathbf{A})\Psi|^2 - \frac{\lambda}{2}(|\Psi|^2 - \rho_0)^2 \right\}, \tag{14.67}$$

Varying this last action gives rise the galilean-invariant Gross-Pitaevskii nonlinear Schrödinger equation for Ψ

$$i(\partial_t - ie^* A_0)\Psi = -\frac{1}{2m^*}(\nabla - ie^* \mathbf{A})^2 \Psi + \lambda(|\Psi|^2 - \rho_0^*)\Psi. \tag{14.68}$$

As we know from Chapter 11, this equation shows that the electron gas behaves as a charged fluid. We saw there that such a fluid exhibits the Meissner effect and other superconducting phenomena.

15
Lattice Field Theory

The Feynman diagram expansion of relativistic field theories gives rise to integrals that diverge at short distances or equivalently at large momenta. To have a well-defined theory we need to *regulate* these divergences. There are many ways to do this, but one that has both practical and conceptual advantages is to put the theory on a discrete space-time lattice. In this chapter we will explore this.

15.1 Boson Fields

Put a real variable $\varphi(\mathbf{n})$ at each point $\mathbf{n} = (n_1, \ldots, n_d) \in \mathbf{Z}^d$ on a d-dimensional square lattice whose points are a distance a apart.

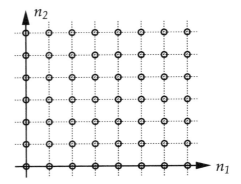

Fig 1. A square lattice.

We consider the integral

$$\mathcal{Z} = \int d[\varphi] \exp\left\{ -\sum_{\mathbf{n} \in \mathbf{Z}^d} \sum_{i=1}^{d} a^d \left[\frac{1}{2} \frac{(\varphi(\mathbf{n}+\mathbf{i}) - \varphi(\mathbf{n}))^2}{a^2} + \frac{1}{2} m^2 \varphi^2(\mathbf{n}) \right] \right\} \quad (15.1)$$

In this expression $d[\varphi]$ is shorthand for $\prod_{\mathbf{n}} d\varphi(\mathbf{n})$ and $\mathbf{i}$ is the vector $(\ldots, 1, \ldots)$ with the 1 in the ith place. One should think of $\mathcal{Z}$ as the partition function of some statistical system of springs and masses. The exponent is clearly intended to be an approximation to

$$L = \int d^d x \left\{ \frac{1}{2} (\partial \varphi)^2 + \frac{1}{2} m^2 \varphi^2 \right\}. \quad (15.2)$$

We are really interested in the Green functions

$$\langle \hat{\varphi}(\mathbf{n}_1) \ldots \hat{\varphi}(\mathbf{n}_n) \rangle = \frac{1}{\mathcal{Z}} \int d[\varphi] \{\varphi(\mathbf{n}_1) \ldots \varphi(\mathbf{n}_n)\} e^{\left\{ -\sum_{\mathbf{n}} \sum_{i=1}^{d} (\ldots) \right\}}. \quad (15.3)$$

We can evaluate these by the simple but rather unilluminating procedure of Fourier transforming $\varphi(\mathbf{n})$. Let us temporarily set $a = 1$, so that it does not clutter our expressions. We can always restore it later by dimensional analysis. We now write $\varphi(\mathbf{n})$ in terms of its Fourier components:

$$\varphi(\mathbf{n}) = \int_{-\pi}^{\pi} \frac{d^d k}{(2\pi)^d} \tilde{\varphi}(\mathbf{k}) e^{i \mathbf{n} \cdot \mathbf{k}}. \quad (15.4)$$

Because of the discrete cubic lattice the momenta $\mathbf{k}$ are restricted to lie in the Brillouin zone.

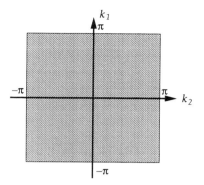

Fig 2. The Brillouin zone.

The lattice Fourier transform uses the identities

$$\int_{-\pi}^{\pi} \frac{d^d k}{(2\pi)^d} e^{i(\mathbf{n}-\mathbf{n}')\cdot \mathbf{k}} = \delta_{\mathbf{n}\mathbf{n}'} \tag{15.5}$$

and

$$\sum_{\mathbf{n}} e^{i\mathbf{n}\cdot(\mathbf{k}-\mathbf{k}')} = \sum_{\mathbf{m}} (2\pi)^d \delta^d(\mathbf{k} - \mathbf{k}' + 2\pi \mathbf{m}), \tag{15.6}$$

which give

$$\tilde{\varphi}(\mathbf{k}) = \sum_{\mathbf{n}} e^{-i\mathbf{n}\cdot \mathbf{k}} \varphi(\mathbf{n}). \tag{15.7}$$

In terms of $\tilde{\varphi}(\mathbf{k})$, and setting $a = 1$, we have

$$\mathcal{Z} = \int d[\tilde{\varphi}] e^{-\int \frac{d^d k}{(2\pi)^d} \frac{1}{2}\left[m^2 + 2\sum_1^d (1-\cos k_i)\right]|\tilde{\varphi}(\mathbf{k})|^2}. \tag{15.8}$$

The two-point function is given by the inverse of the matrix defining the quadratic form in the exponent, and is

$$\langle \tilde{\varphi}(\mathbf{k}_1)\tilde{\varphi}(\mathbf{k}_2)\rangle = \delta^d(\mathbf{k}_1 - \mathbf{k}_2)\frac{(2\pi)^d}{m^2 + 2\sum_1^d(1 - \cos k_i)}. \tag{15.9}$$

In configuration space this becomes

$$\langle \hat{\varphi}(\mathbf{n}_1)\hat{\varphi}(\mathbf{n}_2)\rangle = \int \frac{d^d k}{(2\pi)^d} \frac{e^{i(\mathbf{n}-\mathbf{n}')\cdot \mathbf{k}}}{m^2 + 2\sum_1^d(1 - \cos k_i)}. \tag{15.10}$$

Note that

$$2(1 - \cos k) = k^2 + O(k^4), \tag{15.11}$$

so the small $\mathbf{k}$ behavior of the integrand is similar to the continuum expression.

188   15. Lattice Field Theory

To gain more insight into the lattice propagator we rearrange the exponent in (15.1) so that

$$Z = \int d[\varphi] \exp\left\{ +\sum_{\mathbf{n},i} \varphi(\mathbf{n}+\mathbf{i})\varphi(\mathbf{n}) - \sum_{\mathbf{n}} \frac{1}{2}(2d+m^2)\varphi^2(\mathbf{n}) \right\}. \quad (15.12)$$

Rescale $\varphi \to \sqrt{2d+m^2}\,\varphi$ and absorb the jacobian into the measure. Then

$$Z = \int d\left[\frac{\varphi}{\sqrt{2\pi}}\right] \exp\left\{ \frac{1}{2d+m^2} \sum_{\mathbf{n},i} \varphi(\mathbf{n}+\mathbf{i})\varphi(\mathbf{n}) - \frac{1}{2}\sum_{\mathbf{n}} \varphi^2(\mathbf{n}) \right\}. \quad (15.13)$$

If m^2 is large, it makes sense to expand out the first term in the exponent and obtain a typical "strong-coupling" or "high-temperature" expansion

$$Z = \int d\left[\frac{\varphi}{\sqrt{2\pi}}\right] \left(1 + \frac{1}{2d+m^2} \sum_{\mathbf{n},i} \varphi(\mathbf{n}+\mathbf{i})\varphi(\mathbf{n}) + \ldots \right) e^{-\frac{1}{2}\sum_{\mathbf{n}} \varphi^2(\mathbf{n})}. \quad (15.14)$$

We use the formulae

$$\int \frac{d\varphi}{\sqrt{2\pi}} e^{-\frac{1}{2}\varphi^2} = 1,$$

$$\int \frac{d\varphi}{\sqrt{2\pi}} \varphi^{2n+1} e^{-\frac{1}{2}\varphi^2} = 0,$$

$$\int \frac{d\varphi}{\sqrt{2\pi}} \varphi^2 e^{-\frac{1}{2}\varphi^2} = 1,$$

$$\int \frac{d\varphi}{\sqrt{2\pi}} \varphi^4 e^{-\frac{1}{2}\varphi^2} = 3, \quad \text{etc.} \quad (15.15)$$

to organize the expansion into digrams that form closed loops such as

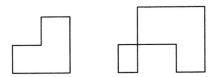

with a weight $1/(m^2+2d)$ for each link used. The factor of 3 in the φ^4 integral means that a diagram with a single crossing such as

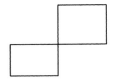

has a weight +3. This can be understood as the sum of three routings of the paths at the crossing point

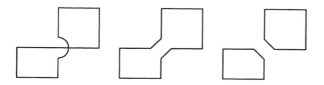

All other combinatorial factors arising from the expansion of the exponent in (15.14) and the subsequent integrations combine to make the expansion of $\mathcal{Z}$ the partition function of a gas (or solution) of closed noninteracting loops. To prove that the $n!$'s really do conspire in this way requires a study of the theory of lattice random walks. Before we address this topic, take a moment to note that when we calculate Green functions, the $\varphi(\mathbf{n})$ act as sources where paths may start and stop.

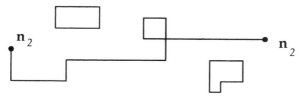

Fig 3. A graph included in the expansion of $\langle \hat{\varphi}(\mathbf{n}_1)\hat{\varphi}(\mathbf{n}_2)\rangle$.

As usual the factor of $1/\mathcal{Z}$ serves to remove the gas of disconnected vacuum loops.

15.2 Random Walks

Let us temporarily forget about field theory and construct a theory of random walks on a hypercubic lattice. Suppose

$$\Gamma(\mathbf{n}_2, \mathbf{n}_1, t) \equiv \Gamma(\mathbf{n}_2 - \mathbf{n}_1, 0, t) \tag{15.16}$$

denotes the number of possible paths from $\mathbf{n}_1$ to $\mathbf{n}_2$ that are t steps long (t being an integer.). It takes only a moment to see that $\Gamma(\mathbf{n}, 0, t)$ is the coefficient of $x_1^{n_1} x_2^{n_2} \ldots x_d^{n_d}$ in

$$\left(x_1 + \frac{1}{x_1} + x_2 + \frac{1}{x_2} + \ldots + x_d + \frac{1}{x_d}\right)^t. \tag{15.17}$$

If we put $x_i = e^{ik_i}$, we can extract this coefficient via a Fourier integral

$$\Gamma(\mathbf{n}, 0, t) = \int_{-\pi}^{\pi} \frac{d^d k}{(2\pi)^d} e^{-i\mathbf{k}\cdot\mathbf{n}} (2\cos k_1 + 2\cos k_2 + \ldots 2\cos k_d)^t. \tag{15.18}$$

190   15. Lattice Field Theory

If we then define

$$G(\mathbf{n}, 0, \mu) = \sum_{t=0}^{\infty} e^{-\mu t} \Gamma(\mathbf{n}, 0, t), \tag{15.19}$$

we see that $G(\mathbf{n}, 0, \mu)$ is essentially the propagator (15.10)

$$G(\mathbf{n}, 0, \mu) = \int_{-\pi}^{\pi} \frac{d^d k}{(2\pi)^d} \frac{e^{-i\mathbf{k}\cdot\mathbf{n}}}{1 - e^{-\mu}\sum_1^d (2\cos k_i)}. \tag{15.20}$$

If we set $e^{-\mu} = (m^2 + 2d)^{-1}$, we can rewrite (15.20) as

$$G(\mathbf{n}, 0, \mu) = (m^2 + 2d) \int_{-\pi}^{\pi} \frac{d^d k}{(2\pi)^d} \frac{e^{-i\mathbf{k}\cdot\mathbf{n}}}{m^2 + 2\sum_1^d (1 - \cos k_i)}. \tag{15.21}$$

The factor in front of the integral reflects the rescaling made in (15.13). We have thus shown that the sum over paths from $\mathbf{n}_1$ to $\mathbf{n}_2$ with a weight of $(m^2 + 2d)^{-1}$ for each link does indeed produce the lattice propagator.

If we want to sum over the closed vacuum loops to get the lattice version of the Ouroboros diagrams, we have to be a little more careful to avoid overcounting. If we fix a point on the loop and sum over all paths through that point. We find

$$e^{-\mu t} \Gamma(\mathbf{n}, \mathbf{n}, t) = \int_{-\pi}^{\pi} \frac{d^d k}{(2\pi)^d} e^{-\mu t} \left(2 \sum_1^d \cos k_i\right)^t. \tag{15.22}$$

When now sum over $\mathbf{n}$, we must take care to divide by $2t$. The factor t arises because the same contribution to the partition function will occur when $\mathbf{n}$ is any of the t vertices on the loop. The 2 occurs because the same term can be generated from walks going in either direction round the loop. [This last factor would be absent if we were considering charged fields by using complex $\varphi(\mathbf{n})$. Complex fields require arrows on the links to keep track of orientation.]

The total sum over all configurations of one loop is therefore

$$N \int_{-\pi}^{\pi} \frac{d^d k}{(2\pi)^d} \sum_{t=1}^{\infty} \frac{1}{2t} e^{-\mu t} \left(2 \sum_1^d \cos k_i\right)^t$$

$$= -\frac{N}{2} \int_{-\pi}^{\pi} \frac{d^d k}{(2\pi)^d} \ln\left(1 - e^{-\mu} \sum_1^d (2\cos k_i)\right)$$

$$= const. - \frac{N}{2} \int_{-\pi}^{\pi} \frac{d^d k}{(2\pi)^d} \ln\left(m^2 + 2\sum_1^d (1 - \cos k_i)\right). \tag{15.23}$$

Here, of course, N is the total number of sites on the lattice. Clearly (15.23) is the lattice verion of the connected Ouroboros graph.

Finally if we want to sum over a gas of n loops we must divide by $n!$ since the loops are indistinguishable. Therefore

$$\mathcal{Z} = (const.) \exp\left(-\frac{N}{2} \int \int_{-\pi}^{\pi} \frac{d^d k}{(2\pi)^d} \ln\left(m^2 + 2\sum_1^d (1 - \cos k_i)\right)\right)$$

$$= \det^{-1/2}\left(-\nabla^2 + m^2\right) \tag{15.24}$$

where ∇^2 is the lattice laplacian. This is exactly the expression we would have obtained by performing the gaussian integral.

15.3 Interactions and Bose Condensation

We have seen that a free scalar field thory — i.e., one whose functional integral is purely gaussian — can be interpreted as a gas of noninteracting worldlines. (A physical model would be a gas of noninteracting polymers.). If we introduce a nongaussian $\lambda \varphi^4$ interaction term into the action

$$\mathcal{Z} = \int d[\varphi] e^{\left\{-\sum_{\mathbf{n}} \sum_{i=1}^{d} a^d \left[\frac{1}{2} \frac{(\varphi(\mathbf{n}+\mathbf{i})-\varphi(\mathbf{n}))^2}{a^2} + \frac{1}{2} m^2 \varphi(\mathbf{n}) + \frac{\lambda}{4!} \varphi^4(\mathbf{n})\right]\right\}}, \tag{15.25}$$

then the $\lambda \varphi^4$ term can be used in the large m expansion at points where lines cross. It has the effect (because of the minus sign outside the sum in the exponent) of reducing the contribution to the sum of loops whenever they have intersections. In other words it leads to short range repulsion between loops and between one part of a loop and another. This type of interaction is actually used to model the effects of short-range repulsions in polymer solutions.

If we were to take $\lambda < 0$, then the integral would diverge at large φ. This is reflected in the force between the polymers being attractive. The force does not saturate, and if one has n walks close together, the action per unit length is proportional to

$$S \approx \alpha n - \beta \lambda n^2, \tag{15.26}$$

for some α, β. [The factor n^2 is because there are $n(n-1)/2$ pairs of lines that interact.] This is essentially the same as in the continuum where an n-particle bound state has mass

$$M \approx mn - K\lambda n^2. \tag{15.27}$$

Such a bound state will become tachyonic (i.e. have a negative mass-squared) if n is large enough. When this happens the vacuum will become filled with a tangled mess of worldlines. A polymer precipitates out of solution if the effective interstrand force becomes attractive.

If we now return to the case of $\lambda > 0$, but with $m^2 < 0$, so that spontaneous symmetry breaking occurs, then a similar picture holds. Having $m^2 < 0$ means that

$$2de^{-\mu} > 1. \tag{15.28}$$

At each step of a random walk on the hypercubic lattice one has a choice of $2d$ directions to go. When the condition (15.28) is satisfied, the $e^{-\mu}$ factor is not sufficient to discourage long paths. In other words the configurational entropy of minus $\ln 2d$ per step wins over the positive μ energy-cost per step. Again the vacuum fills with a spaghetti of long paths. This "spaghetti" phase is the worldline picture of Bose condensation.

15.3.1 Rotational Invariance

The lattice breaks rotational invariance. For large values of m^2 this will be reflected in a lack of rotational invariance in the two-point function. How is the rotational invariance restored when m^2 becomes small? Let us look in detail at the two-dimensional case:

$$G(\mathbf{n}, m^2) = \int_{-\pi}^{\pi} \frac{d^2k}{(2\pi)^2} \frac{e^{-i\mathbf{k}\cdot\mathbf{n}}}{m^2 + 2\sum_1^d (1 - \cos k_i)}. \tag{15.29}$$

Suppose $\mathbf{n}$ is becoming large in a particular direction specified by a unit vector $\mathbf{e}$:

$$\mathbf{n} = r\mathbf{e}, \qquad \mathbf{e}^2 = 1. \tag{15.30}$$

We expect G to fall off exponentially at large $|r|$

$$G(r\mathbf{e}, m^2) \approx e^{-\kappa(\mathbf{e})|r|}. \tag{15.31}$$

The problem is to compute $\kappa(\mathbf{e})$, the inverse correlation length in the direction $\mathbf{e}$. We expect κ to be anisotropic at large m^2 but to be isotropic as m^2 becomes small.

Define

$$f(\xi) = \int_{-\infty}^{\infty} dr\, e^{-ir\xi} G(r\mathbf{e}, m^2). \tag{15.32}$$

This quantity should be analytic in a neighborhood of the real ξ axis and the asymptotic behavior at large r of $G(r\mathbf{e}, m^2)$ will be determined by the nearest singularity to the real ξ axis. The nearest singularity will be on the imaginary axis if G does not oscillate. For example, a singularity at $\xi = i\zeta_0$ would make $G \propto e^{-\zeta_0 r}$. This is because

$$\int_{-\infty}^{\infty} e^{ir\xi} \frac{1}{\xi - i\zeta_0} \frac{d\xi}{2\pi i} = e^{-\zeta_0 r}, \quad \text{for } r > 0. \tag{15.33}$$

Now

$$f(\xi) = \pi \int_{-\pi}^{\pi} \frac{d^2k}{(2\pi)^2} \frac{\delta(\mathbf{k}\cdot\mathbf{e} - \xi)}{m^2 + 2\sum_1^d (1 - \cos k_i)}$$

$$= \pi \int_{-\pi}^{\pi} \frac{d^2k}{(2\pi)^2} \frac{\delta(\mathbf{k}\cdot\mathbf{e} - \xi)}{D(\mathbf{k})}. \tag{15.34}$$

For small m^2, the singularity is expected to be caused by zeros of the denominator $D(\mathbf{k})$ pinching the contour of integration. This is certainly what happens in the continuum where

$$f(i\zeta) = \int_{-\infty}^{\infty} dk \frac{1}{k^2 + m^2 - \zeta^2}. \tag{15.35}$$

15.3 Interactions and Bose Condensation

The integrand has poles at $k = \pm i\sqrt{m^2 - \zeta^2}$ and these pinch when $\zeta = m$. The resultant singularity makes the propagator fall off as $e^{-m|r|}$, a result we know already.

To locate the pinch we move the k integral off the real axis by setting $\mathbf{k} = i\mathbf{K}$. (The fact the k integrals have real endpoints at $k = \pm \pi$ is not relevant to finding the pinch.) Then a pinch at $\xi = i\zeta_0$ requires

$$D(\mathbf{K}) = 0,$$
$$\frac{\partial D}{\partial \mathbf{K}} = 0, \quad \text{on} \quad \mathbf{K} \cdot \mathbf{e} = \zeta_0. \tag{15.36}$$

We can impose the constraint by means of a Lagrange multiplier

$$\frac{\partial}{\partial \mathbf{K}} \{D(\mathbf{K}) - \lambda \mathbf{K} \cdot \mathbf{e}\} = 0. \tag{15.37}$$

These equations have a simple geometric interpretation. The set of points $\mathbf{K} \cdot \mathbf{e} = \zeta_0$ is a straight line perpendicular to $\mathbf{e}$ and at a distance ζ_0 from the origin. Equation (15.36) says that, if there is a pinch singularity at at $i\zeta_0$, then this line must be tangent to the curve $D(\mathbf{K}) = 0$. A more direct way to see this is to note that the points of intersection of the contour of integration $\mathbf{K} \cdot \mathbf{e} = \zeta_0$ with $D(\mathbf{K}) = 0$ are the location of the poles of the inegrand in the complex $\mathbf{k}$ plane. Clearly they can only pinch if they are coincident and the line is tangent.

In our case

$$D(\mathbf{K}) = 4 + m^2 - 2(\cosh K_1 + \cosh K_2). \tag{15.38}$$

For large m^2 the curve $D(\mathbf{K}) = 0$ is essentially a square with sides at $\pm \cosh^{-1}(4 + m^2)/2$

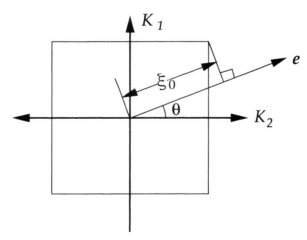

Fig 4. *The construction giving the pinch points.*

A little geometry shows that

$$\zeta_0 \approx (\cos\theta + \sin\theta)\cosh^{-1}(4 + m^2)/2$$

$$\approx (\cos\theta + \sin\theta)\ln(4+m^2). \tag{15.39}$$

Therefore, for large m^2

$$G(|r|\mathbf{e}, m^2) \approx e^{-|r|\zeta_0} = \frac{1}{(4+m^2)^{|n_1|+|n_2|}}. \tag{15.40}$$

This should perhaps not be surprising. At large mass the propagator is dominated by the shortest lattice path between 0 and $\mathbf{n} = (n_1, n_2)$. There are many of these, as the figure shows. They all have length $|n_1| + |n_2|$.

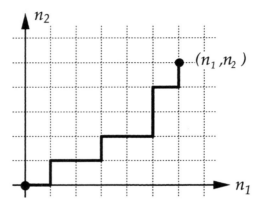

Fig 5. *Any of the shortest paths from 0 to (n_1, n_2) has length $|n_1| + |n_2|$.*

As m^2 becomes smaller the corners of the curve $D(\mathbf{K}) = 0$ round off, tending to circularity at at small m^2.

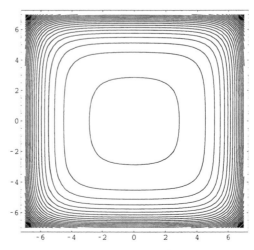

Fig 6. *Contours of $D(\mathbf{K}) = 0$ for varying m^2.*

It is easy to show that along any of the four axes

$$\kappa = \zeta_0 = \cosh^{-1}(1 + \frac{1}{2}m^2), \tag{15.41}$$

while at 45° we have

$$\kappa = \zeta_0 = \sqrt{2}\cosh^{-1}(1 + \frac{1}{4}m^2). \tag{15.42}$$

Both expressions are equal to m when m^2 is small.

Continuum Limit

So far we have been working in lattice units where $a = 1$, m is dimensionless, and the coordinates n_i are integers. We can rewrite expressions such as

$$\langle \hat{\varphi}(\mathbf{n}_2)\hat{\varphi}(\mathbf{n}_1) \rangle = e^{-m|\mathbf{n}_2 - \mathbf{n}_1|} \tag{15.43}$$

as

$$\langle \hat{\varphi}(\mathbf{n}_2)\hat{\varphi}(\mathbf{n}_1) \rangle = e^{-\frac{m}{a}|\mathbf{n}_2 a - \mathbf{n}_1 a|}. \tag{15.44}$$

To take the continuum limit we set $\mathbf{n}a = \mathbf{x}$, $m_{continuum} = ma^{-1}$ and take $m \to 0$ and $a \to 0$ simultaneously so that $m_{continuum}$ remains fixed. Once we have done this we can regard $\mathbf{x}$ as being continuous. Since, for small m, rotational invariance is restored, there will be no remaining evidence that our theory was originally defined on a discrete square lattice.

15.4 Lattice Fermions

We now come to the problem of putting Dirac fermions on a lattice. This turns out to be rather tricky, and there are a number of problems that have not yet been solved in a satisfactory manner.

The most obvious way to discretize the continuum Dirac action

$$S = \int d^d x \left\{ m\bar{\psi}\psi + \bar{\psi}\gamma^i \partial_i \psi \right\} \tag{15.45}$$

leads to the so-called "naive" action

$$S = \sum_{\mathbf{n}} a^d \left\{ m\bar{\psi}(\mathbf{n})\psi(\mathbf{n}) + \frac{1}{2a}\bar{\psi}(\mathbf{n}) \sum_i \gamma^i (\psi(\mathbf{n}+\mathbf{i}) - \psi(\mathbf{n}-\mathbf{i})) \right\}. \tag{15.46}$$

As before it is convenient to set $a = 1$ and rescale the fields by introducing a hopping parameter K so that

$$\psi_{continuum} = \sqrt{2K}\psi_{lattice}, \quad \text{and} \quad m = \frac{1}{2K}. \tag{15.47}$$

196   15. Lattice Field Theory

With these rewrites the action becomes

$$S = \sum_{\mathbf{n}} \left\{ \bar{\psi}(\mathbf{n})\psi(\mathbf{n}) + K\bar{\psi}(\mathbf{n}) \sum_{\mathbf{i}} \gamma^i (\psi(\mathbf{n}+\mathbf{i}) - \psi(\mathbf{n}-\mathbf{i})) \right\}. \tag{15.48}$$

When we integrate over the Grassmann variables $\psi(\mathbf{n})$, $\bar{\psi}(\mathbf{n})$, we need to have a factor of $\bar{\psi}\psi$ at each site in order to get a nonzero answer. To achieve this we can either use the $\bar{\psi}\psi$ mass term at each site, or a pair of adjacent $\bar{\psi}(\mathbf{n})\gamma^i\psi(\mathbf{n}+\mathbf{i})$ hopping terms. In this way we see, at least as long as we do not try to utilize any link twice, that we get a path expansion for the propagator in the form

$$\langle \psi(\mathbf{n}_2)\bar{\psi}(\mathbf{n}_1) \rangle = \sum_{paths} K^L \prod \gamma^i. \tag{15.49}$$

Here the sum is over all paths from $\mathbf{n}_1$ to $\mathbf{n}_2$, L is the length of the path being considered, and $\prod \gamma^i$ is the product of the γ matrices along the path with the understanding that a step in the $-\mathbf{i}$ direction has a factor of $-\gamma^i$. At first one may think that the sum should be over self-avoiding paths or else we would get in trouble with the $\psi^2 = \bar{\psi}^2 = 0$ condition on the Grassmann variables. This is not so, however. There is conspiracy between graphs contributing to the propagator and closed-loop graphs contributing to the vacuum diagrams that allows us to sum over unconstrained paths. To see how this comes about consider the following simple example:

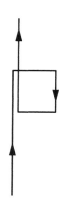

The part of the diagram where two lines share the same link is forbidden in the expansion, but consider also the diagram where the propagator goes straight through and there is a vacuum loop:

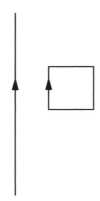

Again we cannot put the loop in contact with the propagator line, but the closed vacuum loop has, as do all closed fermion loops, a relative minus sign compared to the straight-through propagator. Thus adding in both forbidden diagrams changes nothing.

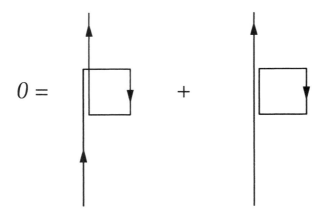

That this conspiracy works in general is easily seen by diagonalizing the action in momentum space:

$$\mathcal{Z} = \int d[\bar{\psi}]d[\psi] \exp\left\{\sum_k \lambda_k \bar{\psi}_k \psi_k\right\} = \prod_k \lambda_k \qquad (15.50)$$

and

$$\langle \psi_k \bar{\psi}_k \rangle = \frac{1}{\mathcal{Z}} \int d[\bar{\psi}]d[\psi] \psi_k \bar{\psi}_k \exp\left\{\sum_k \lambda_k \bar{\psi}_k \psi_k\right\} = \frac{1}{\lambda_k}. \qquad (15.51)$$

One sees that it is the cancellation between the vacuum integral and the other terms that leads to the propagator being the inverse of the matrix in the exponent. We can use a method similar to that for boson random walks to check that the sum (15.49) gives the correct propagator:

$$\sum_{paths} K^L \prod_L \gamma^i = \sum_L \int_{-\pi}^{\pi} \frac{d^d k}{(2\pi)^d} e^{-i\mathbf{k}\cdot\mathbf{n}} \left(\sum_i \gamma^i 2iK \sin k_i\right)^L$$

$$= \int_{-\pi}^{\pi} \frac{d^d k}{(2\pi)^d} e^{-i\mathbf{k}\cdot\mathbf{n}} \frac{1}{1 - \sum_i \gamma^i 2iK \sin k_i}$$

$$= \int_{-\pi}^{\pi} \frac{d^d k}{(2\pi)^d} e^{-i\mathbf{k}\cdot\mathbf{n}} \frac{1 + \sum_i \gamma^i 2iK \sin k_i}{1 + \sum_i 4K^2 \sin^2 k_i}. \quad (15.52)$$

After setting $m = (2K)^{-1}$ and remembering that the lattice ψ field is related to continuum field by $\psi_{continuum} = \sqrt{2K}\psi_{lattice}$ we see that for small $\mathbf{k}$ this looks like the continuum propagator

$$\langle \psi(\mathbf{x})\bar{\psi}(\mathbf{y})\rangle = \int \frac{d^d k}{(2\pi)^d} e^{i\mathbf{k}\cdot(\mathbf{x}-\mathbf{y})} \frac{m + i\slashed{k}}{m^2 + k^2}. \quad (15.53)$$

There is a lurking problem, however. This shows up if we try to evaluate

$$\langle \bar{\psi}_\alpha(\mathbf{x})\psi_\alpha(\mathbf{x})\rangle = \int_{-\pi}^{\pi} \frac{d^d k}{(2\pi)^d} \frac{\text{tr}(1)}{1 + \sum_i 4K^2 \sin^2 k_i} \quad (15.54)$$

as $m = 1/2K \to 0$. As we should expect, we find a large contribution to the integral near $\mathbf{k} = 0$, where the $\sin k \approx k$ becomes small. Unfortunately, because $\sin \pi = 0$, there are equally large contributions from points where any component of $\mathbf{k}$ is close to $\pm\pi$. In d dimensions there are 2^d such points in the Brillouin zone. All make the same contribution to the integral.

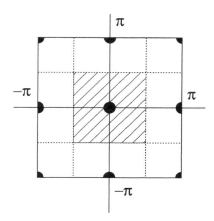

Fig 7. Each of the three dissected squares surrounding the dissected dots contribute the same to $\langle \bar{\psi}_\alpha(\mathbf{x})\psi_\alpha(\mathbf{x})\rangle$ as the shaded region.

We therefore find 2^d times the correct continuum expression. It is as if we started with one species of fermion, but have somehow ended up with 2^d species. The same problem will occur in any quantity we chose to compute. This is the notorious *fermion doubling problem*.

Many approaches have to tried to circumvent this multiplication of fermion species. Two of the most popular are *Wilson fermions*, and *Kogut-Susskind* or *staggered fermions*.

Wilson Fermions

Here we replace the $\pm \gamma^\mu$ factors in the hopping terms by the projection operators $(1 \pm \gamma^\mu)$. Then

$$G = \int_{-\pi}^{\pi} \frac{d^d k}{(2\pi)^d} e^{-i\mathbf{k}\cdot\mathbf{n}} \frac{1 + 2iK \sum_\mu \gamma_\mu \sin k_\mu + 2K \sum_\mu \cos k_\mu}{(1 - 2\sum_\mu \cos k_\mu)^2 + 4K^2 \sum_\mu \sin k_\mu}. \quad (15.55)$$

In this propagator the denominator can only be small near $\mathbf{k} = 0$ and hence there is only one fermion. Unfortunately the $(1+\gamma)$ factor breaks the $K \to -K$ symmetry of the naive action and Wilson fermions are not massless when $K = 0$. This has the consequence that, once interactions are introduced, it is no longer easy to locate the value of K at which the fermions become light, so taking the continuum limit is difficult.

Kogut-Susskind Fermions

These only partially solve the doubling problem, but have an interesting geometric interpretation.

We begin by diagonalizing the action in spin space. To do this set (in four dimensions)

$$\psi_{naive}(n_1, \ldots, n_4) = \gamma_4^{n_4} \gamma_3^{n_3} \gamma_2^{n_2} \gamma_1^{n_1} \psi_{K-S}(n_1, \ldots, n_4), \quad (15.56)$$

then the action becomes

$$S = \sum_{\mathbf{n},i} \left\{ \bar\psi(\mathbf{n})\psi(\mathbf{n}) + (-1)^{\phi(i,\mathbf{n})} \bar\psi(\mathbf{n}) \{\psi(\mathbf{n}+\mathbf{i}) - \psi(\mathbf{n}-\mathbf{i})\} \right\}. \quad (15.57)$$

(I have dropped the $K - S$ label from the fields.) The phases $(-1)^{\phi(i,\mathbf{n})}$ arise from commuting the γ's past one another. For example,

$$\begin{aligned}
&\bar\psi(\mathbf{n})\gamma_1(\psi(\mathbf{n}+1) - \psi(\mathbf{n}-1)) \\
&\to \bar\psi(\mathbf{n})\gamma_1^{n_1}\gamma_2^{n_2}\gamma_3^{n_3}\gamma_4^{n_4}\gamma_1(\gamma_4^{n_4}\gamma_3^{n_3}\gamma_2^{n_2}\gamma_1^{n_1+1}\psi(\mathbf{n}+1) \\
&\qquad - \gamma_4^{n_4}\gamma_3^{n_3}\gamma_2^{n_2}\gamma_1^{n_1-1}\psi(\mathbf{n}-1)) \\
&= (-1)^{n_2+n_3+n_4}\bar\psi(\mathbf{n})\gamma_1(\psi(\mathbf{n}+1) - \psi(\mathbf{n}-1)). \quad (15.58)
\end{aligned}$$

Similarly

$$\begin{aligned}
\phi(2, \mathbf{n}) &= n_3 + n_4, \\
\phi(3, \mathbf{n}) &= n_3, \\
\phi(4, \mathbf{n}) &= 0. \quad (15.59)
\end{aligned}$$

15. Lattice Field Theory

The only remnant of the γ algebra is the fact that

$$\prod_{plaquette} (-1)^\phi = -1. \tag{15.60}$$

This corresponds to the $\gamma_\mu \gamma_\nu \gamma_\mu \gamma_\nu = -1$ if $\mu \neq \nu$. Since the action is now diagonal in the spin components, we may keep only one of the four spin labels and have reduced the number of fermions (in four dimensions) from 16 to 4.

Kogut-Susskind fermions are really a lattice version of "Kahler-Dirac" fermions. These obey a first-order differential equation obtained by taking an alternative square root of the laplacian than that utilized by Dirac. If d is the exterior derivative acting on differential forms (totally antisymmetric covariant tensors), and δ is its adjoint, then $-\nabla^2 = (d + \delta)^2$, so $D = d + \delta$ can be used as an alternative to the Dirac equation. Instead of spinors this operator acts on the totality of differential forms of different degree. In four dimensions there are 16 independent components for totally antisymmetric tensors, and so Kahler-Dirac fermions look like a set of four Dirac fermions. They are not exactly equivalent, however, since they interact differently with a gravitational field.

15.4.1 No Chiral Lattice Fermions

The real world contains *chiral fermions*. In other words neutrinos are left-handed while antineutrinos are right handed. If there are any right-handed neutrinos, they do not participate in weak interactions, and so are invisible to us. A theorem due to Nielsen and Ninomiya, and extended by Friedan, asserts that no discretization of the Dirac equation on a periodic lattice can provide such chiral fermions. These authors' topological arguments demonstate that even if we manage to construct a chiral fermion pole in a propagator at some point in the lattice Brillouin zone, there will be another point at which there are fermions with the opposite chirality. Furthermore, gauge invariance will require us to couple both chiralities equally to any gauge field. Apart from the frustration this result generates for numerical study of the weak interactions, it has comforting theological implications: it is unlikely that our universe is a simultation on some megasupercomputer, so we are not in danger of being terminated when the grant expires.

16
The Renormalization Group

The renormalization group was discovered in relativistic field theories by Stückleberg, and independently by Gell-Mann and Low. Later Leo Kadanoff understood that something similar to the field theory renormalization group was the origin of universality in critical phenomena. It was Ken Wilson who, in a series of papers in the early 1970s, showed how to reconcile the continuum field theory and lattice statistical mechanics viewpoints and made the renormalization group into a practical tool for calculation. It is fair to say that Kadanoff and Wilson changed our entire attitude to quantum field theory. Before them field theory consisted of the Feynman rules for writing down the perturbation expansion, and the renormalization program for extracting finite answers from the divergent Feynman integrals. There was no clear notion of what a field theory *was* outside perturbation theory. Because the perturbation series is at best an asymptotic expansion, and because such expansions do not uniquely determine the quantity they represent, there was no agreement as to what constituted a valid computation of any nonperturbative effect. People of equal skill and good faith could come to very different conclusions. This is not true today.

In this chapter we will examine the renormalization group in the context of statistical mechanics. The lesson we will draw from this is that in order to take the continuum limit of a regulated theory we must locate a point in the parameter space of the theory where the correlation length becomes large. This will happen at a *critical point*, a place where a continuous phase transition occurs. Having located such a point, we zero in on it, rescaling our fields as we do so as to keep the correlation functions finite. Because continuous phase transitions exhibit a large degree of *universality*, the resultant continuum theory will usually be independent of the exact procedure used for regularization.

16.1 Transfer Matrices

The simplest model exhibiting a continuous phase transition is the Ising model of a ferromagnet. The degrees of freedom in this model are classical *spins* σ, each of which takes one of two possible values, $\sigma = 1$ corresponding to spin up (↑), or $\sigma = -1$ corresponding to spin down (↓).

The partition function for the one-dimensional version of the Ising model is

$$\mathcal{Z} = \sum_{\sigma_i = \pm 1} \exp\left\{ J \sum_i \sigma_i \sigma_{i-1} - h \sum_i \sigma_i \right\}. \tag{16.1}$$

Here we sum over all configurations of a line of spins σ_i.

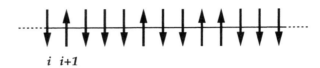

Fig 1. A typical spin configuration.

The positive parameter J controls the coupling between the spins. With J large (corresponding to low temperature), the spins tend to line up. With J small, the typical configurations become random.

The partition function may also be written as

$$\mathcal{Z} = \sum_{\sigma_i} \prod_i (\cosh J + \sigma_i \sigma_{i-1} \sinh J) \, e^{\frac{h}{2}\sigma_i} e^{\frac{h}{2}\sigma_{i-1}}, \tag{16.2}$$

and this form suggests the introduction of a *transfer matrix* $\langle \sigma | T | \sigma' \rangle$, which is defined by

$$T = \begin{bmatrix} e^{J-h} & e^{-J} \\ e^{-J} & e^{J+h} \end{bmatrix} = e^{-\frac{h}{2}\hat{\sigma}_z}(e^J + \hat{\sigma}_x e^{-J})e^{-\frac{h}{2}\hat{\sigma}_z}. \tag{16.3}$$

Here $\hat{\sigma}_{x,z}$ denotes the Pauli σ-matrices. They act as usual on a two-dimensional Hilbert space. We denote the eigenstates of $\hat{\sigma}_z$ by $|\sigma\rangle$, so with

$$\hat{\sigma}_z |\sigma\rangle = \sigma |\sigma\rangle, \qquad \hat{\sigma}_x |\sigma\rangle = |-\sigma\rangle. \tag{16.4}$$

The transfer matrix can also be written

$$T = e^{-\frac{h}{2}\hat{\sigma}_z} e^{\Theta \hat{\sigma}_x} e^{-\frac{h}{2}\hat{\sigma}_z} \sqrt{2 \sinh 2J}, \tag{16.5}$$

where $\tanh \Theta = e^{-2J}$.

The sum over the spin values at any site is now interpreted as the sum over one of the repeated indices in the matrix product of many of the T's. For N spins we have

$$\mathcal{Z} = \sum_{\{\sigma_i\}} \langle \sigma_N | T | \sigma_{N-1} \rangle \langle \sigma_{N-1} | T | \sigma_{N-2} \rangle \ldots \langle \sigma_1 | T | \sigma_0 \rangle. \tag{16.6}$$

How we treat outermost spins, σ_N and σ_0, depends on the boundary conditions we wish to impose. For periodic boundary conditions we set $\sigma_N = \sigma_0$ and sum over them. This corresponds to taking a trace, so

$$\mathcal{Z} = \mathrm{Tr}\left(T^N\right) = \lambda_+^N + \lambda_-^N. \tag{16.7}$$

Here the $\lambda_\pm$ are the two eigenvalues of T:

$$\lambda_\pm = e^J \left(\cosh h \pm \sqrt{\cosh^2 h - 2e^{-2J} \sinh 2J}\right). \tag{16.8}$$

As we approach the thermodyamic limit, $N \to \infty$, the larger eigenvalue λ_+ dominates, and when considering the free energy F defined by

$$\mathcal{Z} = e^{-NF}, \tag{16.9}$$

we may ignore the lesser eigenvalue and write

$$\begin{aligned}F &= -J - \ln\left(\cosh h + \sqrt{\cosh^2 h - 2e^{-2J} \sinh 2J}\right) \\ &= -J - \cosh^{-1}\left(\frac{\cosh h}{\sqrt{2e^{-J} \sinh 2J}}\right).\end{aligned} \tag{16.10}$$

From F we can find many of the properties of the system, for example, the magnetization $M(J, h)$ is given by

$$M(J, h) = \langle \sigma_i \rangle = \frac{1}{\mathcal{Z}} \sum_\sigma \sigma_i e^{-H} = -\frac{\partial F}{\partial h} = \frac{\sinh h}{\sqrt{\cosh^2 h - 2e^{-2J} \sinh 2J}}. \tag{16.11}$$

Here i can be any site, since they are all equivalent. For this one-dimensional model there is therefore no spontaneous magnetization at any finite value of J, i.e., M is a continuous function of h. Only when $J = \infty$, corresponding to zero temperature, do we find a sharp change in $\langle \sigma \rangle$. In this case there are only two contributing configurations. One with all spins up, and energy Nh, and one with all spins down, and energy $-Nh$. These give

$$\langle \sigma \rangle = \tanh Nh \to \mathrm{sgn}\, h \tag{16.12}$$

as N becomes large.

We have so far considered only periodic boundary conditions. If we want other boundary conditions, for example, leftmost spin up and rightmost spin down, then

$$\begin{aligned}\mathcal{Z} &= \langle \uparrow | T^N | \downarrow \rangle \\ &= \langle \uparrow | + \rangle \lambda_+^N \langle + | \downarrow \rangle + \langle \uparrow | - \rangle \lambda_-^N \langle - | \downarrow \rangle \\ &= \sum_\pm \varphi_\pm^*(\uparrow) \varphi_\pm(\downarrow) e^{-N(-\ln \lambda_\pm)}.\end{aligned} \tag{16.13}$$

This is reminiscent of the propagator function for a two-level quantum mechanical system with energies $-\ln \lambda_\pm$, the transfer matix playing the role of of $e^{-\hat{H}}$. The hamiltonian operator and Hilbert-space language are here being used as a tool for studying the sum over configurations — the converse of the path-integral approach to quantum mechanics where sum over configurations (the path integral) was introduced as tool for studying the hamiltonian.

The spin–spin correlation function (for any boundary conditions) is

$$\langle \sigma_i \sigma_{i-m} \rangle = \frac{1}{\mathcal{Z}} \sum_{\{\sigma\}} \sigma_i \sigma_{i-m} e^{-H}, \qquad (16.14)$$

and this is equal to

$$\frac{\langle \sigma_N | T_N T_{N-1} \ldots T_i \hat{\sigma}_z T_{i-1} \ldots \hat{\sigma}_z T_{i-m-1} \ldots T_0 | \sigma_0 \rangle}{\langle \sigma_N | T^N | \sigma_0 \rangle}. \qquad (16.15)$$

If we define a "Heisenberg" operator by $\hat{\sigma}_z(i) = T^{-i} \hat{\sigma}_z T^i$ we may use the Gell-Mann-Low theorem to write the correlation function as a ground-state expectation value

$$\langle +|T\{\hat{\sigma}_z(i)\hat{\sigma}_z(i-m)\}|+\rangle, \qquad (16.16)$$

where the "time" ordering symbol T denotes a product with the larger site indices to the left. We are of course using the fact that as N becomes large

$$|+\rangle \propto T^N |\sigma\rangle \qquad (16.17)$$

independently of what state $|\sigma\rangle$ we chose at the ends.

We can evaluate this expression explicitly when $h = 0$. We use

$$\langle \sigma | T | \sigma' \rangle = \cosh J + \sigma \sigma' \sinh J, \qquad (16.18)$$

togther with

$$\sum_{\pm 1} \sigma = 0, \qquad \sum_{\pm 1} \sigma^2 = 2, \qquad (16.19)$$

to find

$$\langle \sigma_i \sigma_{i-m} \rangle = \frac{(2\cosh J)^{n-|m|}(2\sinh J)^{|m|}}{(2\cosh J)^N} = (\tanh J)^{|m|}. \qquad (16.20)$$

The correlations therefore decay exponentially,

$$\langle \sigma_i \sigma_{i-m} \rangle = e^{-|i-j|/\xi} \qquad (16.21)$$

with $\xi^{-1} = -\ln \tanh J$. (A positive number since $0 < \tanh J < 1$.) The quantity ξ is called the correlation length.

16.1.1 Continuum Limit

Since $\sigma = \pm 1$, the spin configurations can never be regarded as being in any sense smooth continuous fields. The *correlation functions*, however, do have a smooth continuum limit — i.e., a limit in which the underlying graininess of the system becomes invisible. This occurs when we adjust the parameters in the theory so that the correlation functions become very long-ranged compared to the lattice spacing. For our one-dimensional Ising model we have

$$\langle \sigma_i \sigma_j \rangle = e^{-|i-j|(-\ln \tanh J)}. \tag{16.22}$$

If we set $x = ja$ with a the lattice spacing, we can write the two-point function as

$$\langle \sigma(x)\sigma(x') \rangle = e^{-|x-x'|m}. \tag{16.23}$$

Here the "mass" m is given by

$$m = -\frac{1}{a} \ln \tanh J. \tag{16.24}$$

We define the "continuum limit" by taking the lattice spacing $a \to 0$ at the same time as taking $J \to \infty$ in such a way that m remains fixed. Once we have done this we may regard x as continuous. Further, once we have decided to paramaterize the corelations in terms of m rather than J, there is no further dependence on J or a visible in the continuum expression. We think of m as being a "renormalized" quantity, while J is a "bare" parameter associated with the cut-off theory defined in terms of the small length a.

We may vary both J and a without affecting m provided their variations are suitably linked. For large J we have

$$ma \approx 2e^{-2J}, \tag{16.25}$$

so variations in them will not affect the continuum limit provided

$$a\frac{\partial J}{\partial a} = -2. \tag{16.26}$$

This is more profitably expressed in terms of the temperature $T = J^{-1}$ as

$$a\frac{\partial T}{\partial a} = 2T^2 \stackrel{def}{=} \beta(T). \tag{16.27}$$

The last definition serves to give us our first sight of a β-*function*, a function of central importance in renormalization theory. Equation (16.27) is our first example of a *renormalization group equation*.

16.1.2 Two-Dimensional Ising Model

A finite temperature, the one-dimensional Ising model Ising model has only one phase. Onsager[1] obtained the free energy of the two-dimensional version the model where

$$\mathcal{Z} = \sum_{\sigma_n=\pm 1} \exp\left\{ J \sum_{<nm>} \sigma_n \sigma_m - h \sum_n \sigma_n \right\}. \tag{16.28}$$

---

[1] Lars Onsager. Born November 27, 1903, Kristiania, Norway. Died October 5, 1976, Coral Gables, FL.

Here $\langle \mathbf{nm} \rangle$ denotes pairs of adjacent sites. He showed that the free energy was nonanalytic at at $J_{crit} = -\frac{1}{2}\ln(\sqrt{2}-1) \approx .4406$. This turns out to be a critical point. As we increase J toward J_{crit} the correlation length diverges, and for $J > J_{crit}$ the system spontaneously magnetizes. The latter statement means that $\langle\sigma\rangle$ regarded as a function of h has a discontinuity, jumping from $+M$ to $-M$ as we cross the line $h = 0$. This signals that the $\sigma \to -\sigma$ symmetry has been spontaneously broken. Onsager later showed that the magnetization $M(J)$, behaves nonanalytically in the neighborhood of the critical point, varying as

$$M(J) \propto (J - J_{crit})^\beta. \tag{16.29}$$

The exponent β is equal to $\frac{1}{8}$.

The Onsager solution starts from the transfer matrix

$$T = e^{\frac{J}{2}\sum_{i=1}^{N} \hat{\sigma}_z^{(i)} \hat{\sigma}_z^{(i-1)}} e^{\theta \sum_{i=1}^{N} \hat{\sigma}_x^{(i)}} e^{\frac{J}{2}\sum_{i=1}^{N} \hat{\sigma}_z^{(i)} \hat{\sigma}_z^{(i-1)}} (2\sinh 2J)^{\frac{N}{2}}, \tag{16.30}$$

where $\tanh\theta = e^{-2J}$. This couples an entire row of N spins to the next row. The relevant Hilbert space is therefore

$$\mathcal{H} = \bigotimes_{1}^{N} \mathcal{H}_2, \tag{16.31}$$

where $\mathcal{H}_2$ is the usual spin-$\frac{1}{2}$ space on which the individual $\hat{\sigma}^{(i)}$ act. Now the intermediate states $|\sigma\rangle$ are simultaneous eigenstates of all the $\hat{\sigma}_z^{(i)}$

$$\hat{\sigma}_z^{(i)}|\sigma\rangle = \sigma_i|\sigma\rangle, \tag{16.32}$$

where σ_i is value of the ith spin in the current row.

We will not go through the details of Onsager's solution (it works by mapping the problem onto a system of two-dimensional free Majorana fermions), but will restrict ourselves only to those issues that serve to illuminate the renormalization process.

As the coupling J approaches J_{crit} from either side, the correlation length of the spins diverges $\xi \propto |J - J_{crit}|^{-\nu}$ with $\nu = 1$. As the correlation length grows, the spin-spin correlator becomes rotationally symmetric at large separation, i.e.,

$$\langle\sigma_\mathbf{n}\sigma_\mathbf{m}\rangle \propto e^{-|\mathbf{n}-\mathbf{m}|/\xi}, \tag{16.33}$$

where $|\mathbf{n}-\mathbf{m}| = \sqrt{(n_1 - m_1)^2 + (n_2 - m_2)^2}$ is the euclidean distance between the points. Similar properties hold for any of the n-point functions. The Ising model therefore has a continuum limit in this region of J_{crit}. This is a general principle:

Continuuum limit = Critical point.

16.2 Block Spins and Renormalization Group

The notion of renormalization and the renormalization group is best introduced by thinking about spin configurations in the two-dimensional Ising model.

16.2 Block Spins and Renormalization Group

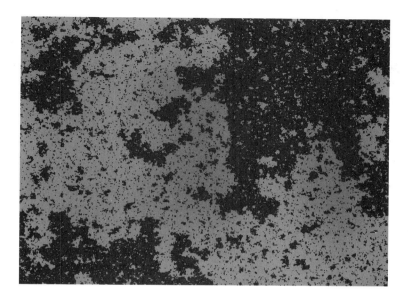

Fig 2. A spin configuration at $T = T_c$.

Fig. 2 shows a typical spin configuration exactly at T_c. Configurations at temperatures just above and just below the critical temperature look very similar. How can we tell them apart?

Suppose that such a configuration is reduced in scale by dividing it into blocks of nine spins, which are then allowed to vote on whether to be up or down.

Fig 3. A block votes to be spin-down ($s = 1 \equiv \uparrow$ is white).

After voting we shrink each block spin down to the size of an individual spin so the new system can be compared with the old one.

After a few iterations of this process a configuration with $T > T_c$ will evolve to complete randomness, while a configuration with $T < T_c$ will evolve to all spins up or all spins down. The critical configuration, however, preserves its character under the block voting procedure. It is a scale invariant *fractal*.

We can formalize the blocking process. Suppose we have a set of spins s and

$$\mathcal{Z} = \sum_{\{\sigma\}} e^{-H(\{\sigma\})}, \quad (16.34)$$

so that the probability distribution for the configuration $\{\sigma\}$ is

$$P(\sigma) = \frac{1}{\mathcal{Z}} e^{-H(\{\sigma\})}. \tag{16.35}$$

We set about *coarse graining* the system by defining general *block spins*. To do this, introduce a conditional probability $P(\{\sigma'\}|\{\sigma\})$. This is the probability of finding the block-spin configuration $\{\sigma'\}$, given that the original spin configuration is $\{\sigma\}$.

For example the 3×3 block voting process introduced above would have

$$P(\{\sigma'\}|\{\sigma\}) = \prod_I \delta\left(\sigma'_I - \text{sgn} \sum_{i \in I} (\sigma_i)\right). \tag{16.36}$$

Here I labels the new block spin made out of the nine original spins. In this case the block spins are a deterministic function of the old spins. Because $P(\{\sigma'\}|\{\sigma\})$ is a probability, we must have

$$\sum_{\{\sigma'\}} P(\{\sigma'\}|\{\sigma\}) = 1. \tag{16.37}$$

We now use $P(\{\sigma'\}|\{\sigma\})$ to write

$$\mathcal{Z} = \sum_{\{\sigma'\}} \sum_{\{\sigma\}} P(\{\sigma'\}|\{\sigma\}) e^{-H(\{\sigma\})} = \sum_{\{\sigma'\}} e^{-H'(\{\sigma'\}) + G} \tag{16.38}$$

[one should fix the division of the exponent into H' and G by requiring $\sum_{\sigma'} H'(\sigma') = 0$, or something similar].

By construction the new hamiltonian H' gives rise to the same partition function as before (after we take into account the constant G). Furthermore, an intelligent definition of the new block-spins as local functions of the old (as is the case of the 3×3 block voting process) preserves all the long-distance physics of the model. The blocking process is best thought of as a "blurring," or reduction in resolution, of our view of the spin system. It is equivalent to the "change of cut-off without changing the physics" introduced in Section 16.1.1. The change in hamiltonian is a discrete version of the beta function introduced in equation (16.27).

After blocking it is convenient to shrink the system by a factor R^{-1} ($\frac{1}{3}$ with the block vote) so that each block spin occupies the same space as one old spin.

Repeating the process we have

$$H(\sigma) \xrightarrow{P} H'(\sigma') \xrightarrow{P} H''(\sigma'') \to \ldots. \tag{16.39}$$

This gives us a discrete renormalization group flow.[2]

Suppose there exists a hamiltonian H^* such that

$$H^* \to H^*. \tag{16.40}$$

---

[2] The use of the word "group" is traditional. It is intended to draw attention to the fact that one may take powers of the block spin operation by applying the operation repeatedly. The blocking process is *not* invertible, however, so the renomalization "group" is only a *semigroup*.

16.2 Block Spins and Renormalization Group

Such an H^* is said to be a *fixed point* of the renormalization group transformation.

The physical significance of a fixed point depends on the way that neighboring hamiltonians behave under renormalization. Suppose we look at a hamiltonian that lies near H^*:

$$H = H^* + \sum_i a_i O^i. \tag{16.41}$$

Here the O^i are some set of operators, or additional interactions. Under the renormalization group flow we will have

$$H \to H^* + \sum_i a_i' O^i, \tag{16.42}$$

which we think of as the map

$$a_i \xrightarrow{P} a_i'. \tag{16.43}$$

Near H^* this flow should be linear $a_i \to a_i' = A_{ij} a_j + O(a^2)$. There is no reason why A_{ij} should be symmetric, but unless we are unlucky it can still be diagonalized. If so, we may assume that the O^i have been chosen so that the matrix A_{ij} is diagonal with entries Λ_i. Then

$$a_i \xrightarrow{P} \Lambda_i a_i \xrightarrow{P} \Lambda_i^2 a_i \xrightarrow{P} \ldots \tag{16.44}$$

(no sum on i intended).

If $|\Lambda_i| < 1$ the coefficient of O^i decreases under the renormalization group flow. We say that O^i is (infrared) *irrelevant*. Conversely, if $|\Lambda_i| > 1$, the coefficient of O^i increases and we say that O^i is a *relevant* perturbation of H^*. When $|\Lambda_i| = 1$, we say that O^i is a *marginal* perturbation. If H^* correponds to a massless free field theory, these terms are equivalent to a *nonrenormalizable, superrenormalizable,* and *renormalizable* perturbation, respectively. Relevent operators take us away from criticality. They correpond to terms such as temperature, mass, magnetic field, etc. Note most simple operators are not eigenvectors of the renormalization group flow. They will mix with one another unless prevented by some symmetry.

Correlation length

Each time we apply P, the system shrinks by $\frac{1}{3}$, or some other fixed ratio R^{-1}, so the correlation length does likewise:

$$\xi \xrightarrow{P} \frac{1}{3}\xi. \tag{16.45}$$

At a fixed point ξ does not change (because H does not), so there ξ must be either 0 or ∞. The latter is the interesting case. It correponds to a critical point for the statistical-mechanics problem, or, equivalently, to a massless field theory.

Suppose

$$H = H^* + aO, \tag{16.46}$$

with O relevant. Thus $a \xrightarrow{P} \Lambda a$ with $\Lambda = R^y$ and $y > 0$. Then we find that ξ will vary with a as

$$\xi = \left(\frac{a}{a_0}\right)^{-\nu}, \qquad \nu = \frac{1}{y}. \tag{16.47}$$

To see this observe that if we start with $a = a_0$ and $\xi = \xi_0$, then after N steps of the renormalization group flow we have $a = R^{Ny} a_0$, $\xi = R^{-N} \xi_0$. Taking logarithms we get

$$Ny \ln R = \ln\left(\frac{a}{a_0}\right),$$

$$-N \ln R = \ln\left(\frac{\xi}{\xi_0}\right). \tag{16.48}$$

Therefore, dividing one equation by the other, we find

$$-y \ln\left(\frac{\xi}{\xi_0}\right) = \ln\left(\frac{a}{a_0}\right), \tag{16.49}$$

or

$$\xi = \xi_0 \left(\frac{a}{a_0}\right)^{-\frac{1}{y}}. \tag{16.50}$$

This is a typical scaling argument.

Note that it is the value of y that is related to the physically measurable exponent ν, not the eigenvalue Λ itself.

There are many possible choices for the blocking operation P. Not all of them will have fixed points, and those that do may have rather different fixed-point hamiltonians H^*. Nonethless, the physical deductions from the existence of the fixed point should be the same. In particular, whether we form blocks out of 3×3 or 5×5 arrays of spins we should find the same value for ν. Similarly, if we have a choice of P's with different length ratios R, we expect that all the eigenvalues will be of the form $\Lambda_i = R^{y_i}$, where the y_i are be independent of R.

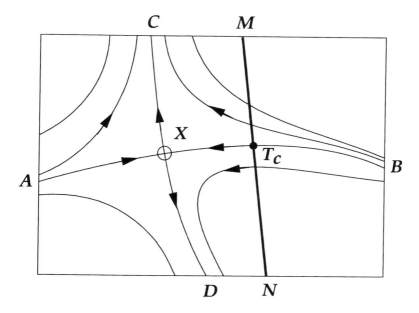

Fig 4. Renormalization group flow in the space of all couplings.

When we have a particular model such as the two-dimensional Ising model with its pair of parameters J, h, the fixed-point hamiltonian H^* will not in general lie in parameter space of the model. Instead, at T_c, the parameter space of the model will intersect the manifold of points that flow into the fixed point, represented by X in Fig 3. This attractive manifold is represented by the curve AXT_cB in the figure. The set of model hamiltonians obtained by varying J is represented by the line MT_cN. At $J^{-1} = T_c$, H differs from H^* only by irrelevant operators. All points on the attractive manifold have $\xi = \infty$ (ξ shrinks as we iterate the renormalization group, but $\xi = \infty$ at X, so it must have been ∞ all along), so the attractive manifold is a *critical surface*. Points not on the critical surface will usually flow to other fixed points where the correlation length is zero (correponding to infinite or zero temperature). Occasionally they may flow to other $\xi = \infty$ fixed points. This last situation give rise to *crossover* phenomena.

This picture of the critical temperature being determined by the intersection of the parameter space of the theory with the critical surface of some fixed point is the origin of *universality*. All theories whose parameter space intersects the capture domain of the fixed point will have exactly the same critical behavior. Thus the liquid gas critical point transition for fluids as diverse as water and helium all have the critical behavior of the three-dimensional Ising model. The lesson for field theory is that many different ways of regularizing or discretizing a field theory will lead to the same continuum theory.

16.2.1 Correlation Functions

For any operator $O(\mathbf{r})$ we can obtain the connected correlation function $G(\mathbf{r}_1 - \mathbf{r}_2) = \langle O(\mathbf{r}_1)O(\mathbf{r}_2)\rangle_{conn}$ by setting $H \to H - \sum_\mathbf{r} h(\mathbf{r})O(\mathbf{r})$ and differentiating

$$G(\mathbf{r}_1 - \mathbf{r}_2; H, h) = \frac{\delta^2 \ln \mathcal{Z}(H, h)}{\delta h(\mathbf{r}_1)\delta h(\mathbf{r}_2)}. \tag{16.51}$$

Now suppose that under the renormalization group we have, for spatially constant h, $H \xrightarrow{P} H'$, $h \xrightarrow{P} h' = \Lambda_O h$ where $\Lambda_O = R^{y_O}$. Then because the renormalization group preserves $\mathcal{Z}$ we have

$$\frac{\delta^2 \ln \mathcal{Z}(H', h')}{\delta h'(\mathbf{r}_1)\delta h'(\mathbf{r}_2)} = \frac{\delta^2 \ln \mathcal{Z}(H, h)}{\delta h'(\mathbf{r}_1)\delta h'(\mathbf{r}_2)}. \tag{16.52}$$

A little thought is required at this point. The left-hand side of (16.52) is simply the correlation function in the block-spin theory. Since the scale has been reduced by R after the block spinning, this is

$$G(\frac{1}{R}(\mathbf{r}_1 - \mathbf{r}_2), H', h'). \tag{16.53}$$

On the right-hand side, changing $h'(\mathbf{r})$ by $\delta h'(\mathbf{r})$ corresponds to altering the coefficient of all the $O(\mathbf{r})$ in the block centered at the point $\mathbf{r}$ by an amount $\delta h = R^{-y_O}\delta h'$. We are thus computing

$$R^{-2y_0} \sum_{(\mathbf{r}_1 \in \text{block } 1)} \sum_{(\mathbf{r}_2 \in \text{block } 2)} G(\mathbf{r}_1 - \mathbf{r}_2; H, h), \tag{16.54}$$

and assuming that $\mathbf{r}_1$ and $\mathbf{r}_2$ are sufficiently far apart that there is no significant change in the correlations across the block. Thus, the right-hand side is equal to

$$R^{2d-2y_0} G(\mathbf{r}_1 - \mathbf{r}_2; H, h). \tag{16.55}$$

We have therefore shown that

$$G(\frac{1}{R}(\mathbf{r}_1 - \mathbf{r}_2), H', h') = R^{2d-2y_0} G(\mathbf{r}_1 - \mathbf{r}_2; H, h). \tag{16.56}$$

Setting $h = 0$, and $H = H^*$, this shows us that, at the fixed point, the correlator has power-law decay

$$G(\mathbf{r}; H^*, 0) = |\mathbf{r}|^{-(2d-2y_0)}. \tag{16.57}$$

This $\mathbf{r}$ dependence suggests that each O behaves as if in some sense it has dimension $[O] = M^{d-y_O}$. We call $d - y_O$ the *scaling dimension* of the operator. We will have $y_0 > 0$, implying that the perturbation represented by O is relevant, if the scaling dimension of the operator is *less than* the dimension of space-time.

The difference between the engineering dimension and the scaling dimension of the operator reflects the self-similar fractal nature of the field configurations at the critical point.

17
Fields and Renormalization

In this chapter we will use the insight gained from the block-spin view of the renormalization group to guide our understanding of continuum Feynman diagram calculations.

17.1 The Free-Field Fixed Point

We begin by considering a scalar field theory in d-dimensional space-time. As usual we have set $\hbar = c = 1$ so that all physical quantities Q have dimensions $[Q] = M^y$ for some y. To find these dimensions y, we begin with the observation that, since the action integral

$$S[\varphi] = \int d^d x \left\{ \frac{1}{2}(\partial \varphi)^2 + V(\varphi) \right\} \tag{17.1}$$

appears in the path integral

$$\mathcal{Z} = \int d[\varphi] e^{-S[\varphi]} \tag{17.2}$$

as the argument of the exponential function, then S must be a *dimensionless* quantity.

Now S contains d powers of the space-time coordinate x coming from the $d^d x$, and -2 powers from the two ∂_{x^μ}'s. The field φ must therefore have dimension

$$[\varphi] = L^{1-\frac{d}{2}}, \tag{17.3}$$

or equivalently

$$[\varphi] = M^{\frac{d}{2}-1}, \tag{17.4}$$

to compensate for these. Since we always use M as the dimensional quantity, we simply say that φ has dimension $\frac{d}{2} - 1$. In particular φ has dimension 1 when $d = 4$, and is dimensionless when $d = 2$.

When $V = 0$ the field φ is free and massless. The correlation length is infinite, and the theory is at a scale-invariant critical point, called the *free-field fixed point*. For the free massless field the $\langle \varphi(x)\varphi(x') \rangle$ correlation function is calculated exactly in the appendix on dimensional regularization. There we find

$$\langle \varphi(x)\varphi(x') \rangle = \frac{\Gamma(\frac{d}{2} - 1)}{4\pi^{\frac{d}{2}}} \left(\frac{1}{x^2}\right)^{\frac{d}{2}-1}. \tag{17.5}$$

The correlator is given by a power law, as it should be at a fixed point, and using the definitions given in the last chapter we see that the *scaling dimension* of φ is also equal to $\frac{d}{2} - 1$. This equality of engineering and scaling dimension is characteristic of the free-field fixed point. The engineering dimension of φ is always equal to $\frac{d}{2} - 1$, but its scaling properties will be different at other fixed points.

In the last chapter we showed an operator is *relevant* when its scaling dimension is less than the dimension of space-time. We can therefore investigate the relevance of various perturbations of the free massless action. At the free-field fixed point, the scaling (and engineering) dimension of the monomial φ^n is $n(\frac{d}{2} - 1)$. Thus, for example, a $\lambda\varphi^4$ interaction will be relevant if and only if

$$2d - 4 < d. \tag{17.6}$$

That is if $d < 4$. In exactly four dimensions such an interaction is *marginal*. In more than four dimensions, it is irrelevant. The later statement means that if we try to construct a continuum $\lambda\varphi^4$ theory in the vicinity of the free-theory fixed point, the $\lambda\varphi^4$ interaction will have no effect on the continuum limit. In the old fashioned terminology such irrelevant interactions were called *nonrenormalizable*. Nonrenormalizable interactions were regarded as a problem because their effects could not be made "finite," or cut-off independent, in any simple way. The modern language says essentially the same thing: when any finite amount of an irrelevant interaction appears in the action of a cut-off theory, it will have no effect on physics at energies far below the cut-off scale. In other words, the interaction will disappear when the cut-off is removed to infinitely high energy. Thus, we cannot have such interactions without explicit cut-off dependence. We no longer regard this as a technical problem though. It is just the way things are.

There is an intuitive explanation as to why a $\lambda\varphi^4$ interaction is irrelevant above four dimensions. Consider two smooth manifolds of dimension D_1 and D_2 embedded in a d dimensional space. Generically they intersect in a $D_1 + D_2 - d$ dimensional manifold. For example, two surfaces in three dimensions typically intersect in a one-dimensional curve ($2 + 2 - 3 = 1$), two lines in two dimensions intersect in a zero-dimensional point ($1 + 1 - 2 = 0$), and two lines in three dimensions usually do not intersect at all. Since $1 + 1 - 3 = -1$, the negative dimension should be interpreted as generic non-intersection. A $\lambda\varphi^4$ term in the lagrangian corresponds to a point-contact interaction between the worldline of two particles, and to have any effect these worldlines must touch. Now, as we have seen in Chapter 15, particle worldlines in euclidean space-time are effectively random walks, and a random walk is a set of fractal dimension two. If we can extend our intersection-dimension counting to fractals, we immediately see that two random walks will usually fail to meet when $d > 4$. Furthermore, in more than four dimensions, a random walker will never revisit a place she has been, and there will be no self interactions either. Thus, unless the world lines are made "fat" by putting them on a lattice, a point-contact interaction can have no effect. This heuristic explanation can be elevated to a rigorous argument.

In fewer than four dimensions random walks do intersect, and the free-theory fixed point is unstable to the $\lambda\varphi^4$ perturbation. The renormalization group trajectories flow to new fixed point at finite $\lambda = \lambda^*$. This *Wilson-Fisher* fixed point controls the long distance behavior of the theory. For a single component φ field, this long-distance behavior coincides with that of the Ising model.

A mass term $\frac{1}{2}m\varphi^2$ is relevant when

$$d - 2 < d. \tag{17.7}$$

This inequality is satisfied identically, so a mass term is *always* a relevant perturbation to the free theory.

In $d = 1$ our field theory reduces to quantum mechanics. Here we usually write $x(t)$ instead of $\varphi(t)$. The engineering dimension of x is, perhaps surprisingly, $-\frac{1}{2}$. (This is because of our $\hbar = c = 1$ conventions.) Similarly $[x^n] = -\frac{1}{2}n$. Thus any polynomial in x is relevant. This is consistent with the fact that quantum mechanics is sensitive to all aspects of the potential $V(x)$ in which a particle moves. The large number of possible relevant interactions also means that in quantum mechanics we have to be much more careful in how we discretize our path integrals. We have already seen an example of this in Chapter 12 where we saw that different discretizations may correspond to different orderings of non-commuting operators in the hamiltonian.

An important case occurs in two dimensions, where φ is dimensionless. Here interaction densities of the form

$$\lambda_{\alpha\beta}(\varphi)\partial_\mu\varphi^\alpha\partial_\mu\varphi^\beta, \tag{17.8}$$

where $\lambda_{\alpha\beta}(\varphi)$ is an arbitrary function of a set of fields φ^α, will be marginal. These give rise to the the *nonlinear σ-models*. We will devote some time to the case that $\delta_{\alpha\beta} + \lambda_{\alpha\beta}(\varphi)$ is a metric on the N-sphere.

17.2 The Gaussian Model

Consider a two-dimensional complex field φ whose magnitude is fixed to be unity. One can think of φ as representing a classical spin field where the spins are forced to

17. Fields and Renormalization

point in the $x-y$ plane. The partition function for such spins with a ferromagnetic interaction is

$$\mathcal{Z} = \int d[\varphi]d[\varphi^*]\delta(|\varphi|^2 - 1)e^{-\frac{1}{2g^2}\int d^2x|\partial\varphi|^2}. \tag{17.9}$$

For a ferromagnet $g^2 = J/T$, where J is the exchange coefficient and T the temperature.

We would like to compute correlation functions such as $\langle \varphi(x)\varphi^*(y) \rangle$ in this theory.

By writing

$$\varphi(x) = e^{i\theta(x)}, \qquad \varphi^*(x) = e^{-i\theta(x)} \tag{17.10}$$

we can simplify the partition function to

$$\mathcal{Z} = \int d[\theta] e^{-\int d^2x \frac{1}{2g^2}(\partial\theta)^2}. \tag{17.11}$$

In this form we are reduced to computing correlators of the exponential operators $e^{\pm i\theta(x)}$ in a free-field theory.

To compute correlators of the form

$$\langle e^{i\theta(x_1)} e^{i\theta(x_2)} \ldots \rangle, \tag{17.12}$$

we use the gaussian integral identity (12.74)

$$\langle e^{i\int d^2x J(x)\theta(x)} \rangle = e^{-\frac{1}{2}\int d^2x d^2y J(x) G(x,y) J(y)}. \tag{17.13}$$

Here $G(x, y)$ is the propagator. For the two-dimensional field theory with action

$$S = \frac{1}{g^2} \int d^2x \left\{ \frac{1}{2}(\partial\theta)^2 + \frac{1}{2}m_0^2\theta^2 \right\}, \tag{17.14}$$

the propagator is

$$G(x, m_0^2) = g^2 \int \frac{d^2k}{(2\pi)^2} \frac{e^{ikx}}{k^2 + m_0^2}$$

$$= g^2 \left(-\frac{1}{2\pi} \ln m_0|x| + const. + O(m_0|x|) \right). \tag{17.15}$$

We want the propagator for the massless, $m_0 = 0$, case. Unfortunately, as $m_0 \to 0$ the k integral diverges at small k. This infrared divergence (associated with long distance) is the origin of the m_0 in the logarithm, and means that we cannot simply set $m_0 = 0$ in $G(x, y)$. We will therefore perform our calculations by assuming that a small mass is present. Provided we are careful in what we choose to calculate, we will discover that the final result will not depend on m_0.

Let us calculate $\langle \varphi(x_1)\varphi^*(x_2) \rangle$. This we do by setting $J(x) = \delta(x - x_1) - \delta(x - x_2)$. Thus,

$$\langle \varphi(x_1)\varphi^*(x_2) \rangle = \langle e^{i\theta(x_1)} e^{-i\theta(x_2)} \rangle$$
$$= \exp\left(G(0) - G(|x_1 - x_2|) \right). \tag{17.16}$$

17.2 The Gaussian Model

We now have to face the problem of interpreting $G(0)$. This expression is divergent, being $\propto \ln 0$. This time the infinity is due to an *ultraviolet* or *short-distance* divergence. One way to proceed is to replace the operator $\varphi(x_1) = \exp\{i\theta(x_1)\}$ by

$$\varphi_a(x_1) = e^{i \int d^2x \rho_a(x-x_1)\theta(x)}, \tag{17.17}$$

where $\rho_a(x)$ is some smearing function localized in a region of radius a about the origin, and obeying

$$\int d^2x \, \rho_a(x) = 1. \tag{17.18}$$

This smeared operator can be thought of as a kind of block spin. Alternatively, we may regularize the entire theory by defining it on a lattice with lattice spacing a. The effect will be the same in either case: the replacement of the ill-defined $\ln m_0 0$ by the finite expression $\ln m_0 a$.

With the cut-off a in place we have

$$\langle \varphi(x_1)\varphi^*(x_2)\rangle = \exp\left(\frac{g^2}{2}(\ln am_0 - \ln m_0|x_1 - x_2|)\right)$$

$$= \left(\frac{a}{|x_1 - x_2|}\right)^{\frac{g^2}{2\pi}}. \tag{17.19}$$

Note how the infrared cut-off m_0 has canceled between the two logarithms. The short distance cut-off a is still present, however.

We may isolate the a dependence by introducing a *renormalization point* or *renormalization mass*, which we will denote by μ. We factor $\langle \varphi(x_1)\varphi^*(x_2)\rangle$ as

$$\left(\frac{a}{|x_1 - x_2|}\right)^{\frac{g^2}{2\pi}} = \left(\frac{1}{\mu|x_1 - x_2|}\right)^{\frac{g^2}{2\pi}} (a\mu)^{\frac{g^2}{2\pi}}, \tag{17.20}$$

and regard the a independent left-hand factor as the two-point function of a *renormalized field*, $\varphi^{(R)}(x)$, where

$$\varphi(x) = \sqrt{Z}\varphi^{(R)}(x) \tag{17.21}$$

with

$$\sqrt{Z} = (a\mu)^{\frac{g^2}{4\pi}} = e^{\frac{g^2}{4\pi} \ln \mu a}. \tag{17.22}$$

The original $\varphi(x)$ is called the *bare*, or unrenormalized, field. The correlation function of the renormalized field is

$$\langle \varphi^{(R)}(x_1)\varphi^{*(R)}(x_2)\rangle = \left(\frac{1}{\mu|x_1 - x_2|}\right)^{\frac{g^2}{2\pi}}. \tag{17.23}$$

This expression no longer depends on the cut-off a. It is therefore "finite," or no longer divergent in the $a \to 0$ limit.

To make it clear that the $\sqrt{Z}$'s are local — being a property of the individual $\varphi(x)$ rather than the overall correlator — we can look a more general correlator

$$\langle \varphi(x_1)\ldots\varphi(x_n)\varphi^*(y_1)\ldots\varphi^*(y_m)\rangle = G_{bare}^{(n,m)}(x_1,\ldots,x_n;y_1,\ldots,y_m). \tag{17.24}$$

The same method as before shows that this is equal to

$$(m_0^2)^{\frac{g^2}{4\pi}(n-m)^2} (a)^{\frac{g^2}{4\pi}(n+m)} \left(\frac{\prod_{i \neq j}^n |x_i - x_j| \prod_{i' \neq j'}^m |y_{i'} - y_{j'}|}{\prod_{i,j} |x_i - y_j|} \right)^{\frac{g^2}{2\pi}}. \quad (17.25)$$

Note that m_0 appears to a positive power. There will therefore be no divergence if we try to take m_0 to zero, but the resultant expression will vanish unless $n = m$. The case $n = m$ is special because then the correlators are invariant under the global $O(2)$ rotation

$$e^{i\theta(x)} \to e^{i\alpha} e^{i\theta(x)}, \quad e^{-i\theta(x)} \to e^{-i\alpha} e^{-i\theta(x)}. \quad (17.26)$$

The mass term $\frac{1}{2} m_0^2 \theta^2$ explicitly breaks this symmetry and tends to line up all the spins in the $\theta = 0$ direction. The vanishing of the $n \neq m$ correlators as $m_0 \to 0$ means that the $O(2)$ symmetry is not spontaneously broken. This is a special case of a general theorem proved in somewhat different forms by Mermin and Wagner and by Coleman. The theorem asserts that no continuous global symmetry can be spontaneously broken in two or fewer dimensions.

Assuming that $n = m$, we may insert factors of μ next to all the a's and $|x - y|$'s without changing anything. Thus,

$$G_{bare}^{n,n} = (\mu a)^{\frac{g^2}{4\pi}(2n)} \left(\frac{\prod_{i \neq j}^n (\mu |x_i - x_j|) \prod_{i' \neq j'}^n (\mu |y_{i'} - y_{j'}|)}{\prod_{i,j} (\mu |x_i - y_j|)} \right)^{\frac{g^2}{2\pi}}. \quad (17.27)$$

The cut-off dependence is again removed by setting $\varphi(x) = \sqrt{Z} \varphi^{(R)}$ with $Z = (\mu a)^{\frac{g^2}{2\pi}}$. Thus,

$$G_R^{n,n} \equiv \langle \varphi^{(R)}(x_1) \ldots \varphi^{(R)}(x_n) \varphi^{*(R)}(y_1) \ldots \varphi^{*(R)}(y_m) \rangle$$
$$= \left(\frac{\prod_{i \neq j}^n (\mu |x_i - x_j|) \prod_{i' \neq j'}^m (\mu |y_{i'} - y_{j'}|)}{\prod_{i,j} (\mu |x_i - y_j|)} \right)^{\frac{g^2}{2\pi}}. \quad (17.28)$$

The renormalized fields $\varphi^{(R)}(x)$:

- Have cut-off independent ("finite") Green functions.
- Depend on the renormalization point μ.
- Have scaling dimension $M^{\frac{g^2}{4\pi}}$.
- Have engineering dimension M^0.

The dependence of the renormalized correlator on μ can be obtained by observing that the correlator of the *bare* fields does not depend on μ. Thus

$$0 = \mu \frac{\partial}{\partial \mu} \langle \varphi(x_1) \ldots \varphi(x_n) \varphi^*(y_1) \ldots \varphi^*(y_n) \rangle$$
$$= Z^n \left(\mu \frac{\partial}{\partial \mu} + 2n\mu \frac{\partial \ln \sqrt{Z}}{\partial \mu} \right) \langle \varphi^{(R)}(x_1) \ldots \varphi^{(R)}(x_n) \varphi^{*(R)}(y_1) \ldots \varphi^{*(R)}(y_m) \rangle.$$
$$(17.29)$$

Since

$$\mu \frac{\partial \ln \sqrt{Z}}{\partial \mu} = \frac{g^2}{4\pi}, \tag{17.30}$$

we find that

$$\left(\mu \frac{\partial}{\partial \mu} + 2n \frac{g^2}{4\pi}\right) G_R^{(n,n)}(x_1, \ldots, x_n; y_1, \ldots, y_n) = 0. \tag{17.31}$$

This is an example of a renormalization group equation (RGE) in the form used in particle physics. It differs from the RGE that we obtained earlier [Equation (16.27)] in that it describes how renormalized quantities vary when the renormalization point is changed while keeping the bare theory fixed. In statistical mechanics it is more common to vary the cut-off a while keeping the renormalized quantities fixed. The two approaches are equivalent.

The utility of such RGEs will become clearer as we proceed.

The expression

$$\gamma(g^2) = \mu \frac{\partial \ln \sqrt{Z}}{\partial \mu} = \frac{g^2}{4\pi} \tag{17.32}$$

is called the *anomalous dimension* of the field. It is the difference between the scaling dimension of the field and its engineering dimension.

17.3 General Method

The general strategy of renormalization is to compute some correlation function

$$G = G_B^{(n)}(x, g_B, \Lambda) \tag{17.33}$$

as a power series expansion in the bare couplings g_B, using a theory with a cut-off at some large momentum $\Lambda = a^{-1}$. Here x is shorthand for all the x_i's in the correlator, and the subscript B on G means that we have computed a correlation function of bare fields. We will need to include all couplings that are *relevant* at the free-field fixed point and are compatable with any symmetries that the theory may possess. We now rewrite this correlator in terms of $G_R^{(n)}$, which contains renormalized fields and renormalized couplings — the latter determined by a parametrization of the theory at some renormalization point μ. We will have

$$G_B^{(n)}(x, g_B, \Lambda) = Z^{\frac{n}{2}}(g_B, \frac{\mu}{\Lambda}) G_R^{(n)}(x, g_R, \mu, \Lambda). \tag{17.34}$$

The idea is that, if $x \gg \Lambda^{-1}$, then all effects of any *irrelevant* couplings in the bare theory will be negligible, and, in particular, $G_R^{(n)}$ will be Λ independent, i.e.,

$$G_B^{(n)}(x, g_B, \Lambda) = Z^{\frac{n}{2}}(g_B, \frac{\mu}{\Lambda}) G_R^{(n)}(x, g_R, \mu). \tag{17.35}$$

From this we can deduce the renormalization group equation. We write the trvial identity

$$0 = \mu \frac{\partial G_B^{(n)}}{\partial \mu}\bigg|_{g_B, \Lambda}, \tag{17.36}$$

as

$$0 = \left(\mu \frac{\partial}{\partial \mu} + \beta(g_R) \frac{\partial}{\partial g_R} + n\gamma(g_R)\right) G_R^{(n)}(x, g_R, \mu), \tag{17.37}$$

where

$$\beta(g_R) = \mu \frac{\partial g_R}{\partial \mu}\bigg|_{g_B, \Lambda}, \qquad \gamma(g_R) = \mu \frac{\partial \ln \sqrt{Z}}{\partial \mu}\bigg|_{g_B, \Lambda}. \tag{17.38}$$

We could equally well write

$$0 = \mu \frac{\partial G_R^{(n)}}{\partial \Lambda}\bigg|_{g_R, \mu}, \tag{17.39}$$

as

$$0 = \left(\Lambda \frac{\partial}{\partial \Lambda} + \tilde{\beta}(g_B) \frac{\partial}{\partial g_B} - n\tilde{\gamma}(g_B)\right) G_B^{(n)}(x, g_B, \Lambda), \tag{17.40}$$

where

$$\tilde{\beta}(g_B) = \Lambda \frac{\partial g_B}{\partial \Lambda}\bigg|_{g_R}, \qquad \tilde{\gamma}(g_B) = \Lambda \frac{\partial \ln \sqrt{Z}}{\partial \Lambda}\bigg|_{g_R}. \tag{17.41}$$

This would be closer to what we did in deriving equation (16.27), and is the way that the RGE often appears in the statistical mechanics literature.

Obviously the β's and γ's do not depend on x. By construction they do not depend on Λ, so they cannot depend on μ either. (They are dimensionless and there would be no way to make a dimensionless expression were they to contain μ.) They are therefore simple power series in the dimensionless coupling constant g_R^2.

17.4 Nonlinear σ Model

As a concrete example of renormalization at work consider the two-dimensional nonlinear $O(N)$ nonlinear σ-model. The name "nonlinear σ-model" comes from particle physics where σ-models provide a framework for studying the consequences of the notion that the three light pions, $\pi^\pm, \pi^0$, are the Goldstone bosons of spontaneously broken chiral symmetry. The $O(N)$ *linear* σ-model has an $O(N)$ symmetric interaction potential

$$V(\varphi_i) = \frac{\lambda}{4!}(|\varphi|^2 - 1)^2, \tag{17.42}$$

where φ is a N component real field $\varphi = (\varphi_1, \ldots, \varphi_N)$. The name "$\sigma$-model" arises because, when expanding about the broken symmetry vacuum $\varphi_0 = (1, 0, \ldots)$, it is traditional to write $\varphi = (\sigma, \pi_1, \ldots, \pi_{N-1})$. The multiplet of $N - 1$ π fields represent the Goldstone boson pions, and the σ field is the remaining massive radial mode. In the "nonlinear" version of this model, we take $\lambda \to \infty$, so $|\varphi|^2$ is forced to take the value unity.

We can think of the fields in the nonlinear model as mappings $\mathbf{R}^2 \to S^{N-1}$. In other words we have an N component field $\mathbf{n} \equiv (n_1, \ldots, n_N)$ subject to the constraint $\mathbf{n}^2 = 1$. One may generalize this notion by considering mappings from $\mathbf{R}^N$ to any manifold M, which is then called the *target space*. We will restrict ourselves to the simplest case, however.

The partition function is given by

$$\mathcal{Z} = \int d[\mathbf{n}] \delta(\mathbf{n}^2 - 1) e^{-\frac{1}{2g^2} \int d^2 x \, (\partial_\mu \mathbf{n}) \cdot (\partial_\mu \mathbf{n})}. \tag{17.43}$$

We are considering the two-dimensional version of the model because it is in this dimension that the coupling constant g^2 is dimensionless (in d dimensions $[g^2] = M^{2-d}$ and the interactions are marginal perturbations of the free-field fixed point).

To derive the perturbation series it is convenient to make the replacement $\mathbf{n} \to g\mathbf{n}$, so we have

$$\mathcal{Z} = \int d[\mathbf{n}] \delta\left(\mathbf{n}^2 - \frac{1}{g^2}\right) e^{-\frac{1}{2} \int d^2 x \, (\partial_\mu \mathbf{n}) \cdot (\partial_\mu \mathbf{n})}. \tag{17.44}$$

To evaluate the functional integral we must select a co-ordinate system on the target space S^{N-1}. We will take

$$n_1 = \sqrt{\frac{1}{g^2} - \pi^2} = \frac{1}{g}\sqrt{1 - g^2 \pi^2},$$

$$n_i = \pi^i \quad N \geq i > 1. \tag{17.45}$$

These co-ordinates only parametrize half of the sphere, and the π fields should be restricted to the region $\pi^2 g^2 < 1$, but these problems do not show up in the perturbation series because this only probes small fluctuations about the uniform configuration.

We can now write

$$\mathcal{Z} = \int d[\pi] \prod_x \sqrt{g(x)} e^{-\frac{1}{2} \int d^2 x \, g_{ij} \partial \pi^i \partial \pi^j}, \tag{17.46}$$

where g_{ij} is the metric tensor on S^{N-1} in our chosen coordinates, and $g = \det(g_{ij})$. The metric tensor is

$$g_{ij} = \delta_{ij} + g^2 \frac{\pi^i \pi^j}{1 - g^2 \pi^2}, \tag{17.47}$$

and
$$\sqrt{g} = \frac{1}{\sqrt{1-g^2\pi^2}}. \tag{17.48}$$

Thus,
$$\mathcal{Z} = \int \frac{d[\pi]}{\sqrt{1-g^2\pi^2}} e^{-\frac{1}{2}\int d^2x \left(\delta_{ij}+g^2 \frac{\pi^i\pi^j}{1-g^2\pi^2}\right)\partial\pi^i\partial\pi^j}. \tag{17.49}$$

To lowest order the action is simply $N-1$ free π fields. Then comes the interaction terms – different expressions at each order in g^2. There is one factor of $1/\sqrt{1-g^2\pi^2}$ at each point x in $\mathbf{R}^2$, so taking note that $\delta^2(0) = \int d^2k/(2\pi)^2$ is the density of points in $\mathbf{R}^2$ we can write

$$\mathcal{Z} = \int d[\pi] e^{-\frac{1}{2}\int d^2x \left\{\left(\delta_{ij}+g^2 \frac{\pi^i\pi^j}{1-g^2\pi^2}\right)\partial\pi^i\partial\pi^j + \delta^2(0)\ln(1-g^2\pi^2)\right\}}. \tag{17.50}$$

The delta function gives an extra point-contact interaction that is essential for preserving the Ward indentities.

In order to avoid infrared problems cause by long-wavelength low-energy fluctuations, it is preferable to compute $O(N)$ invariant correlators such as

$$\langle \mathbf{n}(x) \cdot \mathbf{n}(y) \rangle = \frac{1}{g^2} \langle \sqrt{1-g^2\pi^2(x)}\sqrt{1-g^2\pi^2(y)} \rangle + \langle \pi(x) \cdot \pi(y) \rangle. \tag{17.51}$$

As in the gaussian model, these turn out to be infrared finite without the need to introduce a symmetry breaking field to give mass to the Goldstone modes.

The lowest-order terms are

$$\langle \frac{1}{g^2} \rangle + \frac{1}{g^2}\left(-\frac{g^2}{2}\langle \pi^2(x) \rangle - \frac{g^2}{2}\langle \pi^2(y) \rangle\right) + \langle \pi^i(x)\pi^i(y) \rangle. \tag{17.52}$$

The propagator is

$$\langle \pi^i(x)\pi^j(y) \rangle = \delta_{ij} \int \frac{d^2x}{(2\pi)^2} \frac{e^{ik(x-y)}}{k^2+m_0^2} = -\frac{\delta_{ij}}{2\pi}\ln m_0(x-y) = \delta_{ij}G(x-y). \tag{17.53}$$

Once again m_0 is an infrared cut-off, necessary at intermediate stages of the computation, but dropping out in the final result.

We find
$$g^2 \langle \mathbf{n}(x) \cdot \mathbf{n}(y) \rangle = 1 + (N-1)g^2(G(x-y) - G(0)) + O(g^4) \tag{17.54}$$

and we see that the m_0 regulator cancels between the two terms.

At next order we must keep track of terms from expanding out the square root

$$\sqrt{1-g^2\pi^2} = 1 - \frac{1}{2}g^2\pi^2 - \frac{1}{8}g^4(\pi^2)^2 + \frac{1}{16}g^6(\pi^2)^3 + \ldots \tag{17.55}$$

to order g^4. In addition we begin to see interaction terms coming from the metric. The terms from expanding out the square roots to $O(g^4)$ are

$$\frac{1}{4}g^4 \langle \pi^2(x)\pi^2(y) \rangle - \frac{1}{8}g^4 \langle (\pi^2(x))^2 \rangle - \frac{1}{8}g^4 \langle (\pi^2(y))^2 \rangle. \tag{17.56}$$

17.4 Nonlinear σ Model

These contribute

$$g^4 \left\{ \frac{1}{2}(N-1)G^2(x-y) + \frac{1}{2}G^2(0)\{\frac{1}{2}(N-1)^2 - \frac{1}{2}(N-1)^2 - (N-1)\} \right\}. \tag{17.57}$$

The three polynomials multiplying $G^2(0)$ come from the three Wick-contraction terms

$$\frac{g^4}{4} \langle \pi^i(x)\pi^i(x) \rangle \langle \pi^j(y)\pi^j(y) \rangle = \frac{g^4}{4}(N-1)^2 G^2(0), \tag{17.58}$$

$$-\frac{g^4}{8} \left\{ \langle \pi^i(x)\pi^i(x) \rangle \langle \pi^j(x)\pi^j(x) \rangle + \langle \pi^i(y)\pi^i(y) \rangle \langle \pi^j(y)\pi^j(y) \rangle \right\}$$
$$= -\frac{g^4}{4}(N-1)^2 G^2(0), \tag{17.59}$$

and

$$-\frac{g^4}{4} \left\{ \langle \pi^i(x)\pi^j(x) \rangle \langle \pi^j(x)\pi^i(x) \rangle + \langle \pi^i(y)\pi^j(y) \rangle \langle \pi^j(y)\pi^i(y) \rangle \right\}$$
$$= -\frac{g^4}{4}(N-1)G^2(0). \tag{17.60}$$

Next we have loops with interactions derived from

$$\frac{g^2}{2}\frac{\pi^i \pi^j}{1-g^2\pi^2} \partial \pi^i \partial \pi^j = g^2 \pi^i \pi^j \partial \pi^i \partial \pi^j + O(g^4), \tag{17.61}$$

together with

$$\frac{1}{2}\delta^2(0) \ln(1-g^2\pi^2) = -\frac{1}{2}\delta^2(0)\pi^2 + O(g^4). \tag{17.62}$$

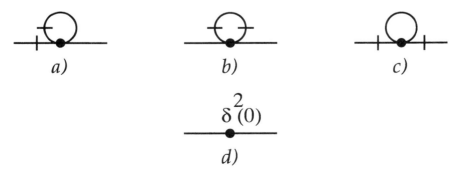

Fig 1. *One-loop diagrams arising from the interaction $g^2\pi^i\pi^j\partial\pi^i\partial\pi^j$. The crosslines indicates which propagators are to be differentiated.*

These may dress any of the $O(g^2)$ propagators.

17. Fields and Renormalization

In k space diagram a) is equal to

$$\frac{k_\mu}{k^2}\left(\int \frac{d^2q}{(2\pi)^2}\frac{q_\mu}{q^2}\right)\frac{1}{k^2}, \tag{17.63}$$

and this is zero by the $q \to -q$ symmetry of the integrand.
Diagram b) is equal to

$$\frac{1}{k^2}\left(\int \frac{d^2q}{(2\pi)^2}\frac{q^2}{q^2}\right)\frac{1}{k^2}, \tag{17.64}$$

and cancels against diagram d) arising from the measure.
This leaves diagram c), which is

$$-g^2\int d^2z\, \partial_z G(z-y)\partial_z G(x-z)(N-1)G(0). \tag{17.65}$$

Integrating by parts, and using $\partial_x^2 G(x-y) = \delta^2(x-y)$, reduces this to

$$g^2(N-1)G(x-y)G(0). \tag{17.66}$$

We insert these diagrams into all the $O(g^2)$ diagrams we have considered so far. The result is that

$$g^2\langle\mathbf{n}(x)\cdot\mathbf{n}(y)\rangle = 1 + g^2(N-1)\{G(x-y) - G(0)\} +$$
$$+ \frac{g^4}{2}(N-1)\{G(x-y) - G(0)\}^2 + O(g^6). \tag{17.67}$$

Once again we find the infrared finite, but short-distance divergent combinations $\{G(x-y) - G(0)\}$.
Using $G(x-y) = -\frac{1}{2\pi}\ln m_0|x-y|$ with $G(0) = G(\Lambda^{-1})$ we have

$$g^2\langle\mathbf{n}(x)\cdot\mathbf{n}(y)\rangle = 1 - \frac{g^2}{2\pi}(N-1)\ln\Lambda|x-y| + \frac{1}{2}\left(\frac{g^2}{2\pi}\right)^2(N-1)\ln^2\Lambda|x-y| + \cdots. \tag{17.68}$$

If we set $N = 2$, we recognize the first two terms in the expansion

$$\exp g^2\{G(x-y) - G(0)\} \propto \frac{1}{|x-y|^{\frac{g^2}{2\pi}}}. \tag{17.69}$$

We now understand the origin of the divergent logarithms in perturbation theory. They are a result of the expansion attempting to modify the power law dependence of the correlator so as to reflect the scaling dimension of the fields and not their engineering dimension. They are a direct consequence of the fractal nature of the typical field configurations.

17.4.1 Renormalizing

We need to absorb Λ into a renormalized field $\mathbf{n} = \sqrt{Z}\mathbf{n}_R$ and a renormalized coupling constant $g_R^2(g^2)$. First we get rid of the initial factor of g^{-2} by replacing our current definition of $\mathbf{n}$ by the original one.

To lowest order we have

$$\langle \mathbf{n}(x) \cdot \mathbf{n}(y) \rangle = 1 - \frac{g^2}{2\pi}(N-1)\ln\Lambda|x-y| + O(g^4)$$

$$= \left(1 + \frac{g_R^2}{2\pi}(N-1)\ln\left(\frac{\mu}{\Lambda}\right)\right)\left(1 - \frac{g_R^2}{2\pi}(N-1)\ln\mu|x-y|\right) + O(g^4). \quad (17.70)$$

To this order $g^2 = g_R^2 + O(g_R^4)$, and

$$Z = 1 + \frac{g_R^2}{2\pi}(N-1)\ln\left(\frac{\mu}{\Lambda}\right) + O(g_R^4). \quad (17.71)$$

Let us define Z at arbitrary order by requiring that $G_R \equiv Z^{-1}G_B$ be unity at $|x| = \mu^{-1}$. Therefore,

$$Z(g_B^2) = 1 - (N-1)\frac{g_B^2}{2\pi}\ln\left(\frac{\Lambda}{\mu}\right) + \frac{(N-1)}{2}\frac{g_B^4}{(2\pi)^2}\ln^2\left(\frac{\Lambda}{\mu}\right). \quad (17.72)$$

We will define g_R^2 by

$$\left.\frac{\partial G_R}{\partial \ln|x|}\right|_{x=\mu^{-1}} = -(N-1)\frac{g_R^2}{2\pi}. \quad (17.73)$$

The left-hand side of this expression is

$$Z^{-1}(g_B^2)\left\{-(N-1)\frac{g_B^2}{2\pi} + (N-1)\frac{g_B^4}{(2\pi)^2}\ln\left(\frac{\Lambda}{\mu}\right)\right\}, \quad (17.74)$$

so

$$g_R^2 = Z^{-1}(g_B^2)\left\{1 - \frac{g_B^2}{2\pi}\ln\left(\frac{\Lambda}{\mu}\right)\right\}g_B^2 + O(g_B^6)$$

$$= \left(1 + (N-1)\frac{g_B^2}{2\pi}\ln\left(\frac{\Lambda}{\mu}\right)\right)\left\{1 - \frac{g_B^2}{2\pi}\ln\left(\frac{\Lambda}{\mu}\right)\right\}g_B^2 + O(g_B^6)$$

$$= \left(1 + (N-2)\frac{g_B^2}{2\pi}\ln\left(\frac{\Lambda}{\mu}\right)\right)g_B^2 + O(g_B^6). \quad (17.75)$$

Reverting this series gives

$$g_B^2 = \left(1 - (N-2)\frac{g_R^2}{2\pi}\ln\left(\frac{\Lambda}{\mu}\right)\right)g_R^2 + O(g_R^6). \quad (17.76)$$

We now substitute for g_B^2 in $Z(g_B^2)$ and find

$$Z = 1 - (N-1)\frac{g_R^2}{2\pi}\ln\left(\frac{\Lambda}{\mu}\right) + (N-1)(N-2)\frac{g_R^4}{(2\pi)^2}\ln^2\left(\frac{\Lambda}{\mu}\right)$$

$$+ \frac{(N-1)}{2} \frac{g_R^4}{(2\pi)^2} \ln^2\left(\frac{\Lambda}{\mu}\right) + O(g_R^6), \tag{17.77}$$

giving

$$Z = 1 - (N-1)\frac{g_R^2}{2\pi} \ln\left(\frac{\Lambda}{\mu}\right) + (N-1)(N-\frac{3}{2})\frac{g_R^4}{(2\pi)^2} \ln^2\left(\frac{\Lambda}{\mu}\right) + O(g_R^6). \tag{17.78}$$

Now we use these results to find $G_R = Z^{-1} G_B$. We need to use

$$(1 + ax + bx^2 + \cdots)^{-1} = 1 - ax + (a^2 - b)x^2 + \cdots. \tag{17.79}$$

We find

$$G_R = \left(1 + (N-1)\frac{g_R^2}{2\pi} \ln\left(\frac{\Lambda}{\mu}\right) + \frac{1}{2}(N-1)\frac{g_R^4}{(2\pi)^2} \ln^2\left(\frac{\Lambda}{\mu}\right)\right) \times$$

$$\times \left(1 - (N-1)\frac{g_R^2}{2\pi} \ln \Lambda|x| + \frac{1}{2}(N-1)\frac{g_R^4}{(2\pi)^2} \ln^2 \Lambda|x| + \right.$$

$$\left. (N-1)(N-2)\frac{g_R^4}{(2\pi)^2} \ln \Lambda|x| \ln\left(\frac{\Lambda}{\mu}\right)\right) + O(g_R^6). \tag{17.80}$$

After the dust has settled, we end up with

$$G_R = 1 - (N-1)\frac{g_R^2}{2\pi} \ln \mu|x| + \frac{1}{2}(N-1)\frac{g_R^4}{(2\pi)^2} \ln^2 \mu|x| + O(g_R^6). \tag{17.81}$$

Gazing at this expression one has the sense that *montes parturiunt, nascetur ridiculus mus*.[1] All that seems to have happened is that g_B^2 has been replaced by g_R^2 and Λ by μ. This simplicity is a consequence of our choice of the definition of g_R^2. A similar replacement will not work at higher order.

We can now find the renormalization group equation coefficients either by demanding that G_R obey

$$\left(\mu \frac{\partial}{\partial \mu} + \beta(g_R^2) \frac{\partial}{\partial g_R^2} + 2\gamma(g_R^2)\right) G_R(\mu, x, g_R^2) = 0, \tag{17.82}$$

or from the definitions of β, γ in terms of Z.

To persue the former course we assume that

$$\beta(g_R^2) = b_1 g_R^4 + O(g_R^6),$$
$$\gamma(g_R^2) = a_1 g_R^2 + a_2 g_R^4 (O(g_R^6)). \tag{17.83}$$

Now substitute Equation (17.81) into (17.82) and require that the coefficient of each power of g_R^2 should vanish. We find for the coefficient of g_R^2

$$-(N-1)\frac{g_R^2}{2\pi} + 2a_1 g_R^2 = 0, \tag{17.84}$$

---

[1] Quintus Horatius Flaccus (Horace), *Ars Poetica* 139.

so
$$a_1 = \frac{1}{2}(N-1)\frac{1}{2\pi}. \tag{17.85}$$

For the coefficient of g_R^4 we have

$$(N-1)\frac{g_R^4}{(2\pi)^2}\ln\mu|x| - b_1\frac{g_R^4}{2\pi}\ln\mu|x| - (N-1)a_1\frac{g_R^4}{2\pi}\ln\mu|x| + a_2 g_R^4 = 0. \tag{17.86}$$

Thus,
$$b_1 = -(N-2)\frac{1}{2\pi}, \quad a_2 = 0. \tag{17.87}$$

Therefore,
$$\beta(g_R^2) = -(N-2)\frac{g_R^4}{2\pi}$$
$$\gamma(g_r^2) = (N-1)\frac{g_R^2}{2\pi} + 0.g_R^4 + O(g_R^6). \tag{17.88}$$

We can confirm these results by a direct computation from the definitions

$$\beta(g_R^2) = \mu\frac{\partial g_R^2}{\partial \mu}\bigg|_{g_B^2,\Lambda}, \tag{17.89}$$

$$\gamma(g_R^2) = \mu\frac{\partial \ln\sqrt{Z}}{\partial \mu}\bigg|_{g_B^2,\Lambda}. \tag{17.90}$$

Recall that
$$g_R^2 = Z^{-1}(g_B^2)\left\{1 - \frac{g_B^2}{2\pi}\ln\left(\frac{\Lambda}{\mu}\right)\right\} g_B^2 = \left(1 + (N-2)\frac{g_B^2}{2\pi}\ln\left(\frac{\Lambda}{\mu}\right)\right) g_B^2, \tag{17.91}$$

so
$$\mu\frac{\partial g_R^2}{\partial\mu}\bigg|_{g_B^2,\Lambda} = -(N-2)\frac{g_B^4}{2\pi} + O(g_B^6) = -(N-2)\frac{g_R^4}{2\pi} + O(g_R^6). \tag{17.92}$$

Similarly we have
$$Z(g_B^2) = 1 - (N-1)\frac{g_B^2}{2\pi}\ln\left(\frac{\Lambda}{\mu}\right) + \frac{(N-1)}{2}\frac{g_B^4}{(2\pi)^2}\ln^2\left(\frac{\Lambda}{\mu}\right), \tag{17.93}$$

so
$$\mu\frac{\partial Z}{\partial\mu}\bigg|_{g_B^2,\Lambda} = (N-1)\frac{g_B^2}{2\pi} - (N-1)\frac{g_B^4}{(2\pi)}\ln\left(\frac{\Lambda}{\mu}\right). \tag{17.94}$$

We need to premultiply this with Z^{-1} and express everything in terms of g_R^2 so we get

$$\gamma(g_R^2) = \frac{1}{2}Z^{-1}\mu\frac{\partial Z}{\partial\mu}\bigg|_{g_B^2,\Lambda}$$

$$= \frac{1}{2}\left(1 + (N-1)\frac{g_R^2}{2\pi}\ln\left(\frac{\Lambda}{\mu}\right) + \frac{(N-1)}{2}\frac{g_R^4}{(2\pi)^2}\ln^2\left(\frac{\Lambda}{\mu}\right)\right)$$

$$\times\left((N-1)\frac{g_R^2}{2\pi} - (N-1)\frac{g_R^4}{(2\pi)^2}\ln\left(\frac{\Lambda}{\mu}\right) - (N-1)(N-2)\frac{g_R^4}{(2\pi)^2}\ln^2\left(\frac{\Lambda}{\mu}\right)\right)$$

$$= \frac{1}{2}(N-1)\frac{g_R^2}{2\pi} + \frac{1}{2}((N-1)^2 - (N-1) - (N-1)(N-2))\frac{g_R^4}{(2\pi)^2}\ln\left(\frac{\Lambda}{\mu}\right)$$

$$= \frac{1}{2}(N-1)\frac{g_R^2}{2\pi} + 0.g_R^4 + O(g_R^6). \tag{17.95}$$

17.4.2 Solution of the RGE

Why are we spending so much time discussing the renormalization group? The utility of the RGE stems from the fact that the n-th term in the perturbation expansion for $G_R(x)$ contains terms like $(g_R^2)^n(\ln\mu|x|)^n$. As x moves away from the renormalization point where $\mu|x| = 1$ the logarithms become large and make the higher order corrections nonnegligible even when g_R^2 itself is small. The RGE allows us to control these logarithms, at least for certain ranges of x. To see how this works we need to think about the solution of the RGE. The RGE is a first-order partial differential equation and may be solved by the *method of characteristics*. We write the RGE

$$\left(\mu\frac{\partial}{\partial\mu} + \beta(g_R^2)\frac{\partial}{\partial g_R^2} + 2\gamma(g_R^2)\right)G_r(\mu, x, g_R^2) = 0, \tag{17.96}$$

in the form

$$\left[\frac{d}{dt} + 2\gamma(g^2(t))\right]G_R(\mu(t)x, g^2(t)) = 0. \tag{17.97}$$

Here $\mu(t)$ and $g_R^2(t)$ obey the equations

$$\frac{d\mu(t)}{dt} = \mu, \quad \frac{dg_R^2}{dt} = \beta(g_R^2(t)). \tag{17.98}$$

The curve $(\mu(t), g_R^2(t))$ is called a *characteristic* of the partial differential equation. There is one through each point in the $\mu(t)$, $g_R^2(t)$ plane. Integrating (17.97) along one of these characteristics gives us

$$G(\mu x_0 e^{(t-t_0)}, g_R^2(t)) = G(\mu x_0, g_R^2(t_0))e^{-\int_{t_0}^t \gamma(g_R^2(t'))dt'}, \tag{17.99}$$

where t and $g_R^2(t)$ are connected by

$$(t - t_0) = \int_{g_R^2(t_0)}^{g_R^2(t)} \frac{dg^2}{\beta(g^2)}. \tag{17.100}$$

We can interpret this equation pictorially:

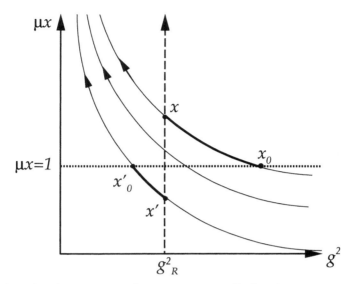

Fig 2. The RGE characteristics for an asymptotically free theory.

The diagram shows the characteristic curves in the μ, g_R^2 plane in the case that the β-function is negative. Since G_R is a function of the product μx, we can just as well regard the μ as being x in this plot. We would like to use the RGE to find $G_R(\mu x, g_R^2)$ for varying x and a fixed small value of g_R^2 (the vertical dashed line). Since we defined the scale of G_R by the condition that it be unity when $\mu x = 1$ (the horizontal dotted line), this condition provides our initial data. By integrating (17.97) along the characteristic from x to the point x_0, where $\mu x_0 = 1$, we find $G_R(\mu x)$. From the curves one can see that the larger μx becomes, the larger the values of g_R^2 at which we need to know $\gamma(g_R^2)$. Since perturbation theory only gives us $\gamma(g_R^2)$ for small g_R^2, we loose control over the calculation in the large x region. On the other hand, if we seek $G_R(\mu x)$ for values where μx is less than unity (x' say), we need to know $\gamma(g_R^2)$ only in the region between x' and x_0', where it is always small. We can therefore find $G_R(\mu x)$ very accurately for small x. Theories where the β-function is negative, and, consequently, the short distance behavior is governed by perturbation theory, are said to be *asymptotically free* at short distance.

17.5 Renormalizing $\lambda \varphi^4$

Take as our action

$$S = \int d^4x \left\{ \frac{1}{2}(\partial \varphi_B)^2 + \frac{1}{2} m_B^2 \varphi_B^2 + \frac{1}{4!} \lambda_B \varphi_B^4 \right\}. \tag{17.101}$$

We want to make all the correlators finite by absorbing the "infinities," or cut-off Λ-dependent terms, into various renormalization constants. For theory with a finite renormalized mass we can, for example, set

$$\lambda_B = Z_1(\frac{m_R^2}{\Lambda^2}, \lambda_R)\lambda_R,$$

$$\varphi_B = \sqrt{Z}(\frac{m_R^2}{\Lambda^2}, \lambda_R)\varphi_R,$$

$$m_B^2 = \Lambda^2 Z_2(\frac{m_R^2}{\Lambda^2}, \lambda_R). \tag{17.102}$$

If we want the final theory to be massless, then we need to introduce a renormalization point μ and set

$$\lambda_B = Z_1(\frac{\mu}{\Lambda}, \lambda_R)\lambda_R,$$

$$\varphi_B = \sqrt{Z}(\frac{\mu}{\Lambda}, \lambda_R)\varphi_R,$$

$$m_B^2 = \Lambda^2 Z_2(\frac{\mu}{\Lambda}, \lambda_R). \tag{17.103}$$

Here we will work with the massless theory.

To one loop the propagator is

$$= \frac{1}{p^2} - \frac{\lambda_B}{2}\left(\frac{1}{p^2}\right)^2 \int \frac{d^4p}{(2\pi)^4} \frac{1}{p^2 + m_B^2}. \tag{17.104}$$

Let us choose

$$m_B^2 = -\frac{\lambda_B}{2} \int \frac{d^4p}{(2\pi)^4} \frac{1}{p^2}, \tag{17.105}$$

so that to this order the theory is massless.

To this (one-loop) order no wavefunction renormalization is required, so $Z = 1$.

The bare vertex is equal to $-\lambda_B$. At one loop order there are three vertex correction diagrams, so the total vertex is

17.5 Renormalizing $\lambda\varphi^4$

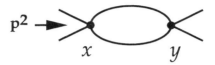

Look at one of the vertex correction diagrams

It is easiest to evaluate this Feynman diagram in real space, where the relevant integral is

$$\int d^4x\, G^2(x,y)e^{ip(x-y)} = \frac{1}{(4\pi^2)^2}\int d^4x\, e^{ipx}\frac{1}{|x^2|^2}. \quad (17.106)$$

We could (and soon will) evaluate this by dimensional regularization, but for now we simply note that the integral is effectively

$$\frac{1}{(4\pi^2)^2}\int \frac{d^4x}{|x^2|^2} \quad (17.107)$$

with an large separation cut-off provided by the oscillating exponential at $x-y \approx p^{-1}$. We also need to include a short-distance cut-off at $(x-y) = a = \Lambda^{-1}$. Thus,

$$\int d^4x\, G^2(x,y)e^{ip(x-y)} \approx \frac{2\pi^2}{(4\pi^2)^2}\int_{\Lambda^{-1}}^{|p|^{-1}}\frac{r^3 dr}{r^4} = \frac{1}{8\pi^2}\ln\left(\frac{\Lambda}{|p|}\right). \quad (17.108)$$

The unrenormalized four-point function is therefore

$$\Gamma^{(4)}(s,t,u) = \lambda_B + \lambda_B^2 \frac{1}{16\pi^2}\left(\ln\frac{\sqrt{s}}{\Lambda} + \ln\frac{\sqrt{t}}{\Lambda} + \ln\frac{\sqrt{u}}{\Lambda}\right) + O(\lambda_B^3). \quad (17.109)$$

Here s, t, u are the usual kinematic invariants $s = (p_1+p_2)^2$ etc.

As mentioned above, to one-loop order, we do not need any wave-function renormalization, and $Z = 1$. To make the theory finite we need only express λ_B in terms of a renormalized coupling constant λ_R. We will define λ_R by equating it to the four-point vertex evaluated with a symmetric combination of momenta flowing into the vertex. We take four momenta p_1, p_2, p_3, p_4 with $p_1+p_2+p_3+p_4 = 0$, $p_i \cdot p_j = \frac{\mu^2}{4}(4\delta_{ij}-1)$. Thus, $(p_i+p_j)^2 = \mu^2$. Then,

$$\lambda_R = \lambda_B + \lambda_B^2 \frac{3}{16\pi^2}\ln\left(\frac{\Lambda}{\mu}\right) + \cdots. \quad (17.110)$$

Therefore,

$$\beta(\lambda_R) = \mu \left.\frac{\partial \lambda_R}{\partial \mu}\right|_{\Lambda,\lambda_B} = +\frac{3}{16\pi^2}\lambda_R^2. \qquad (17.111)$$

The plus sign means that $\lambda\varphi^4$ is not asymptotically free.

The renormalized scattering amplitude at other values of the incoming momenta is found by substituting for λ_B in terms of λ_R as usual. To one-loop order the reversion of the series is trivial, and we have

$$\lambda_B = \lambda_R - \lambda_R^2 \frac{3}{16\pi^2} \ln\left(\frac{\Lambda}{\mu}\right) + O(\lambda_R^3). \qquad (17.112)$$

On inserting this into (17.109) the cut-off-dependent terms cancel, and we are left with

$$\Gamma^{(4)}(s,t,u) = \lambda_R \left\{1 + \frac{1}{16\pi^2}\lambda_R(\ln\frac{\sqrt{s}}{\mu} + \ln\frac{\sqrt{t}}{\mu} + \ln\frac{\sqrt{u}}{\mu})\right\} + O(\lambda_R^3) \qquad (17.113)$$

which is "finite."

18
Large N Expansions

In this chapter we will explore the critical properties of both the linear and non-linear versions of the $O(N)$ σ-model. By exploiting the simplifications that occur when N is large, we will show that, despite their formal differences, the two models are describing the same physics. This is an example of universality.

In this chapter we will use dimensional regularization. This makes the computations much more compact. This computational efficiency does come at some cost however. Dimensional regularization is so effective at sweeping divergences under a rug that one completely looses sight of the role of the cut-off in defining the theory.

18.1 $O(N)$ Linear σ-Model

We begin by using dimensional regularization to compute the β-function for the $O(N)$ linear σ-model. This is just $\lambda \varphi^4$ theory with $O(N)$ symmetry. We will take the interaction to be

$$V(\varphi) = \frac{\lambda_B}{4!} \left(\sum_{i=1}^{N} (\varphi_a)^2 \right)^2. \tag{18.1}$$

To determine the renormalized interaction, it is simplest to compute the diagrams for the scattering of two distinguishable particles, created by φ_a and φ_b with the index $a \neq b$. The lowest-order diagram is

18. Large N Expansions

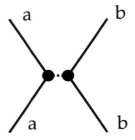

This contributes

$$S_{ab \to ab} = -\frac{\lambda_B}{3}. \tag{18.2}$$

Note that the amplitude is only one third of that for the indistinguashable particle case. This is because the dotted line representing the interaction can only appear in the t channel. For indistinguishable particles we would have to add the s and u channel versions of this diagram.

The one-loop diagrams, together with their symmetry factors are

18.1 $O(N)$ Linear σ-Model

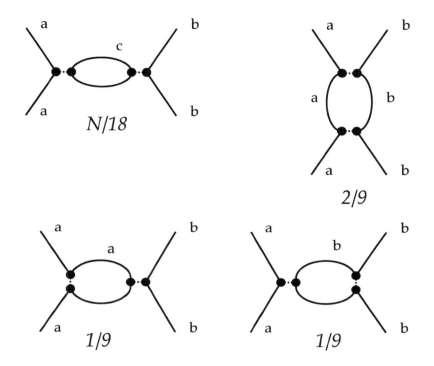

Thus,

$$S_{ab\to ab} = -\frac{\lambda_B}{3} + \lambda_B^2 \frac{N}{18}\Pi(t) + \lambda_B^2 \frac{2}{9}\Pi(s) + \lambda_B^2 \frac{2}{9}\Pi(t). \tag{18.3}$$

Each of the loops corresponds to

$$\Pi(p^2) = \int d^d x\, e^{ipx} (G(x,0))^2, \tag{18.4}$$

where p is the appropriate momentum.

The bubble $\Pi(p^2)$ is evaluated via dimensional regularization using the expression

$$G(x,0) = \frac{1}{4\pi^{d/2}} \Gamma\left(\frac{d}{2} - 1\right) \frac{1}{(x^2)^{d/2-1}}. \tag{18.5}$$

We find

$$\Pi(p) = \frac{1}{16\pi^{d/2}} \left(\Gamma\left(\frac{d}{2}-1\right)\right)^2 \Gamma\left(2 - \frac{d}{2}\right) \left(\frac{p^2}{4}\right)^{d/2-2}. \tag{18.6}$$

18. Large N Expansions

Near four dimensions this is

$$\Pi(p) \approx \frac{1}{16\pi^2}\left\{\frac{1}{2-\frac{d}{2}} - \ln\frac{p^2}{4}\right\} = \frac{1}{8\pi^2}\left\{\frac{1}{4-d} - \ln|p| + const.\right\}. \quad (18.7)$$

The occurrence of $\ln|p|$ seems slightly disconcerting because the argument of any mathematical function must be dimensionless. The mystery is solved by noting that λ_B is no longer dimensionless once we are away from $d = 4$. It is convenient to keep the renormalized coupling constant dimensionless, so we will set

$$\lambda_B = \mu^{4-d} Z_1(\lambda_R, d)\lambda_R. \quad (18.8)$$

Here μ has dimensions of mass and will play the role of renormalization point in the dimensional regularization approach.

The one-loop scattering amplitude is therefore

$$-\frac{\lambda_B}{3} + \frac{\lambda_B^2}{8\pi^2}\left\{\frac{N+8}{18}\frac{1}{4-d}\right\} + \frac{\lambda_B^2}{8\pi^2}\left\{\frac{N+4}{18}\ln\sqrt{t} + \frac{4}{18}\ln\sqrt{s}\right\}, \quad (18.9)$$

where s, t are the usual kinematic invariants.

To cancel the poles and render the theory finite we define

$$\lambda_B = \mu^{4-d}\lambda_R\left\{1 + \lambda_R\frac{N+8}{48\pi^2}\frac{1}{4-d} + O(\lambda_R^2)\right\}. \quad (18.10)$$

This gives for the scattering amplitude

$$-\frac{\lambda_R}{3} + +\frac{\lambda_R^2}{8\pi^2}\left\{\frac{N+4}{18}\ln\left(\frac{\sqrt{t}}{\mu}\right) + \frac{4}{18}\ln\left(\frac{\sqrt{s}}{\mu}\right)\right\} + O(\lambda_R^3). \quad (18.11)$$

We extract $\beta(\lambda_R)$ from equation (18.10) by observing that

$$0 = \mu\left.\frac{\partial\lambda_B}{\partial\mu}\right|_{\lambda_B} = (4-d)\lambda_B + \beta(\lambda_R)\frac{\partial\lambda_B}{\partial\lambda_R}. \quad (18.12)$$

Thus,

$$0 = (4-d)\mu^{4-d}\lambda_R\left(1 + \lambda_R\frac{N+8}{48\pi^2}\frac{1}{4-d}\right) + \beta(\lambda_R)\mu^{4-d}\left(1 + \lambda_R\frac{N+8}{24\pi^2}\frac{1}{4-d}\right), \quad (18.13)$$

or

$$0 = (4-d)\lambda_R\left(1 - \lambda_R\frac{N+8}{48\pi^2}\frac{1}{4-d}\right) + \beta(\lambda_R). \quad (18.14)$$

Finally

$$\beta(\lambda_R) = (d-4)\lambda_R + \frac{N+8}{48\pi^2}\lambda_R^2 + O(\lambda_R^3). \quad (18.15)$$

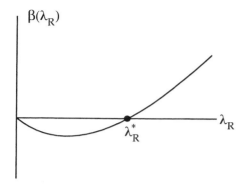

Fig 1. The function $\beta(\lambda_R)$ in $4-\epsilon$ dimensions.

If we plot $\beta(\lambda_R)$ in dimension $d = 4 - \epsilon$, we find that there is an infrared stable fixed point at $\lambda_R = O(\epsilon)$. This is the *Wilson-Fisher* fixed point. It governs the critical behavior of the $O(N)$ model in dimensions $2 < d < 4$. The critical exponents will differ from their gaussian theory values by $O(\epsilon)$. The perturbation expansion in powers of λ_R is equivalent to the ϵ-*expansion*, which gives the critical exponents as a power series in ϵ. In obtaining this expansion, dimensional regularization combines the role of the regulator and the expansion parameter. This makes the computations very efficient, but obscures the physics.[1]

There is no real physical meaning to the theory in $4-\epsilon$ dimensions when ϵ is not an integer. The expansion is simply a trick to enable us to compute exponents for $d = 3$, where $\epsilon = 1$. It is tempting, however, to pretend that there is a continuous range of theories with interesting physics for $0 \leq \epsilon \leq 2$. In the next couple of sections we will succumb to this temptation.

18.2 Large N Expansions

In this section we will consider $O(N)$ symmetric models for a range of dimensions between two and four. It turns out that when N is very large, we can essentially solve these models in closed form.[2] We will study this limit for both the $O(N)$ nonlinear σ-model that we considered earlier, and the linear version of the $O(N)$ σ-model.

---

[1] For a cut-off based approach to the ϵ-expansion I recommend the article by Brézin, Le Guillou, and Zinn-Justin in volume 6 of the series *Phase Transitions and Critical Phenomena*, C. Domb, M. S. Green, eds. (Academic Press 1976).

[2] The discussion of the large N limit given here closely follows that in E. Brézin, J. Zinn-Justin, Physical Review B **14** (1976) pp.3110-3120.

18. Large N Expansions

As before, the partition function of the nonlinear σ-model is

$$\mathcal{Z} = \int d[\varphi] \prod_x \delta(|\varphi(x)|^2 - 1) e^{-\frac{1}{2g_B^2} \int d^d x (\partial \varphi)^2}, \tag{18.16}$$

where φ is an N component real field. We impose the constraint $|\varphi(x)|^2 = 1$ by writing the delta function as

$$\delta(|\varphi(x)|^2 - 1) = \int_C \frac{d\alpha(x)}{2\pi} e^{-\alpha(x)(|\varphi(x)|^2 - 1)}, \tag{18.17}$$

where C is a contour running up the imaginary axis. There is one variable $\alpha(x)$ for each space-time point x, so α becomes a new "Lagrange multiplier" field. The partition function is now

$$\mathcal{Z} = \int d[\varphi] d[\alpha] e^{-\frac{1}{2g_B^2} \int d^d x \{(\partial \varphi)^2 + i 2 g_B^2 \alpha (|\varphi|^2 - 1)\}}. \tag{18.18}$$

The integral over $\varphi(x)$ is now a gaussian, so we can do it and obtain

$$\mathcal{Z} = \int d[\alpha] e^{-\frac{N}{2} \ln \det(-\partial^2 + 2 g_B^2 \alpha) + i \int d^d x \alpha(x)}. \tag{18.19}$$

The "large N" limit is one in which N is taken to infinity, while the product $g_B^2 N$ is kept fixed. If we define $m^2(x) = 2 g_B^2 \alpha(x)$, we can now write the partition function as

$$\mathcal{Z} = \int d[m^2] e^{-\frac{N}{2} \ln \det(-\partial^2 + m^2) + N \int d^d x \frac{1}{g_B^2 N} m^2}. \tag{18.20}$$

Here each of the terms in the exponent has a factor of N, and this means that when N is large, the $m^2(x)$ functional integral may be evaluated by the method of steepest descent. In effect we simply seek a stationary point of the exponent and evaluate the integral by replacing the exponent by its value at this stationary point. Making the plausible assumption that $m^2(x)$ is independent of position at this point, the condition for stationarity is

$$\frac{\partial}{\partial m^2} \left\{ -\frac{N}{2} \int \frac{d^d k}{(2\pi)^d} \ln(k^2 + m^2) + N \frac{1}{g_B^2 N} \right\} = 0. \tag{18.21}$$

This is

$$1 = N g_B^2 \int \frac{d^d k}{(2\pi)^d} \frac{1}{k^2 + m^2}. \tag{18.22}$$

In this form the equation is essentially the same "gap equation," that we met with when discussing the BCS theory of superconductors.

The dimensionally regularized integral is

$$\int \frac{d^d k}{(2\pi)^d} \frac{1}{k^2 + m^2} = \frac{1}{2^d \pi^{d/2}} \Gamma(1 - \frac{d}{2}) (m^2)^{d/2 - 1}, \tag{18.23}$$

so

$$1 = N g_B^2 \left\{ \frac{1}{2^d \pi^{d/2}} \Gamma(1 - \frac{d}{2}) (m^2)^{d/2 - 1} \right\}. \tag{18.24}$$

18.2 Large N Expansions

Let us define

$$A(d) = \frac{1}{2^d \pi^{d/2}} \Gamma(1 - \frac{d}{2}) \equiv \frac{1}{2^d \pi^{d/2}} \frac{2\Gamma(2 - \frac{d}{2})}{2 - d}, \quad (18.25)$$

so we can write the gap equation as

$$\frac{1}{g_B^2} = N A(d) (m^2)^{d/2-1}. \quad (18.26)$$

The gamma function diverges as $d \to 2$, so to have a finite expression in this limit we define g_R^2 by setting

$$\frac{1}{g_B^2} = \frac{(\mu^2)^{d/2-1}}{g_R^2} + N A(d)(\mu^2)^{d/2-1}. \quad (18.27)$$

Note that we have introduced a dimensional parameter μ (having $[\mu] = M$) so that g_R^2 remains dimensionless in any space-time dimension. The relation between g_R^2 and g_B^2 can also be written as

$$g_B^2 = \mu^{2-d} g_R^2 \left(1 + g_R^2 N A(d)\right)^{-1} = \mu^{2-d} g_R^2 Z_1. \quad (18.28)$$

In terms of g_R^2 the renormlized gap equation is

$$1 = N g_R^2 A(d) \left\{ \left(\frac{m^2}{\mu^2}\right)^{\frac{d}{2}-1} - 1 \right\}. \quad (18.29)$$

Near $d = 2$ we have

$$A(d) \approx \frac{1}{2\pi} \frac{1}{2-d} + O(1), \quad (18.30)$$

so as we approach $d = 2$ the coefficients in gap equation remain finite, and the equation becomes

$$1 = g_R^2 N \frac{1}{2\pi} \frac{1}{2-d} \left\{ 1 + (\frac{d}{2} - 1) \ln \frac{m^2}{\mu^2} + O((\frac{d}{2} - 1)^2) - 1 \right\}$$

$$\to -g_R^2 N \frac{1}{\pi} \ln \frac{m^2}{\mu^2}. \quad (18.31)$$

In other words

$$m^2 = \mu^2 e^{-\frac{\pi}{g_R^2 N}}. \quad (18.32)$$

The quantity m^2 is a common mass shared by all N of the φ fields. In two dimensions, therefore, the model has no spontaneous symmetry breaking, and no massless Goldstone bosons. Although the bare theory contained no dimensional parameters, the dynamically induced mass has borrowed the renormalization point mass to set the scale.

In more than two dimensions there is a phase transition to a spontaneously broken phase. To see this observe that the quantity $A(d)$ is *negative* in the range of space-time dimensions $2 < d < 4$. For d in this range there is a critical value $(g_R^*)^2$ at which

$$1 = -N(g_R^*)^2 A(d). \tag{18.33}$$

For $g_R^2 < (g_R^*)^2$ there is no solution to the gap equation with positive m^2. It is not quite obvious, but rather than consider imaginary values of m in this region we should simply set $m^2 = 0$, so all the φ particles are massless. We will soon see that the $O(N)$ symmetry is broken spontaneously when $m^2 = 0$, and the massless particles are the Goldstone bosons. Note that as $d \to 2$, $(g_R^*)^2 = O(d-2)$.

Let us compute the $\beta(g_R^2)$ function from Equation (18.28). We use

$$0 = \mu \left.\frac{\partial g_B^2}{\partial \mu}\right|_{g_B^2} = (4-d)g_B^2 + \beta(g_R^2)\frac{\partial g_B^2}{\partial g_R^2}. \tag{18.34}$$

Using $(1 - \frac{d}{2})\Gamma(1 - \frac{d}{2}) = \Gamma(2 - \frac{d}{2})$ we find

$$\beta(g_R^2) = (d-2)g_R^2(1 + g_R^2 N A(d))$$
$$= (d-2)g_R^2 - g_R^4 N \frac{1}{2^d \pi^{d/2}} 2\Gamma(2 - \frac{d}{2})$$
$$\approx (d-2)g_R^2 - \frac{N}{2\pi} g_R^4. \tag{18.35}$$

The approximation in the last line being valid near $d = 2$.

We should compare this with the expression

$$\beta(g_R^2) = -\frac{(N-2)}{2\pi} g_R^4 + O(g_R^6) \tag{18.36}$$

that we obtained in our direct $d = 2$ perturbative calculation. We see that our large N calculation does indeed capture the large N behavior of at least the leading term in $\beta(g_R^2)$.

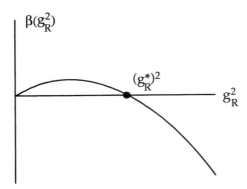

Fig 2. The function $\beta(g_R^2)$ in $2 + \epsilon$ dimensions.

Note that the β-function vanishes at the critical value $(g^*)_R^2$. The critical point is therefore a fixed point of our renormalization group flow, and g_R^2 the deviation of the action from the fixed point is a relevant operator.

The induced mass m^2 is really a function of the bare coupling g_B^2 and d only. Consequently it should not change when we vary μ and g_R^2 keeping g_B^2 constant. This means that m^2 satisfies a renormalization group equation of the form

$$\left(\mu\frac{\partial}{\partial\mu} + \beta(g_R^2)\frac{\partial}{\partial g_R^2}\right)m^2 = 0. \tag{18.37}$$

Substituting for m^2 the expression (18.32), we find that it does indeed satisfy this equation.

18.2.1 Linear vs. Nonlinear σ-Models

In order to analyze the $m^2 = 0$ region of the nonlinear σ-model, we can examine the equation of state. Thinking of the model as one for a ferromagnet, and writing $\varphi = (\sigma, \pi_1, \pi_2, \ldots \pi_{N-1})$, this is the equation linking the magnetization $\langle\sigma\rangle$ to the external symmetry breaking magnetic field h. We include the h field by writing

$$\mathcal{Z}(h) = \int d[\varphi]\delta(|\varphi|^2 - 1)e^{-\frac{1}{2g_B^2}\int d^dx\{(\partial\varphi)^2 - h\sigma\}}. \tag{18.38}$$

We again introduce the field $\alpha(x)$ to enforce the constraint, but now we only integrate over the $N-1$ π fields to get

$$\mathcal{Z} = \int d[\sigma]d[\alpha]e^{-\frac{1}{g_B^2}\int d^dx\{\frac{1}{2}(\partial\sigma)^2 + \frac{1}{2}\alpha g_B^2\sigma^2 - h\sigma - \frac{1}{2}\alpha g_B^2\}}$$
$$\times e^{-\frac{1}{2}(N-1)\ln\det(-\partial^2 + \alpha g_B^2)}. \tag{18.39}$$

Now we argue that we can perform both of the integrals by replacing the exponents by their values at the stationary points. For the α integral, the justification is as before — the large factor of $(N-1)$ outside the ln det. For the σ integral we recall that $g_B^2 N$ is $O(1)$, so g_B^2 is small. This justifies a stationary phase treatment of the σ field.

The equations of motion are:
1) from varying σ

$$\sigma\alpha g_B^2 - h = 0, \tag{18.40}$$

2) from varying α

$$-\frac{(N-1)}{2}\left\{\frac{1}{2^d\pi^{d/2}}\Gamma(1 - \frac{d}{2})(m^2)^{\frac{d}{2}-1}\right\} + \frac{(1-\sigma^2)}{2g_B^2} = 0. \tag{18.41}$$

(we have set $m^2 = \alpha g_B^2$ as before).

To renormalize we set

$$\frac{1}{g_B^2} = \frac{(\mu^2)^{d/2-1}}{g_R^2} + (N-1)A(d)(\mu^2)^{d/2-1}. \tag{18.42}$$

Because we are working in the large N limit, there is no significance to the appearance of $N-1$ instead of N in this equation and the previous one for m^2. We simply cannot distinguish between N and $N-1$. As before, we can write

$$g_B^2 = \mu^{2-d}\left(1 + g_R^2 N A(d)\right)^{-1} = \mu^{2-d} Z_1 g_R^2. \tag{18.43}$$

We also need to define a renormalized field.

$$\sigma_B = \sqrt{Z}\sigma_R,$$
$$h_B = Z_1 Z^{-\frac{1}{2}} h_R. \tag{18.44}$$

We get finite expressions if we take $Z = Z_1$.

We find the equation of state to be

$$(1 - \sigma_R^2) = (N-1)A(d)g_R^2\left(\mu^{2-d}\left(\frac{h_R}{\sigma_R}\right)^{\frac{d}{2}-1} - 1\right). \tag{18.45}$$

This equation is finite as $d \to 2$. When $g_R^2 > (g^*)_R^2$ and $h = 0$, we should write m^2 in place of the combination h_R/σ_R, and the equation of state reduces to the previous gap equation. When $g_R^2 < (g^*)_R^2$ and $h = 0$, on the other hand, there is now still a solution, but it is one with nonzero σ_R^2.

Let us now compare this with the equation of state for the *linear* σ-model, in four or fewer dimensions. Again writing $\varphi = (\sigma, \pi_1, \pi_2, \ldots \pi_{N-1})$, the linear σ-model has partition function

$$\mathcal{Z} = \int d[\pi]d[\sigma]e^{-S(\pi,\sigma,H)}, \tag{18.46}$$

where

$$S = \int d^d x \left\{\frac{1}{2}(\partial\sigma)^2 + \frac{1}{2}(\partial\pi)^2 - H\sigma + \frac{\lambda_B}{4!}(\sigma^2 + \pi^2)^2 + \frac{1}{2}m_B^2(\sigma^2 + \pi^2)\right\}. \tag{18.47}$$

Here we introduce a field α to decouple the interaction, and write

$$\mathcal{Z} = \int d[\pi]d[\sigma]d[\alpha]e^{-\tilde{S}(\pi,\sigma,\alpha,H)}, \tag{18.48}$$

with the new action

$$\tilde{S} = \int d^d x \left\{\frac{1}{2}(\partial\sigma)^2 + \frac{1}{2}(\partial\pi)^2 - H\sigma + \frac{\alpha}{2}(\sigma^2 + \pi^2) - \frac{3}{2\lambda_B}(\alpha - m_B^2)^2\right\}. \tag{18.49}$$

Integrating over π gives us

$$\mathcal{Z} = \int d[\sigma]d[\alpha]e^{-\int d^d x\left\{\frac{1}{2}(\partial\sigma)^2 - H\sigma + \frac{\alpha}{2}\sigma^2 - \frac{3}{2\lambda_B}(\alpha - m_B^2)^2\right\}}$$
$$\times e^{-\frac{1}{2}(N-1)\ln\det(-\partial^2 + \alpha)}. \tag{18.50}$$

Again the large value of N allows us to evaluate the functional integrals by replacing the integrands with their values at stationary points. The equations determining the stationary points are:

18.2 Large N Expansions

1) from the variation of σ

$$\alpha\sigma - H = 0, \tag{18.51}$$

2) from the variation of α

$$\frac{1}{2}\sigma^2 - \frac{3}{\lambda_B}(\alpha - m_B^2) + \frac{(N-1)}{2}\int \frac{d^d k}{(2\pi)^d} \frac{1}{k^2 + \alpha} = 0. \tag{18.52}$$

To renormalize we break the integral up as

$$\int \frac{d^d k}{(2\pi)^d}\left\{\frac{1}{k^2+\alpha} - \frac{1}{k^2} + \frac{\alpha}{(k^2+\alpha)^2}\right\}$$
$$+ \int \frac{d^d k}{(2\pi)^d}\frac{1}{k^2} - \int \frac{d^d k}{(2\pi)^d}\frac{\alpha}{(k^2+\mu^2)^2}, \tag{18.53}$$

and write

$$\frac{m_B^2}{\lambda_B} = \mu^{d-4}\frac{m_R^2}{\lambda_R} - \frac{(N-1)}{6}\int \frac{d^d k}{(2\pi)^d}\frac{1}{k^2}. \tag{18.54}$$

Integrals such as

$$\int \frac{d^d k}{(2\pi)^d}\frac{1}{k^2} \tag{18.55}$$

are zero in dimensional regularization, so effectively

$$\frac{m_B^2}{\lambda_B} = \mu^{d-4}\frac{m_R^2}{\lambda_R}. \tag{18.56}$$

This would *not* be true in a general regularization scheme.

We define the renormalized coupling by setting

$$\frac{1}{\lambda_B} = \frac{\mu^{d-4}}{\lambda_R} - \frac{(N-1)}{6}\int \frac{d^d k}{(2\pi)^d}\frac{1}{(k^2+\alpha)^2}$$
$$= \frac{\mu^{d-4}}{\lambda_R} - \frac{(N-1)\,\Gamma(2-\frac{d}{2})}{6\cdot 2^d \pi^{d/2}}\mu^{d-4}. \tag{18.57}$$

Thus the renormalized equation of state is

$$\sigma_R^2 + \frac{6m_R^2\mu^{d-4}}{\lambda_R} = (N-1)\frac{2\Gamma(2-\frac{d}{2})}{2^d\pi^{d/2}(2-d)}\left(\frac{H_R}{\sigma_R}\right)^{\frac{d}{2}-1}$$
$$+ \mu^{d-4}\left(\frac{H_R}{\sigma_R}\right)\left\{\frac{6}{\lambda_R} - (N-1)\frac{\Gamma(2-\frac{d}{2})}{2^d\pi^{d/2}}\right\}. \tag{18.58}$$

This does not look exactly like the expression we obtained from the nonlinear model — but note that the term linear in H/σ is smaller that the term $(H/\sigma)^{\frac{d}{2}-1}$ in the critical region where H/σ is small, at least provided $d < 4$. If we ignore the subdominant term, then the two equations of state are very similar.

18. Large N Expansions

Let us compute $\beta(\lambda_R)$. Define

$$B(d) = \frac{(N-1)}{6} \frac{\Gamma(2 - \frac{d}{2})}{2^d \pi^{d/2}}, \tag{18.59}$$

so we can rewrite

$$\frac{1}{\lambda_B} = \frac{\mu^{d-4}}{\lambda_R} - \frac{(N-1)}{6} \frac{\Gamma(2 - \frac{d}{2})}{2^d \pi^{d/2}} \mu^{d-4} \tag{18.60}$$

as

$$\lambda_B = \mu^{4-d} \lambda_R (1 - \lambda_R B(d))^{-1}. \tag{18.61}$$

From this we find that

$$\beta(\lambda_R) = \mu \left. \frac{\partial \lambda_R}{\partial \mu} \right|_{\lambda_B} = (d-4) \lambda_R (1 - \lambda_R B(d))$$

$$\approx (d-4) \lambda_R + (N-1) \frac{\lambda_R^2}{48 \pi^2} + O((d-4)^2). \tag{18.62}$$

The last line should be compared with the expression

$$\beta(\lambda_R) = (d-4) \lambda_R + (N+8) \frac{\lambda_R^2}{48 \pi^2} + O(\lambda_R^3) \tag{18.63}$$

that one gets for the $O(N)$ model without making the large N approximation. Once again we see that our method of treating the large N limit is reproducing the standard perturbative computation in their common domain of applicability.

There is a fixed point at

$$\lambda^* = \frac{1}{B(d)}. \tag{18.64}$$

For d near 4 this is at

$$\lambda^* \approx \frac{24 \pi^2 (4-d)}{N}. \tag{18.65}$$

More important, when we set $\lambda_R = \lambda^*$ in (18.58), the subdominant linear term in the equation of state vanishes. In this case we have

$$\sigma_R^2 + \frac{6 m_R^2 \mu^{d-4}}{\lambda_R} = -(N-1) A(d) \left(\frac{H_R}{\sigma_R} \right)^{\frac{d}{2}-1}. \tag{18.66}$$

Let us multiply this equation by μ^{2-d} to make it dimensionless, and introduce the dimensionless parameter g^2 via the equation

$$\frac{1}{g^2} = -(N-1) A(d) - \frac{6 m_R^2 \mu^{-2}}{\lambda_R}. \tag{18.67}$$

The linear σ model equation of state becomes

$$(\sigma_R^2 g^2 \mu^{2-d} - 1) = -g^2 (N-1) A(d) \left\{ \mu^{2-d} \left(\frac{H_R}{\sigma_R} \right)^{\frac{d}{2}-1} - 1 \right\}. \tag{18.68}$$

With a trivial redefinition of the σ_R and H_R fields (to take into account that the field in the nonlinear model is dimensionless, while the σ field in the linear model has dimension $[\sigma] = M^{\frac{d}{2}-1}$), the linear and nonlinear model equations of state become identical.

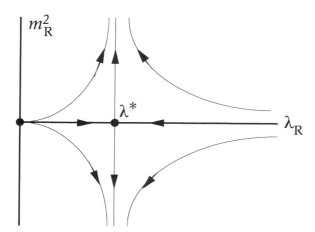

Fig 3. The flows near λ^* in $4 - \epsilon$ dimensions.

The reason why the two models have the same equation of state is brought out by the above figure. When we computed the β function of the linear model in $4 - \epsilon$ dimensions, we found a fixed point λ^*, which was infrared *stable*, and so did not correspond to phase transition occurring as we varied λ_R through it. In the nonlinear model, on the other hand, we found an infrared *unstable* fixed point, which divided the broken from the unbroken symmetry phases.

In both these cases we were looking at the *same* fixed point, but taking different slices through its neigborhood. The linear model fixed point *is* infrared unstable if we allow m_R^2 to vary instead of λ_R. Further the symmetry breaks as m_R^2 goes from positive to negative. Varying g_R^2 in the nonlinear model must therefore be equivalent to varying m_R^2 along the vertical trajectory through λ_R^* in the linear model.

We see that both the linear and the nonlinear versions of the large N σ-model describe exactly the same physics. Despite having different dimensions in which they are perturbatively renormalizable, and having different numbers of degrees of freedom, once we set $\lambda_R = \lambda^*$ to remove all the irrelevant perturbations to the fixed point, their physical predictions are identical.

This is expected to be a general feature of field theories. Any two theories which lie in the catchment area of the same fixed point will have the same continuum limit. Experience leads us to believe that these catchment areas are determined only by the symmetries of the theory and by the space-time dimension.

Appendix A
Relativistic State Normalization

In this appendix I will explain the origin of the relativistically invariant inner product on the single-particle states. I will restrict the derivation to 1+1 dimensions so we can do the arithmetic easily and thus see the essentials.

Passive Lorentz transformations from the x frame to an x' frame moving in the $+x^3$ direction may be written as

$$\begin{pmatrix} (x')^3 \\ (x')^0 \end{pmatrix} = \begin{pmatrix} \cosh s & -\sinh s \\ -\sinh s & \cosh s \end{pmatrix} \begin{pmatrix} x^3 \\ x^0 \end{pmatrix}, \quad (A.1)$$

where, $v = \tanh s$, is the velocity the x' frame as seen from the x frame. The parameter s, called the *rapidity*, is very convenient because, unlike v, it behaves additively under composition of boosts:

$$\begin{pmatrix} \cosh(s+s') & -\sinh(s+s') \\ -\sinh(s+s') & \cosh(s+s') \end{pmatrix}$$
$$= \begin{pmatrix} \cosh s & -\sinh s \\ -\sinh s & \cosh s \end{pmatrix} \begin{pmatrix} \cosh s' & -\sinh s' \\ -\sinh s' & \cosh s' \end{pmatrix}, \quad (A.2)$$

while

$$v_{12} = \frac{v_1 + v_2}{1 + v_1 v_2}. \quad (A.3)$$

The contravariant components of the momentum 4-vector transform similarly, and under an *active* boost:

$$\begin{pmatrix} (p')^3 \\ E' \end{pmatrix} = \begin{pmatrix} \cosh s & \sinh s \\ \sinh s & \cosh s \end{pmatrix} \begin{pmatrix} p^3 \\ E \end{pmatrix} = L(s) \begin{pmatrix} p^3 \\ E \end{pmatrix}. \quad (A.4)$$

A. Relativistic State Normalization

If a particle at rest ($p^3 = 0$, $E = m$) is boosted to rapidity s we have

$$(p')^3 = m \sinh s, \tag{A.5}$$

$$E' = m \cosh s, \tag{A.6}$$

so $(E')^2 - (p'^3)^2 = m^2$.

In relativistic quantum field theory, the Lorentz group is a *symmetry* of the system and must be implementable by a unitary representation of the group acting on the Hilbert space. This means that there is a unitary $U(L(s))$ obeying the group composition law

$$U(L(s_1))U(L(s_2)) = U(L(s_1)L(s_2)) = U(L(s_1 + s_2)). \tag{A.7}$$

Suppose we normalize our single-particle states so that for small (non-relativistic) momenta, **p**, we have

$$\langle \mathbf{p}|\mathbf{p}'\rangle = (2\pi)^3 2m\delta^3(\mathbf{p} - \mathbf{p}'). \tag{A.8}$$

(Except for the $2m$ this is the normalization of a plane wave $\psi = e^{i\mathbf{k}\cdot\mathbf{x}}$.) Now we obtain the state with 3-momentum **p** from that of a particle at rest by application of the boost $L(\mathbf{p})$ that takes $\mathbf{0} \to \mathbf{p}$ i.e.:

$$|\mathbf{p}\rangle = U(L(\mathbf{p}))|0\rangle. \tag{A.9}$$

For states boosted in the x^3 direction $L(\mathbf{p})$ is just $L(s)$ with $p^3 = m \sinh s$. Let us compute the inner product of these states: for small s we can use the previous answer and find

$$\langle \mathbf{p}|\mathbf{p}'\rangle = (2\pi)^3 2m\delta^3(\mathbf{p} - \mathbf{p}') = (2\pi)^3 2\delta(s - s') \tag{A.10}$$

by exploiting the identity $\delta(ax) = a^{-1}\delta(x)$. For larger values of s we can use the group composition law to reduce the calculation to the previous one:

$$\langle \mathbf{p}|\mathbf{p}'\rangle = \langle 0|U^{\dagger}(L(s))U(L(s'))|0\rangle = \langle 0|U(L(s' - s))|0\rangle. \tag{A.11}$$

Since the product is only going to be non-zero when the two momenta coincide, we may safely assume that $s - s'$ is small, and use the non-relativistic result to evaluate this as

$$(2\pi)^3 2\delta(s - s'); \tag{A.12}$$

an answer that is clearly invariant under a simultaneous boost of both states as this would just add the same constant to s and s'. Now remember that, if $f(0) = 0$, we have $\delta(f(x)) = (f'(0))^{-1}\delta(x)$, so we can rewrite the delta function in terms of the momenta as

$$(2\pi)^3 2\delta(s - s') = (2\pi)^3 2E_\mathbf{p}\delta^3(\mathbf{p} - \mathbf{p}'). \tag{A.13}$$

So finally

$$\langle \mathbf{p}|\mathbf{p}'\rangle = (2\pi)^3 2E_\mathbf{p}\delta^3(\mathbf{p} - \mathbf{p}'). \tag{A.14}$$

Appendix B
The General Commutator

Let us evaluate the quantity

$$i\Delta(x - x') = [\varphi(x), \varphi^\dagger(x')] \tag{B.1}$$

for general times x_0 and x'_0. We find

$$i\Delta(x - x') = \int \frac{d^3k}{(2\pi)^3} \frac{1}{2E_\mathbf{k}} \left(e^{-ik(x-x')} - e^{ik(x-x')} \right)$$
$$= \int \frac{d^4k}{(2\pi)^4} 2\pi \, \text{sgn} \, k_0 \, \delta(k^2 - m^2) e^{-ik(x-x')}. \tag{B.2}$$

We see that Δ obeys a source-free Klein-Gordon equation $(\partial^2 + m^2)\Delta = 0$ with Cauchy data $\Delta = 0$, and $\partial_{x_0}\Delta\big|_{x_0=0} = \delta^3(\mathbf{x} - \mathbf{x}')$ on the initial hypersurface $x_0 = 0$.

It should be clear that Δ is invariant under a proper (non-time-reversing) Lorentz tranformation and has the property $\Delta(-x) = -\Delta(x)$. Since we can send x to $-x$ by a proper Lorentz transformation provided x is *spacelike*, we deduce that $\Delta(x) = 0$ if $x^2 < 0$.

The vanishing of the commutator in regions where x and x' cannot be connected by any subluminal signal is consistent with the Copenhagen interpretation of quantum mechanics, which traces the non-commutativity of the operators corresponding to observable to disturbance of the process of measurement of one observable by the measurement by the other.

B. The General Commutator   249

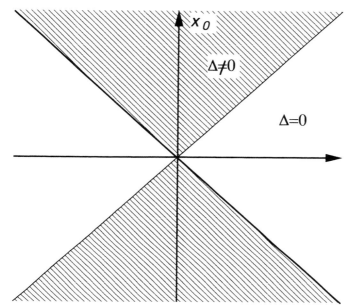

Fig 1. Δ is nonzero only in the shaded area.

Appendix C
Dimensional Regularization

C.1 Analytic Continuation and Integrals

In order to understand how analytic continuation can provide finite expressions for divergent integrals, consider the familiar case of the gamma function, which for Re $x > 0$ is defined by the integral

$$\Gamma(x) = \int_0^\infty dt\, t^{x-1} e^{-t}. \tag{C.1}$$

For other values of x we define $\Gamma(x)$ by analytically continuing this expression via the identity

$$\Gamma(x) = \frac{1}{x}\Gamma(x+1). \tag{C.2}$$

This is established by integrating (C.1) by parts in the region where the integral is convergent. Let us now consider what happens if we assume $-1 < \operatorname{Re} x < 0$. From

$$\frac{d}{dt}(t^x e^{-t}) = xt^{x-1} e^{-t} - t^x e^{-t} \tag{C.3}$$

we have

$$\left[t^x e^{-t}\right]_\epsilon^\infty = x\int_\epsilon^\infty dt\, t^{x-1} e^{-t} - \int_\epsilon^\infty dt\, t^x e^{-t}. \tag{C.4}$$

C.1 Analytic Continuation and Integrals

Here we have cut off the integral at the lower limit so as to avoid the divergence near $t = 0$. Evaluating the left-hand side and dividing by x we find

$$-\frac{1}{x}\epsilon^x = \int_\epsilon^\infty dt\, t^{x-1} e^{-t} - \frac{1}{x} \int_\epsilon^\infty dt\, t^x e^{-t}. \tag{C.5}$$

Since, for this range of x,

$$-\frac{1}{x}\epsilon^x = \int_\epsilon^\infty dt\, t^{x-1}, \tag{C.6}$$

we can rewrite (C.5) as

$$\frac{1}{x} \int_\epsilon^\infty dt\, t^x e^{-t} = \int_\epsilon^\infty dt\, t^{x-1} \left(e^{-t} - 1\right). \tag{C.7}$$

The integral on the right-hand side of this last expression is convergent as $\epsilon \to 0$, so we may safely take the limit and find

$$\frac{1}{x}\Gamma(x+1) = \int_0^\infty dt\, t^{x-1} \left(e^{-t} - 1\right). \tag{C.8}$$

Since the left-hand side is equal to $\Gamma(x)$, we have shown that

$$\Gamma(x) = \int_0^\infty dt\, t^{x-1} \left(e^{-t} - 1\right), \qquad -1 < \operatorname{Re} x < 0. \tag{C.9}$$

Similarly, if $-2 < \operatorname{Re} x < -1$, we have

$$\Gamma(x) = \int_0^\infty dt\, t^{x-1} \left(e^{-t} - 1 + t\right). \tag{C.10}$$

Thus the analytic continuation of the original integral is given by a new integral in which we have subtracted exactly as many terms from the Taylor expansion of e^{-t} as are needed to just make the integral convergent.

The same process occurs when we dimensionally continue a Feynman integral in the space-time dimension d. Because $d^d k \propto k^{d-1} dk$, the analytically continued integrals have had the first few terms in the Taylor expansion of the integrand in terms of the momenta entering the diagram subtracted.

Dimensional regularization has a habit of setting certain classes of integrals to zero. Look at, for example,

$$\int d^d k \frac{1}{k^2 + M^2}. \tag{C.11}$$

From (5.47) we find that this is equal to

$$\Gamma(1 - d/2) \frac{1}{2^d \pi^{d/2}} \left(M^2\right)^{d/2-1}. \tag{C.12}$$

If we set M to zero, this becomes identically zero in any dimension $d > 2$, and so by analytic continuation, for any d — even though we are surely integrating a positive quantity and the integral is divergent at large k. The paradox is resolved once one realizes that the analytically continued expression is not longer given by the original integral, but by that integral with various subtractions.

C.2 Propagators

Here are some useful formulae for propagators in the euclidean region.

First the massless case. Consider the Fourier transform of a general power of k^2:

$$\int \frac{d^d k}{(2\pi)^d} \frac{e^{ikx}}{|k^2|^\alpha} = \frac{1}{\Gamma(\alpha)} \int_0^\infty ds \int \frac{d^d k}{(2\pi)^d} s^{\alpha-1} e^{-sk^2+ikx}$$

$$= \frac{1}{\Gamma(\alpha)} \int_0^\infty ds \left(\sqrt{\frac{\pi}{s}}\right)^d \frac{e^{-x^2/4s}}{(2\pi)^d} s^{\alpha-1}$$

$$= \frac{1}{2^d \pi^{d/2}} \frac{1}{\Gamma(\alpha)} \int_0^\infty ds\, s^{d/2-\alpha-1} e^{-sx^2/4}$$

$$= \frac{1}{2^d \pi^{d/2}} \frac{\Gamma(d/2-\alpha)}{\Gamma(\alpha)} \left(\frac{4}{x^2}\right)^{d/2-\alpha}. \qquad (C.13)$$

The right-hand side of this expression has to be interpreted with caution. If we set $\alpha = -1$, we are taking the Fourier transform of k^2, and this should be $-\nabla^2$ — a differential operator! We must regard the right hand side as a distribution, i.e., always convolve it against some smooth test function $\phi(x)$. It is an amusing exercise to see how the divergence of the $\Gamma(\alpha)$ for α a negative integer, and the appearance of the Taylor coefficients in the analytic continuation, conspire for this convolution to give $\nabla^2 \phi(x)$.

To obtain the x-space propagator for a massless scalar field, we set $\alpha = 1$. Then we find

$$\int \frac{d^d k}{(2\pi)^d} \frac{e^{ikx}}{|k^2|} = \frac{1}{4\pi} \left(\frac{1}{\frac{d}{2}-1} - \ln|x|^2 + \ldots\right), \quad d \approx 2 \qquad (C.14)$$

$$= \frac{1}{4\pi^2} \frac{1}{|x|^2}, \quad d \approx 4. \qquad (C.15)$$

The pole near $d = 2$ is due to an infrared divergence.

For the massive case we evaluate

$$\int \frac{d^d k}{(2\pi)^d} \frac{e^{ikx}}{k^2 + m^2} = \frac{1}{2^d \pi^{d/2}} \int_0^\infty ds\, s^{-d/2} e^{-x^2/4s - m^2 s}. \qquad (C.16)$$

This is a Bessel function. Setting $s = e^t |x|/2m$ we can write it as

$$\frac{1}{2^d \pi^{d/2}} \left|\frac{x}{2m}\right|^{1-d/2} \int_{-\infty}^\infty dt\, e^{(1-\frac{1}{2}d)t - m|x|\cosh t}. \qquad (C.17)$$

Now

$$\int_{-\infty}^\infty e^{(1-\frac{1}{2}d)t - m|x|\cosh t} dt$$

$$= \int_{-\infty}^\infty \cosh \frac{(1-\frac{1}{2}d)t}{2} e^{-m|x|\cosh t} dt = K_{|1-d/2|}(m|x|), \qquad (C.18)$$

where the modified Bessel function, $K_n(x)$, has the large x behavior

$$K_n(x) \approx \sqrt{\frac{\pi}{2x}} e^{-x} \quad \text{as} \quad x \to \infty. \tag{C.19}$$

Appendix D
Spinors and the Principle of the Sextant

In this appendix I will discuss the basic properties of the spinor representations of the rotation group, $SO(N)$, and the orthogonal group, $O(N)$. I will mostly consider the case N *even*, because the odd-dimensional rotation groups are slightly more complicated. Spinors were discovered by Cartan[1] in 1913.

D.1 Constructing the γ-Matrices

We begin by showing that for any N there exists a set of $2N$ matrices obeying

$$\{\gamma_n, \gamma_m\} = 2\delta_{mn}, \tag{D.1}$$

and that they act on a 2^N-dimensional space.

Let $\hat{a}_i, \hat{a}_i^\dagger$, $i = 1, N$ be a set of fermion creation and annihilation operators so $\{\hat{a}_i, \hat{a}_j^\dagger\} = \delta_{ij}$ etc. These anticommutation relations have an obvious 2^N-dimensional Fock-space representation spanned by $(\hat{a}_1^\dagger)^{n_1}(\hat{a}_2^\dagger)^{n_2}\ldots(\hat{a}_N^\dagger)^{n_N}|0\rangle$, where the n_i take the values 0 or 1.

Define

$$\gamma_{2n} = a_n + \hat{a}_n^\dagger, \tag{D.2}$$

$$\gamma_{2n+1} = \frac{1}{i}(a_n - \hat{a}_n^\dagger), \tag{D.3}$$

---

[1] Élie-Joseph Cartan. Born April 9, 1869, Dolomieu, France. Died May 6, 1951, Paris.

then the γ_n are hermitian and have the desired anticommutation relations

$$\{\gamma_n, \gamma_m\} = 2\delta_{mn}. \tag{D.4}$$

D.2 Basic Theorem

Given the matrices γ_μ, $\mu = 1, 2N$. Our next task is to establish the following result:

Theorem:
$\forall R \in SO(2N), \exists U(R) \in SU(2^N)$ *such that*

$$U^\dagger(R)\gamma_\mu U(R) = R_{\mu\nu}\gamma_\nu. \tag{D.5}$$

To motivate the following construction consider the sextant.

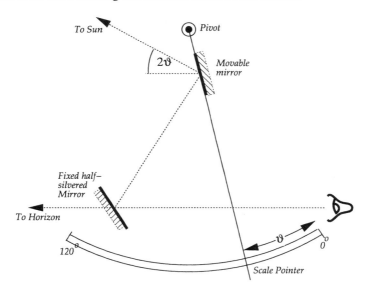

Fig. 1 The sextant.

This familiar instrument is used to measure the altitude of the sun above the horizon while standing on the pitching deck of a ship at sea. A theodolite or similar device would be rendered useless by the ship's motion. The sextant exploits the fact that successive reflection in two mirrors inclined at an angle θ to one another serves to rotate the image through an angle 2θ about the line of intersection of the mirror planes. This is used to superimpose the image of the sun onto the image of the horizon, where it stays even if the instrument is rocked back and forth. This same trick is used in constructing the spinor representations of the rotation group $SO(2N)$:

256   D. Spinors and the Principle of the Sextant

Given a vector **x** with components x^μ, form the object $x = x^\mu \gamma_\mu$. Now, if **n** is a unit vector, then

$$(-\gamma_\nu n^\nu)(x^\mu \gamma_\mu)(\gamma_\sigma n^\sigma) = (x^\mu - 2(\mathbf{n} \cdot \mathbf{x})(n^\mu)) \gamma_\mu \tag{D.6}$$

is the **x** vector reflected in the plane perpendicular to **n**. So

$$-(\gamma_1 \cos\theta/2 + \gamma_2 \sin\theta/2)(-\gamma_1)x(\gamma_1)(\gamma_1 \cos\theta/2 + \gamma_2 \sin\theta/2) \tag{D.7}$$

performs two succesive reflections, first in the "1" plane, and then in a plane at an angle $\theta/2$ to it. In other words, as you may establish from the Clifford algebra,

$$(\cos\theta/2 - \gamma_1\gamma_2 \sin\theta/2)x(\cos\theta/2 + \gamma_1\gamma_2 \sin\theta/2) \tag{D.8}$$

$$= \gamma_1(\cos\theta x^1 - \sin\theta x^2) + \gamma_2(\sin\theta x^1 + \cos\theta x^2) + \gamma_3 x^3 \cdots. \tag{D.9}$$

If we use $(\gamma_1\gamma_2)^2 = -1$, we can write this as

$$e^{-i\frac{1}{4i}[\gamma_1,\gamma_2]\theta} x^\nu \gamma_\nu e^{i\frac{1}{4i}[\gamma_1,\gamma_2]\theta} = x^\nu R_{\nu\mu}\gamma_\mu, \tag{D.10}$$

where R is the $N \times N$ rotation matrix for a rotation through angle θ in the 1-2 plane. It is clear that this construction allows *any* rotation to be performed. The use of mirrors for creating and combining rotations is discussed the the appendix to Misner, Thorn, and Wheeler's weighty tome *Gravitation*.

D.3   Chirality

The 2^N-dimensional space $\mathcal{H}$ is not actually *irreducible* under $SO(2N)$ — the generators of $SO(2N)$,

$$\Gamma_{ij} = \frac{1}{4i}[\gamma_i, \gamma_j] \tag{D.11}$$

keep the particle number, $F = \sum \hat{a}_n^\dagger \hat{a}_n$, either odd or even, so the total space decomposes into the direct sum of two invariant subspaces, the spaces of left or right *chirality*:

$$\mathcal{H} = \mathcal{H}_{odd} \oplus \mathcal{H}_{even}. \tag{D.12}$$

The operator that distinguishes the two spaces, $(-1)^F$, anticommutes with all the γ_μ, commutes with the Γ_{ij}, and can be written $\gamma_{2N+1} = -(i)^N \prod_\mu \gamma_\mu$. This is the higher-dimensional analog of $\gamma_5 = \gamma^0 \gamma^1 \gamma^2 \gamma^3$. Under the bigger group, $O(2N)$, which includes reflections, $\mathcal{H}$ *is* irreducible as one γ_n always changes the "particle" number from odd to even and so takes you from $\mathcal{H}_{odd}$ to $\mathcal{H}_{even}$ etc.

By including γ_{2N+1} with the other gamma matrices one can perform rotations in $2N + 1$ dimensions — but now there is no chirality, and the extra gamma matrix is rather an odd man out. Because of the orphan γ_{2N+1}, the Dynkin diagram of the odd dimensional rotation groups is different — belonging to the Cartan sequence B_n, rather than D_n — see H. Georgi, *Lie Algebras in Particle Physics* for a physicist's account of all this.

D.4 $Spin(2N)$, $Pin(2N)$, and $SU(N) \subset SO(2N)$

The bilinears $\hat{a}_i^\dagger \hat{a}_j$ span the Lie algebra of $U(N)$, which also acts on the 2^N-dimensional space $\mathcal{H}$— but in such a way as to preserve the number of "particles." The spaces $\mathcal{H}_{odd/even}$ therefore further decompose into a total of $N+1$ subspaces, each invariant under $U(N)$ and $SU(N)$. Now $SU(N)$ is a subgroup of $SO(2N)$ and this decomposition is an example the familar breakup of a representation space when we restrict to a subgroup.

It is clear that the group obtained by exponentiating the $\frac{1}{4i}[\gamma_\mu, \gamma_\nu]$ is a subgroup of $SU(2^N)$ and has the same Lie algebra as the rotatation group — but as the familiar example of $O(3)$ shows, it is not isomomorphic to it, forming instead a *double cover* of $SO(2N)$, called $Spin(2N)$. If we include the reflections, we get the group $Pin(2N)$, which is a double cover of $O(2N)$.

Appendix E
Indefinite Metric

In this appendix I give more details of the indefinite metric Hilbert space that appears when we quantize the electromagnetic field using the Lorentz gauge. We will continue to work with a single mode, which we take to be a plane wave propagating in the z direction. We will suppress the continuum normalization for the duration of this appendix, and assume that

$$[\hat{a}_i, \hat{a}_i^\dagger] = 1, \quad i = 1, 2, 3, \tag{E.1}$$

while

$$[\hat{a}_0, \hat{a}_0^\dagger] = -1. \tag{E.2}$$

These operators act on a "Hilbert space" with an indefinite inner product $\langle a|b\rangle$. In particular, the $\hat{a}_0^\dagger$ create states with negative norm-squared. Because of the indefinite nature of the product, the equation $\langle \chi|\chi\rangle = 0$ does not imply that $|\chi\rangle = 0$. The inner product is, however, *nondegenerate* — meaning that any state $|\chi\rangle$ that obeys $\langle \psi|\chi\rangle = 0$ for all $|\psi\rangle \in \mathcal{H}$ must be zero.

It is important that the inner product be nondegenerate. Remember that the notion of the adjoint, or hermitian conjugate, of an operator depends on the choice of inner product. Suppose $A : V \to V$, is an operator on a complex inner-product space V. To define the adjoint $A^\dagger$ we consider the linear functional $f : V \to \mathbf{C}$

given, for each $y \in V$, by[1]

$$f(x) = (y, Ax), \tag{E.3}$$

where (x, y) denotes the (not necessarily positive definite) inner product. If there exists an element $z \in V$ such that

$$f(x) = (z, x), \tag{E.4}$$

then we say that y is in the domain of $A^\dagger$, and set

$$z = A^\dagger y. \tag{E.5}$$

If the inner product were degenerate, then z would not be uniquely defined, and we would have no way to define $A^\dagger$.

For our single mode the physical state condition

$$(\partial^\mu A_\mu)^+ |physical\rangle = 0 \tag{E.6}$$

reads

$$(\hat{a}_0 + \hat{a}_3)|physical\rangle = 0. \tag{E.7}$$

We cannot further restrict the states by also requiring that

$$(\hat{a}_0^\dagger + \hat{a}_3^\dagger)|physical\rangle \stackrel{wrong!}{=} 0, \tag{E.8}$$

not because this is inconsistent with (E.7), but because there will turn out to be no states satisfying $(\hat{a}_0^\dagger + \hat{a}_3^\dagger)|\psi\rangle = 0$. We want to understand how only forcing half of it to vanish nonetheless leads to the full expression $\partial^\mu A_\mu$ being in some sense zero.

Define the operators

$$\hat{a} = \frac{1}{\sqrt{2}}(\hat{a}_3 + \hat{a}_0), \quad \hat{a}^\dagger = \frac{1}{\sqrt{2}}(\hat{a}_3^\dagger + \hat{a}_0^\dagger),$$

$$\hat{b} = \frac{1}{\sqrt{2}}(\hat{a}_3 - \hat{a}_0), \quad \hat{b}^\dagger = \frac{1}{\sqrt{2}}(\hat{a}_3^\dagger - \hat{a}_0^\dagger). \tag{E.9}$$

We find that

$$[\hat{a}, \hat{a}^\dagger] = [\hat{b}, \hat{b}^\dagger] = 0, \quad [\hat{a}, \hat{b}^\dagger] = [\hat{b}, \hat{a}^\dagger] = 1, \tag{E.10}$$

while $[\hat{a}, \hat{a}]$, $[\hat{a}, \hat{b}]$ etc. vanish. We assume that there is a vacuum state, $|0\rangle$, that obeys $\hat{a}|0\rangle = \hat{b}|0\rangle = \hat{a}_{1,2}|0\rangle = 0$ and has $\langle 0|0\rangle = 1$. We then apply powers of $\hat{a}_{1,2}^\dagger$, $\hat{a}^\dagger$ and $\hat{b}^\dagger$ to $|0\rangle$ to build a Fock-space representation.

The physical subspace is

$$\mathcal{H}_{physical} = \{|\psi\rangle \in \mathcal{H} \mid \hat{a}|\psi\rangle = 0\}. \tag{E.11}$$

---

[1] I have to make this point in standard vector space notation. The Dirac notation is too slick in that the symbol $\langle \psi|A|\chi\rangle$ does not distinguish between its two possible meanings: $(|\psi\rangle, A|\chi\rangle)$ or $(A^\dagger|\psi\rangle, |\chi\rangle)$.

It decomposes into a direct sum

$$\mathcal{H}_{physical} = \mathcal{H}_{transverse} \oplus \mathcal{H}_{null}, \tag{E.12}$$

where $\mathcal{H}_{transverse}$ is spanned by states obtained by applying powers of $\hat{a}^{\dagger}_{1,2}$ to the vacuum. $\mathcal{H}_{null}$ is spanned by states obtained by applying powers of $\hat{a}^{\dagger}_{1,2}$ and *at least one* power of $\hat{a}^{\dagger}$ to the vacuum, but involves *no* powers of $\hat{b}^{\dagger}$. (No state created by $\hat{b}^{\dagger}$'s can be killed by $\hat{a}$.)

Any state $|\chi\rangle \in \mathcal{H}_{null}$ is orthogonal to all the other states in $\mathcal{H}_{physical}$ (including itself). In other words

$$\langle physical | null \rangle = 0. \tag{E.13}$$

In computing overlaps, therefore, these states are effectively zero. As far as their physical predictions are concerned, two states that differ by a null state can be identified. Note that $\hat{a}^{\dagger}$ acting on any physical state yields one of these null states, and is thus effectively the zero operator.

Any operator corresponding to a measurable quantity must map the physical subspace into itself. It is easy to see that the hamiltonian

$$\hat{H} = (\hat{a}^{\dagger}_1 \hat{a}_1 + \hat{a}^{\dagger}_2 \hat{a}_2 + \hat{a}^{\dagger}_3 \hat{a}_3 - \hat{a}^{\dagger}_0 \hat{a}_0) \tag{E.14}$$

has this property, as do the components of the field strength tensor.

Appendix F
Phonons and Momentum

Suppose a sound wave in a fluid is described by the velocity potential

$$\theta = A\frac{1}{k}\sin(kx - \omega t). \tag{F.1}$$

The velocity field, $v = \nabla\theta$, is then

$$v = A\cos(kx - \omega t). \tag{F.2}$$

Using the continuity equation $\partial_x \rho v + \partial_t \rho = 0$, setting $\rho = \rho_0 + \delta\rho$, and approximating $\rho v \approx \rho_0 v$, we find that

$$\delta\rho = \frac{1}{\omega}\rho_0 k A \cos(kx - \omega t) + O(A^2). \tag{F.3}$$

The time average of the momentum density ρv is therefore

$$\langle \rho v \rangle = \langle \delta\rho v \rangle = \frac{k\rho_0}{\omega}\frac{1}{2}A^2, \tag{F.4}$$

to $O(A^3)$ accuracy.

Similarly the time average of the kinetic-energy density is

$$\langle \frac{1}{2}\rho_0 v^2 \rangle = \frac{1}{4}\rho_0 A^2. \tag{F.5}$$

Since for any combination of harmonic oscillators the time average of the potential energy is equal to that of the kinetic energy, the total energy density is

$$\langle E \rangle = \frac{1}{2}\rho_0 A^2. \tag{F.6}$$

The wave is composed of phonons of energy $\hbar\omega$, so the number-density of phonons is

$$\langle n \rangle = \frac{1}{\hbar\omega}\langle E \rangle = \frac{1}{2\hbar\omega}\rho_0 A^2. \tag{F.7}$$

The momentum-per-phonon is therefore

$$\frac{1}{\langle n \rangle}\langle \rho v \rangle = \frac{k\rho_0}{\omega}\frac{1}{2}A^2 \left(\frac{1}{2\hbar\omega}\rho_0 A^2\right)^{-1} = \hbar k. \tag{F.8}$$

This is what we would expect from general principles.

Now $\langle \rho v \rangle$ plays a dual role as the momentum density and as the mass-current. A nonzero average for the former therefore implies a steady drift of particles in the direction of wave propagation, in addition to the back-and-forth motion in the wave. We can confirm this by interpreting $v = A\cos(kx - \omega t)$ in a lagrangian sense, i.e. as the equation of motion of an individual particle of fluid. The particle trajectory $x(t)$ is the solution of the equation

$$\frac{dx}{dt} = A\cos(kx - \omega t). \tag{F.9}$$

Since the quantity x appears both in the derivative and in the cosine, this is a nonlinear equation. We solve it perturbatively by setting

$$x(t) = x_0 + A x_1(t) + A^2 x_2(t) + \ldots. \tag{F.10}$$

We find that

$$x_1(t) = -\frac{1}{\omega}\sin(kx_0 - \omega t). \tag{F.11}$$

Subsituting this into (F.9) we find

$$\frac{dx_2}{dt} = \frac{k}{\omega}\sin^2(kx_0 - \omega t). \tag{F.12}$$

Thus $\dot{x}_2$ has a non-vanishing time average, $k/(2\omega)$, leading to a secular drift that is consistent with (F.4).

We also see why there is no net newtonian momentum associated with phonons is a crystal. Referring to Chapter 1 we recall that the atomic displacements in a crystal are given by

$$\eta_n = A\cos(k(na) - \omega t), \tag{F.13}$$

so the crystal equivalent of (F.9) is

$$\frac{d\eta_n}{dt} = A\cos(k(na) - \omega t). \tag{F.14}$$

We see that η does not appear on the right-hand side of this equation. It is therefore a linear equation and gives rise to no net particle drift.

In reality things are a little more complicated than this. The equations of fluid dynamics are nonlinear, and the sound wave (F.1) is a solution only of the linearized equations. The difference between the sound wave in the fluid and the crystal is of second order in A, and both waves could have corrections from second-order effects in their equations of motion. Indeed, in sound fields in ordinary fluids, the mass flow due to the non-zero value of $\langle \delta\rho \mathbf{v} \rangle$ is often masked by larger bulk motions of the fluid. These steady flows, known as *acoustic streaming* and first studied by Rayleigh, are generated by friction effects, which transfer momentum from the sound wave to the background flow. Even in the absence of dissipation there will typically be $O(A^2)$ counterflow corrections to the mean flow, and these can carry as much momentum as the wave itself. In the two-fluid picture, these counterflows, although caused by the sound wave, are not counted as contributing to its momentum but are instead included the supercurrent where they serve to enforce $\nabla \cdot (\rho_s \mathbf{v}_s + \rho_n \mathbf{v}_n) = 0$.

Appendix G
Determinants in Quantum Mechanics

The determinant of the operator appearing in the quadratic part of an action contains all the one-loop effects, and is sometimes called the one-loop effective action. The evaluation of these determinants is therefore an important topic. In general the determinant can only be evaluated approximately, but in the case of quantum mechanics (one-dimensional field theory) there are some interesting results that often allow determinants to evaluated exactly.

Consider an operator of Schrödinger type:
$$T = -\partial_x^2 + V(x), \tag{G.1}$$
on the interval $[a, b]$, and seek to compute
$$D(\lambda) = \text{Det}\,(T - \lambda). \tag{G.2}$$

$D(\lambda)$ should be be some regularized version of the product $\prod_i (\lambda_i - \lambda)$ and so should be an entire function of λ having a zero whenever λ is an eigenvalue of T. Now the eigenvalues of T depend on the boundary conditions imposed on the eigenfunctions at a and b, and so the determinant will also depend on these boundary conditions. The simplest case occurs when we impose *Dirichlet* boundary conditions, i.e. require $y(a) = y(b) = 0$.

As a simple example consider the case of a harmonic oscillator with frequency ω. Here
$$D = \text{Det}\,(-\partial_x^2 + \omega^2) \tag{G.3}$$
on the interval $[0, T]$. The eigenfunctions are
$$y_n(x) \propto \sin \frac{n\pi x}{T}. \tag{G.4}$$

The eigenvalues are
$$\lambda_n = \frac{\pi^2 n^2}{T^2} + \omega^2, \qquad n = 1, 2, 3 \ldots. \tag{G.5}$$
Thus
$$\text{Det}\,(-\partial_x^2 + \omega^2) \to \prod_1^\infty \left(\frac{\pi^2 n^2}{T^2} + \omega^2\right). \tag{G.6}$$

This is divergent, but the ratio of two determinants with different values of ω^2 is finite. In particular,
$$\frac{\text{Det}\,(-\partial_x^2 + \omega^2)}{\text{Det}\,(-\partial_x^2)} = \prod_1^\infty \left(1 + \frac{\omega^2 T^2}{\pi^2 n^2}\right) = \frac{1}{\omega T} \sinh \omega T. \tag{G.7}$$

Note the curious fact that
$$\frac{\text{Det}\,(-\partial_x^2 + \omega^2)}{\text{Det}\,(-\partial_x^2)} = \frac{1}{T} \cdot \frac{1}{\omega} \sinh \omega T = \frac{y_\omega(T)}{y_0(T)}, \tag{G.8}$$
where
$$(-\partial_x^2 + \omega^2) y_\omega(x) = 0. \qquad y_\omega(0) = 0, \quad y'_\omega(0) = 1. \tag{G.9}$$

This is generally true. To see this consider a solution $y(x)$ to the initial value problem
$$(T - \lambda) y_\lambda = 0, \qquad y_\lambda(a) = 0, \quad y'_\lambda(0) = 1. \tag{G.10}$$

Now λ will be an eigenvalue of T if and only if $y_\lambda(b) = 0$. The zeros of $y_\lambda(b) = 0$ therefore coincide with those of $D(\lambda)$. Furthermore, a theorem of Poincaré in the theory of ordinary differential equations ensures that $y_\lambda(x)$ is, as the determinant should be, an entire function of λ. It is fairly easy to show that they have the same behaviors near $\lambda \to \infty$, so their ratio will be an entire bounded function, and so a constant. Thus, $D(\lambda) = C y_\lambda(b)$ for some (divergent) constant C. WE have therefore shown that:

Theorem:
Let T_1 be the differential operator $-\partial_x^2 + V_1(x)$ with domain $\mathcal{D}(T_V) = \{y \in L^2[a, b] | y(a) = y(b) = 0\}$, and T_2 the similar operator with potential V_2. Define the functions $y_1(x; \lambda)$, $y_2(x; \lambda)$ by
$$(T_1 - \lambda) y_1 = 0, \qquad y_1(a; \lambda) = 0, \quad y'_1(a; \lambda) = 1 \tag{G.11}$$
and similarly $y_2(x; \lambda)$. Then
$$\frac{\text{Det}\,(T_1 - \lambda)}{\text{Det}\,(T_2 - \lambda)} = \frac{y_1(1; \lambda)}{y_2(1; \lambda)}. \tag{G.12}$$

It is worth providing an alternative proof. This works by replacing the differential operator T by a discrete approximation on a lattice of spacing a. We replace the function $y(x)$ by a vector y_i and the derivatives by the simplest finite difference
$$(T - \lambda) y \to \tilde{T} y_i = -\frac{1}{a^2}(y_{i+1} - 2y_i + y_{i-1}) + (V_i - \lambda) y_i. \tag{G.13}$$

The differential equation $(T - \lambda)y = 0$ then become a three-term recurrence relation, which, like the original differential operator, has two linearly independent solutions. The condition $y'(0) = 1$ is here $y_1 = a$. On an interval with n discretization points we denote the determinant of the $n \times n$ finite matrix

$$\tilde{T}_n = \frac{1}{a^2} \begin{pmatrix} 2 + a^2(V_n - \lambda) & -1 & 0 & \cdots \\ -1 & 2 + a^2(V_{n-1} - \lambda) & -1 & \cdots \\ 0 & -1 & 2 + a^2(V_{n-2} - \lambda) & \cdots \\ \vdots & \vdots & \vdots & \ddots \end{pmatrix}, \quad \text{(G.14)}$$

representing $\tilde{T}$ by $D_{n+1}(\lambda)$.

Expanding the determinant about the first row we find

$$-\frac{1}{a^2}(D_{i+1} - 2D_i + D_{i-1}) + (V_i - \lambda)D_i = 0. \quad \text{(G.15)}$$

The determinant therefore satisfies the same recurrence relation as the solution to the discrete Dirichlet problem. The quantity that should vanish for λ to be an eigenvalue of the discrete equation is the solution of the recurrence relation one step beyond the end of the lattice (this is the origin of the shift $n \to n+1$ in the index on the determinant). As the lattice spacing is reduced, the solutions of the discrete problem will approach those of the differential equation, and this proves the theorem.

Index

Abrikosov vortices, 134
absorbtive part, 51
analyticity, 48
angle-resolved photoemission, 102
annihilation operators, 5, 30
anomalous dimension, 219
anticommuting c-numbers, 81
antiderivation, 176
antiparticle, 77, 78
asymptotically free theory, 229
axial gauge, 84

Bare field, 70
barotropic flow, 134
BCS theory, 179
Berezin integrals, 171
Bernoulli equation, 134
β-function, 205
block spins, 208
Bogoliubov transformation, 123
Bohm-Staver relation, 115
Boltzmann equation, 103
Borcher's class, 71
Bose condensation, 119
Breit-Wigner peak, 56
Brueckner (or Bethe) Goldstone theorem, 41

Campbell-Baker-Hausdorff, 20, 23
canonical conjugate field, 3
canonical energy-momentum tensor, 22
Casimir energy, 10
causality, 51
characteristic, differential equation of, 228
chirality, 76
circulation, 133
classical spins σ, 202
Clifford algebra, 73
cluster decomposition property, 121
coherent state, 177
complex fields, 17
Compton wavelength, 12
conjugacy class, 138
continuum limit = critical point, 206
contraction, 30, 82
convex hull, 163
Cooper pairs, 134, 183
correlators, 63
cosmological constant, 10
CPT theorem, 54
creation operators, 5, 30
critical point, 206
critical surface, 211
cross-section, 44

268   Index

crossing symmetry, 53
cumulants, 159
current algebra, 21

D'Alembert's paradox, 132
Debye frequency, 116
Debye-Hükel screening, 106, 108
decomposable state, 99
derivation, 176
dimensional regularization, 59
Dirac delta function, 7
Dirac equation, 72
Dirac sea, 78, 81, 100, 141
dispersion relation, 66
Dyson approach, 27

Effective action, 181
effective potential, 161, 162
endpoint singularity, 60
ϵ-expansion, 237
equal-time commutator, 3, 9
euclidean region, 59
Euler equation, 134

f-sum rule, 67
Fermi gas, 102
Fermi liquid, 102
Fermi sea, 100
Fermi statistics, 89
Fermi surface, 102, 114, 116
Fermi's "Golden Rule", 45
fermion, 78, 79
Feynman diagrams, 97
Feynman gauge, 86
Feynman path integral, 143, 145
Feynman propagator, 27, 29
Feynman rules, 37, 42, 88
Feynman's trick, 58
Feynman-Kac formula, 147, 155
fixed point, 209
    free-field fixed point, 214
Fock Space, 7
Fourier transforms, 8
Fréchet derivative, 8
fractal, 207
Fredholm determinant, 154
Fresnel integral, 145, 146
Friedel oscillations, 108
Furrey's theorem, 89

Galilean transformation, 129
gauge invariance, 83, 148
gauge transformation, 83
Gaussian integral, 153
Gell-Mann Low Theorem, 62, 204
generating functional, 159
ghosts, 86
global transformations, 20
Goldstone boson, 169
Goldstone bosons, 167
Goldstone's theorem, 158, 167
grading, 176
grand canonical partition function, 136
grand potential, 136, 137, 142
Grassmann variables, 81, 171, 174, 177
Green function, 29, 51
Gross-Pitaevskii equation, 131
ground-state energy, 10
Gupta-Bleuler formalism, 85

Harmonic crystal, 2
healing length, 133
Heisenberg field, 62
Heisenberg operator, 4
Heisenberg picture, 3, 144
helicity, 76
Higgs-Kibble mechanism, 132, 170
hole, 78
holes, 100
hot dark matter, 77

Imaginary part, 48, 55
in and out states, 68
inner product, 17
interaction picture
    interaction picture state, 26
interaction vertex, 38
interpolating field, 69, 71
irreducible vertex, 93
irrelevant operator, 26, 102
irrotational motion, 132
isospin, 18
isospin algebra, 18

Jellium model, 107

Kahler-Dirac, 200
kinematic invariants, 43
Kirchhoff's rules, 42

Klein-Gordon equation, 13
Kogut-Susskind fermions, 199

Lagrangian, 2
lagrangian density, 10
Landau criterion, 128, 129
Landau damping, 111
Landau-Cutkoski cutting formulae, 51
large N limit, 238
Lehmann-Källén, 64, 69
lifetime, 47, 55
Lindhard function, 106, 108, 111
liquid ^{4}He, 117
local transformations, 20
London penetration depth, 132
long range-order, 140
Lorentz gauge, 85, 86
Lorentz invariance, 12
Lorentz-invariant Phase Space (LIPS), 15, 45
lorentzian inner product, 14
LSZ (Lehman, Symanzik and Zimmerman, 67–69

Mössbauer effect, 34, 35
Madelung transformation, 132
marginal operator, 209, 214
mass-shell, 28
Matsubara frequencies, 140
Maxwell construction, 163
Meissner effect, 132, 170
method of characteristics, 228
midpoint rule, 149
momentum transfer, 43
mutually local operators, 71
Møller scattering, 90
Møller wave operators, 63

Natural units, 12
negative frequency part, 30
Noether's theorem, 18
nonrenormalizable interaction, 26, 209, 214
normal component, 131
normal products, 27, 30
normal threshold, 53

Occupation numbers, 7

Off-diagonal Long-range Order (ODLRO), 120
off-shell, 28
one-particle density matrix, 102
one-particle irreducible diagrams (1PI), 164
optical theorem, 48
order parameter, 120
Ouroboros graphs, 56–58, 165, 190

Partition functions, 136
permutation, 138
Perron-Frobenius theorem, 119
Pfaffian, 173
phase velocity, 129
phonons, 7
physical states, 85, 87
pinch, 193
pinch singularity, 60, 193
plasma frequency, 109, 110
plasma oscillations, 108
plasmon, 109
Poisson distribution, 35
polarization vectors, 88
polarized light, 88
positive frequency, 16
positive-energy, 17
principle of least action, 8
projection operators, 80
pseudomomentum, 24, 35

Quantum electrodynamics, 83
quantum pressure, 134
quasiparticles, 102

Rapidity, 246
Rayleigh-Schrödinger perturbation theory, 49, 50, 89
relevant operator, 209, 214
renormalizable interaction, 209
renormalization group equation, 205
renormalized field, 70
resolution of unity, 15
resonance, 44, 54, 55
roton minimum, 128

S (for scattering) matrix, 37
s, t, and u channels, 44
scalar field, 13

scaling dimension, 212, 214
Schrödinger-picture, 26, 144
Schwarz reflection principle, 53
Schwinger's trick, 58
second quantization, 97
self-energy diagram, 32, 41
Slater-determinant, 99
sound waves, 124
spectral weight $\rho(m^2)$, 65
spin, 13
spinor field, 13, 77, 254
spontaneous symmetry breaking, 10
sum rule, 65
superfield, 176
superfluid mass density, 131
superfluid velocity, 131
superpotential, 176
superrenormalizable interaction, 209
supersymmetry, 174
symmetry factor, 39, 42

Target space, 221
tensor, 97
tensor product, 6
Thomas-Fermi approximation, 107
Thomas-Reiche-Kuhn sum rule, 67
time-dependent perturbation theory, 26
time-independent gauge transformation, 84
time-ordered exponential, 27
time-ordered products, 27

time-ordered vacuum expectation values (TVEV's), 71
time-ordering symbol T, 27
top form, 172
trace-class, 154
two fluid model, 131

Unitarity, 48
universality, 211

Vacuum bubbles, 39, 41
vacuum persistence amplitude, 32
Vlasov equation, 104

Wake, ship's, 129
Ward identity, 92, 94
wave function renormalization constant, 41, 65
weak operator topology, 69
Weyl basis, 76
Weyl equation, 76
Weyl fermions, 76
Wick rotation, 56, 57, 59, 141
Wick's theorem, 97
Wilson fermions, 199
Wilson-Fisher fixed point, 215

Yukawa potential, 32, 33

Zero sound, 113, 114

Graduate Texts in Contemporary Physics

(continued from page ii)

A.M. Zagoskin: **Quantum Theory of Many-Body Systems: Techniques and Applications**